An Introduction to
THE MECHANICS
OF SOLIDS

Stresses and Deformation in Bars

A. S. HALL
Professor of Civil Engineering
University of New South Wales

JOHN WILEY & SONS AUSTRALASIA PTY LTD
Sydney   New York   London   Toronto

Preface

Great changes have recently come about in the more advanced phases of structural analysis, where matrix methods have almost completely displaced the older techniques. Besides being more powerful, the newer methods are clearer and more logical. The majority of structures analysed are still of necessity composed of one-dimensional components such as beams and columns, and the engineer's beam theory, far from being obsolescent, is now the basis of a body of theory more weighty than ever before.

This subject used to be misnamed "Strength of Materials". The current fashion is to call it "Mechanics of Solids" which is also a misnomer. I have tried to mitigate the fault by a sub-title. This book is intended for use by undergraduate students as a first course in structural mechanics. An effort has been made to use symbols, sign conventions and concepts that will require a minimum of adjustment in later work. The concepts of flexibility and stiffness simplify the present work as well as leading on to later phases.

The force–deformation characteristics of one-dimensional members are explicitly developed from those of a small element. The way in which the member characteristics can be used in the analysis of more extensive structures is briefly mentioned in Chapter 9 where the analysis of small pin-jointed trusses is discussed. In this chapter the principles of Virtual Forces and Virtual Displacements are introduced and their mutual relationship is explained.

It is assumed that the student has taken, or is concurrently taking, a course in materials science. A thorough knowledge of the structural properties of materials is essential. Although brief reference is made to these properties in section 2.5 this is in no way intended to be a resumé of materials science; the intention is only to indicate the sense in which certain terms are used in the present book. A knowledge of geometry and of mechanics is also assumed. Although the student will probably have studied the statics of bars previously, it will be desirable to look at Appendix A before commencing the book proper. The sign conventions used in Appendix A are those which are now required for later work, and not those which have been used traditionally.

Quite a number of problems and worked examples make use of materials whose stress–strain relation is not linear. In the first place, the use of such materials is becoming more common as also is the use of such non-linear relationships in certain theories of analysis, so that some mention of non-linearity at an early stage seems desirable. In the second place, I believe the restriction of problems to those involving only a linear relationship enables the student to get by without really appreciating the logical steps of analysis. Inevitably, the problems involving non-linearity are a little more difficult, but it is hoped that a better grasp of the principles will compensate for this.

Many students find difficulty with shear stresses and shear deformations. This subject has been dealt with a little more fully than usual. The emphasis has been laid on the shear flow concept rather than shear stress. Besides being of considerable practical use, the shear flow concept avoids erroneous assumptions required for shear stress calculations.

Chapter 10, which introduces the concept of instability, is also written from a slightly different standpoint. In most books the eigen-value problem, as exemplified by the Euler column problem, is given prominence. Study of this rather special class of instability phenomena does not lead to a good engineering appreciation of instability, but should rather be introduced after the general problem is properly understood. Elementary concepts of dynamic analysis are introduced in Chapter 11 so that the student shall not think exclusively in terms of static analysis.

The book is written for engineering students, and, in consequence, the explanations are given in physical terms as far as possible. The problems and their solutions are expressed in terms of engineering concepts and not in those of applied mathematics. Although the engineer of today requires a sounder grasp of fundamentals than previously, it is still essential for him to relate his theory to physical reality.

I have received a great deal of assistance from the Civil Engineering staff at the University of New South Wales, both in reading the manuscript and in checking the answers to problems. I would make special acknowledgement to Professor A. J. Carmichael, Professor F. E. Archer, Professor R. W. Traill-Nash, Professor R. W. Woodhead, Mr I. J. Somervaille, Mr K. A. Faulkes, Dr R. F. Warner, Dr A. P. Kabaila, Mr R. Frisch-Fay and Dr G. M. Folie. There are possibly still many errors in the text and problems, and I should be pleased to receive a note of any that are noticed.

A.S.H.

Sydney 1969

Contents

Notation

A = area of cross-section of bar
A_y = shear area for shear parallel to y (Chapter 5)
A_z = shear area for shear parallel to z (Chapter 5)
A_0 = area enclosed by wall of a thin-walled tube (section 6.6)
a = amplitude of vibration

B = flange width of channel section
b = width of a rectangular section

C = couple

D = diameter of circular section
D = total depth of channel section
d = diameter of circular section
d = depth of rectangular section

E = Young's modulus of elasticity
E_t = tangent modulus of elasticity
e = deformation
e = eccentricity of loading of a strut

F = force
f_t = tangent flexibility coefficient
f = flexibility coefficient

G = shear modulus of elasticity
g = function giving warp of cross-section (Chapter 5)
g = gravitational acceleration

H = step function
h = height of fall (Chapter 11)

I = moment of inertia
I_p = polar moment of inertia

I_{yy} = moment of inertia about the y axis
I_{yz} = product of inertia about axes y and z
I_{zz} = moment of inertia about the z axis

J = torsion constant

K = stiffness coefficient
K = volumetric modulus of elasticity
k = stiffness coefficient
k_t = tangent stiffness coefficient

L = length of bar
l = length of bar
l = length round wall of tube

M = bending moment
M_y = bending moment around the y axis
M_z = bending moment around the z axis
M_p = plastic moment
m = mass

N = axial force
n = ratio of moduli of elasticity
n = number of coils in a helical spring
n = frequency of vibration

P = external force
$p = \sqrt{k/m}$ = circular frequency
$\bar{p}$ = circular frequency of forcing function

Q = first moment of area
q = shear flow

R = radius
R = reaction
r = radius
r = ratio of amplitudes of vibration

S = shear force
S_y = shear force component in the y direction
S_z = shear force component in the z direction

T = twisting moment (torque)
T = period of vibration
T = temperature rise
t = thickness of web, flange or tube
t = time (Chapter 11)

U = energy
u = displacement in the x direction

V = volume
v = displacement in the y direction (e.g. deflection of a beam)

W = external load
w = intensity of distributed loading

X, Y, Z = planes normal to axes x, y and z respectively
X, Y, Z = body forces in the directions x, y and z respectively
x, y, z = co-ordinate axes
$\bar{x}, \bar{y}, \bar{z}$ = alternative co-ordinate axes

Y = intercept on a line parallel to the y axis (section 4.7)

Z = section modulus

α = coefficient of linear thermal expansion
α = a shear factor (section 5.7)
α = helix angle (section 6.4)
α = phase angle (Chapter 11)
α = deviation

β = coefficient for torsion constant (section 6.7)

γ = shear strain

ϵ = normal strain
ϵ_y = yield strain
$\epsilon_1, \epsilon_2, \epsilon_3$ = principal strains

ϕ = angle of twist
$\phi = \sqrt{F/m}$ = coefficient of forcing function (Chapter 11)

Λ_1, Λ_2 = Lamé's constants

$\lambda_1, \lambda_2 = $ coefficient for torsion stresses (section 6.7)
$\lambda = $ a coefficient (Chapter 11)
$\lambda = \sqrt{P/EI}$ (Chapter 10)

$\mu = $ Poisson's ratio

$\rho = $ curvature (of a bent beam)

$\sigma = $ normal component of stress
$\sigma_y = $ yield stress of bar in uniaxial stress
$\sigma_1, \sigma_2, \sigma_3 = $ principal stresses

$\tau = $ shear stress
$\tau_y = $ yield stress in shear
$\tau_1, \tau_2, \tau_3 = $ principal shear stresses
$\tau_{Xy} = $ shear stress on a plane normal to X and acting in the y direction

$\theta = $ angle with x axis

$\Omega = \int \dfrac{dl}{t} = $ function related to thin-walled tubes (section 6.6)

CHAPTER 1

Definitions

1.1 Introduction

In the course of designing a machine or a building, an engineer is called upon to consider many of its characteristics—economy, durability, appearance, strength and rigidity. Of these aspects, the present book is concerned with strength and rigidity. The importance of strength in any component is at once evident. Rigidity, by which we mean resistance to change of shape or size, is no less important. In the moving parts of a machine, any change of dimensions beyond a specified tolerance will probably prevent the functioning of the machine. In static structures, such as buildings and bridges, excessive distortion also cannot be tolerated. In such structures, fracture is uncommon and complete collapse is usually a matter of extremely large deformation.

A building or machine, which we shall call in general a *structure*, is subjected during its life to a number of external influences which cause it to deform. Chief among these are externally applied loads, both those it is being specifically designed to support or withstand and also others of a more accidental nature such as earthquake forces or forces resulting from misuse. Change of temperature will also affect the shape and size of the structure, and change of humidity may do so as well in certain circumstances.

It is in order to predict the behaviour of his proposed structure that the engineer undertakes an analysis before he proceeds with construction. As well as trying to estimate its rigidity, that is to say its resistance to change of shape or size, he also computes the stresses at what appear to be critical locations. The values of the stresses often, but not always, provide an indication of the margin of safety against collapse or fracture. It is not unusual for an engineer to regard the calculation of stresses as his primary objective, relying on his experience to conclude that if the stresses lie within certain limits (known as *working stresses*) his structure is safe both against collapse and against undue deformation. It should be remembered, however, that if the structure is of an unusual design such a conclusion may not be justified, and specific calculation of both rigidity and stress may be necessary.

How is this behaviour to be predicted? Briefly, the behaviour of the whole structure is deduced from the behaviour of its components. The process is carried out in several stages, and each stage consists of the determination of the behaviour of a component in terms of the properties of a still smaller element. Ideally, we should like to be able to predict the behaviour of a bridge, for example, in terms of the molecular structure of the materials of which it is composed. This ideal, however, is yet to be realized, and at the present time certain links in the chain of synthesis are supplied by experimental relations rather than by deductive reasoning.

Certainly, the resistance of a material to deformation under load and other influences is governed fundamentally by its atomic and molecular structure. However, in most substances the behaviour of elements of finite size is influenced to some extent by manufacturing processes which introduce random effects. The behaviour of such finite-sized elements is therefore still determined experimentally. The material deformation depends on many factors—the load to which it is subjected, whether the load is steady or fluctuating, temperature, the duration of the loading or heating, and so on. The relationships of the deformation to these other variables are expressed by the *structural* properties of the material. Mention will be made of these properties in section 2.5.

From a knowledge of the properties of the material we can derive the deformation characteristics of simple structural components. We may consider first such components as beams, columns, shafts in machinery, tie-rods and springs. These have one feature in common—they are all long compared with their cross-section dimensions, and for brevity they will be referred to in this book either as *bars* or *members*.

This book is concerned with the derivation of the deformation characteristics of bars, which are perhaps the most common of all structural components. Once these properties have been established we may regard the bar as itself an element in a more extensive structure built up from such bars. The analysis of such bar structures, or *frames*, forms a very extensive subject, but a brief introduction to the principles of such analysis is given in Chapter 9.

The study of a complex structure may thus be divided into a number of phases; (*a*) from microstructure to material properties (this phase is examined in materials science), (*b*) from material properties to simple structural components, and (*c*) from simple components to groups of components.

Before discussing the methods of analysis it is necessary to define certain fundamental quantities.

1.2   Force, Stress, Stress-resultant

Several different kinds of forces are recognized. First of all there are *external forces*, which are forces imposed on the body. They may be imposed at the surface by some other body or they may be imposed directly on each individual particle by a force field such as a gravitational or magnetic field. The former are called *surface forces*, while the latter are known as *body forces*.

Then there are *internal forces*, which act between two different portions of the body under consideration. If we imagine any plane within the material, then the force transmitted across this plane is an internal force. An example of forces transmitted across a plane within the material is given in Fig. 1.1 (a). The end forces P tend to cause the bar to fracture across a plane such as $A–A$ and if the material resists this tendency, a force is transmitted across the plane. In this instance the internal force results from an external force acting on the material. Internal forces can occur independently of external forces. For instance, if the central portion of a plate is heated it tends to expand but is restrained by the surrounding material. In consequence, the central part will be in compression while the outer part will be in tension. These effects will balance one another.

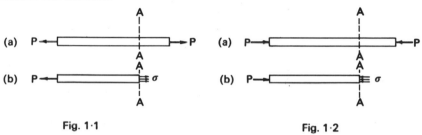

Fig. 1·1   Fig. 1·2

The usual units of force in engineering are the pound and the kip (1 kip = 1000 lb). The ton is sometimes used, but this unit leads to confusion as the ton is sometimes defined as 2240 lb and sometimes as 2000 lb (the short ton).

When a plane within the material is acted upon by an internal force, the intensity of the force is called the *stress* on the plane. Stress is measured usually in pounds per square inch (psi) or kips per square inch (ksi) and sometimes in tons per square inch (tsi). The dimensions of stress are force $\times$ length^{-2}.

A stress produced by a force normal to a given plane is called *normal* stress or *direct* stress. If the force, or component of force is parallel to the plane, the stress produced is called *tangential* stress or *shear* stress.

Consider first the normal stresses, which will be denoted by σ. Fig. 1.1 (a) shows a straight bar subjected to opposing end forces, P, which tend to increase its length. Such forces are called *tensile* forces. If we cut the bar by a cutting plane $A–A$ normal to the axis of the bar, we can consider the equilibrium of one part in isolation. A partial structure isolated in this way is called a *free-body*. Fig. 1.1 (b) shows the free-body diagram of the left-hand portion. If the free-body has no acceleration the end force P is balanced by the resultant of the stresses on the section *A–A*. These are normal stresses, and since they act *outward* from the free-body they are called *tensile* stresses. Fig. 1.2 shows the same bar subjected to *compressive* end forces. The stresses on the cross-section *A–A* are again normal but since they now act *inward* upon the free-body they are called *compressive* stresses. We see that normal stresses can be either tensile or compressive.

If the force is distributed evenly over the plane, the value of the stress is simply the force divided by the area of the plane. Fig. 1.3 (a) shows a tensile

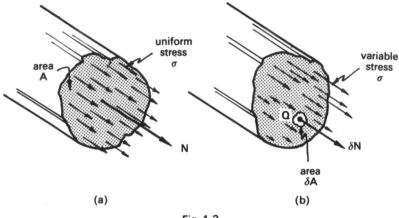

Fig. 1·3

force *N* acting on an area *A* and producing a stress *σ* of uniform intensity. In this case

$$\sigma = N/A \tag{1.1}$$

However, uniform stress rarely occurs in practice, and where it varies from point to point the stress at any point *Q* (Fig. 1.3 (b)) is defined in terms of the small force *δN* acting on an element of area *δA* surrounding *Q*. The limit of the ratio of *δN* to *δA* is taken to be the stress. Thus, at *Q* the stress is

$$\sigma = \operatorname*{\mathcal{L}}_{\delta A \to 0} \frac{\delta N}{\delta A} = \frac{dN}{dA} \tag{1.2}$$

The relationship between force and stress (equation 1.2) can be expressed in the form

$$N = \int \sigma \, dA \tag{1.3}$$

This integral can be represented graphically if at each point of the surface *A* we erect an ordinate, normal to the surface, representing to scale the value of the stress at that point (Fig. 1.4).

These ordinates define a surface *EFGH* (Fig. 1.4) which, in the case of variable stress, will not be parallel to the surface *ABCD* on which the stress acts. The force on any elemental area dA is σdA and this is represented by the volume of a prism (Fig. 1.4) of base dA and length σ. From equation (1.3) it follows that the total force N is represented by the volume of the solid bounded by the plane *ABCD* and the surface *EFGH*.

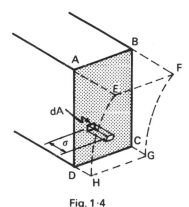

Fig. 1·4

Such a solid is often referred to as a "stress block". It is easy to show that the resultant force acts at a point on *ABCD* opposite the centre of gravity of the stress block. In many problems the stress varies according to a simple law and the volume of the stress block can easily be calculated.

Example 1.1. A rectangle *ABCD*, 4.5' in × 6 in, lies in the *yz* plane (Fig. 1.5). It sustains a stress acting in the *x* direction the intensity of which (in psi) is $\sigma = 80y^2$. Find the magnitude and position of the resultant force on the rectangle.

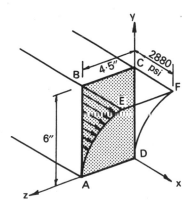

Fig. 1·5

6 Definitions

Solution. (*a*) On an element of area $dy \times dz$ the force is

$$dN = \sigma \, dy \, dz = 80y^2 \, dy \, dz$$

The total force is therefore given by

$$N = \int_0^6 \int_0^{4.5} 80y^2 \, dz \, dy$$

$$= \int_0^6 360y^2 \, dy$$

$$= 25{,}920 \text{ lb}$$

The total moment of the stresses about the z axis is

$$M = \int dN \cdot y = \int_0^6 \int_0^{4.5} 80y^3 \, dz \, dy$$

$$= 116{,}640 \text{ lb-in}$$

The distance of the resultant from the z axis is therefore

$$\bar{y} = \frac{116{,}640}{25{,}920} = 4.5 \text{ in}$$

In this calculation we have merely found the volume and located the centre of gravity of the stress block *ABCDEF* (Fig. 1.5). If these values are known, integration is unnecessary.

(*b*) An alternative solution is possible if we already know the geometrical properties of the stress block (which we usually do). The relevant properties of the parabolic area *ABE* (Fig. 1.5) are given in Table B.1 of Appendix B.

The area of the parabolic segment *ABE* (Fig. 1.5) is $\frac{1}{3} \times 6$ in $\times$ 2880 psi = 5760. Hence the volume of the block is

$$N = 5760 \times 4.5 = 25{,}920 \text{ lb}$$

The centre of gravity of the block is at the same height as the centroid of *ABE* in this case. That is to say, $\bar{y} = \frac{3}{4} \times 6 = 4.5$ in.

The stress block method is useful when the area upon which the stress acts is irregular or discontinuous.

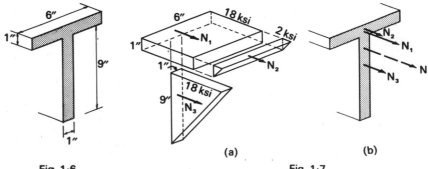

Fig. 1·6　　　　　　　　　　　　　(a)　　　　　　　　(b)　　　Fig. 1·7

Example 1.2. A T-shaped area (Fig. 1.6) is acted upon by a normal stress which varies linearly from 20 ksi at the top of the T to zero at the bottom. Find the magnitude and position of the resultant of these stresses.

Solution. The stress block in this example is discontinuous. It can be divided up into sub-blocks for each of which the volume and centre of gravity are easily computed. One way of doing this is shown diagrammatically in Fig. 1.7 (a). The individual resultants are easily found and are shown in Fig. 1.7 (b). From the shapes of the blocks it is evident that N_1 is $\frac{1}{2}$ in below the top, N_2 is $\frac{1}{3}$ in below the top, and N_3 is 6 in above the bottom. The resultant, N, of the three parallel forces N_1, N_2 and N_3 is 195 kips at 1.95 in below the top.

We now consider shear stresses, which will be denoted by τ. Fig. 1.8 (a) shows a bar projecting from a supporting wall and carrying a transverse end force P. (A member supported in this way is called a *cantilever*.) As before, we cut the bar with a cutting plane A–A and consider the free-body diagram of one part (Fig. 1.8 (b)). As far as vertical equilibrium is concerned, the force P is balanced by the resultant S of the shear stresses on plane A–A. (The free-body is not in rotational equilibrium so evidently there are other stresses on A–A.)

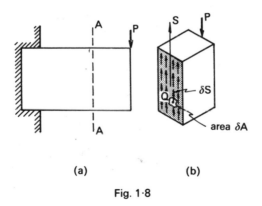

(a)   (b)

Fig. 1·8

As in the case of direct stress, the shear stress is also defined as the *intensity* of force at a particular point. If a small shear force δS acts on an element of area δA surrounding a point Q, the shear stress, τ, at Q is

$$\tau = \operatorname*{\mathcal{L}}_{\delta A \to 0} \frac{\delta S}{\delta A} = \frac{dS}{dA} \tag{1.4}$$

If we divide the total shear force by the total area we obtain a value S/A which is the *average* shear stress. Since shear stresses are rarely distributed uniformly over an area, the value S/A will not in general give the actual value of the shear stress at a given point caused by the force S.

The relationship between force and stress (equation 1.4) can also be expressed in the form

$$dS = \tau \, dA \qquad (1.5)$$

The various elemental forces dS at different points will lie in the plane on which they are acting. They will not necessarily act in the same direction. Consequently, before the resultant can be found, each small force dS must be resolved into convenient components. These components can then be summed by integration in the usual way. The stress block concept is seldom useful for finding the resultant of shear stresses.

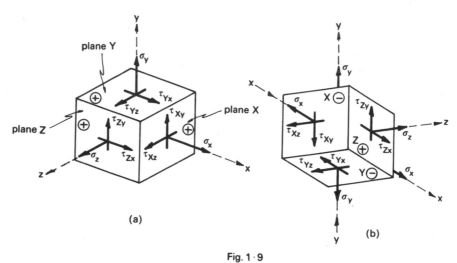

(a)

(b)

Fig. 1·9

Positive and negative senses are associated both with normal stresses and with shear stresses. Fig. 1.9 shows a small block of material with faces normal to the x, y, z axes.

The faces normal to x, y and z will be denoted by X, Y and Z respectively. On each face three components of stress are acting, one normal and two shear components. The stresses may be distinguished by the use of two subscripts, the first identifying the plane and the second denoting the direction of the stress component. Thus σ_{Yy} is the stress on the Y plane in the y direction; τ_{Zx} is the stress on the Z plane in the x direction, and so on. In the case of normal stresses, σ, the double subscript is not strictly necessary since the two letters are inevitably the same. So σ_{xx} is often written simply as σ_x.

The three faces from which the axes are directed outward are called faces of *positive incidence*, and in Fig. 1.9 (a) the block is turned so that these faces are visible. When the stress components on these faces act in the positive direction of x, y or z they are said to be positive.

In Fig. 1.9 (b) the block is turned to show two faces of negative incidence (the axis is directed inward). On these faces the stress components are positive if they *disagree* with the directions of the axes. The normal stresses are positive if tensile, negative if compressive, but the.sign of shear stresses must refer specifically to the direction of the co-ordinate axes.

The single force (or couple) which represents the resultant of the stresses acting on the cross-section of a bar is called a *stress-resultant*. It would be possible to represent the resultant of the normal stresses as a single force normal to the cross-section. For various reasons it is often more convenient to divide this into a force and a couple, the force usually being situated at the centroid of the section. The force is called the *axial force* and the couple is the *bending moment*.

In a similar way, the resultant of the shear stresses on the section is usually divided for convenience into a force (the *shear force*) and a couple (the *twisting moment*).

The signs of stress-resultants are defined in the same way as the signs of stresses. This matter is discussed in some detail in Appendix A. Briefly, we may consider the free body either to one side or the other side of a given cross-section. On one free-body the x axis will have positive incidence at the cut section, while on the other free-body it will have negative incidence. On the face of positive incidence a stress-resultant is said to be positive if it agrees with the co-ordinate axes, and negative otherwise. On the face of negative incidence, the reverse situation applies.

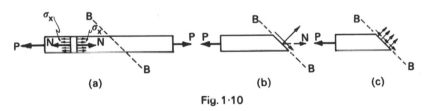

(a)   (b)   (c)

Fig. 1·10

The essential difference between external forces on the one hand, and stresses and stress-resultants on the other, must be clearly understood. In Fig. 1.10 (a) the two forces P are *external forces*. The force at the left-hand end is negative (opposite to x) while that at the right-hand end is positive (in the x direction). The external forces give rise to stresses, and a stress-resultant (axial force), which are shown acting on an element of the bar in Fig. 1.10 (a). Irrespective of whether we consider σ and N on the left-hand face of this element or σ and N on the right-hand face, we conclude that both σ and N are positive.

At any point in a body the nature of the stress will depend, among other things, upon the direction of the cutting plane. In Fig. 1.1 we saw that when a bar is under axial tension the stresses on a normal cross-section are direct tensile stresses.

However, suppose that the bar is cut by an oblique plane $B-B$ (Fig. 1.10 (a)). The end force P is still balanced by a longitudinal force N, equal to P, on plane $B-B$. This force can be resolved into normal and tangential components (Fig. 1.10 (b)) which in turn produce normal and tangential stresses (Fig. 1.10 (c)). Thus, on an oblique plane such as $B-B$, the longitudinal force N gives rise to both normal and tangential stresses. This will be discussed further in Chapter 7.

1.3 Displacement, Deformation, Strain

If all points in a given body undergo the same displacement, then the body remains the same size and shape. It has merely moved to a different location. It has been given a *rigid body displacement*.

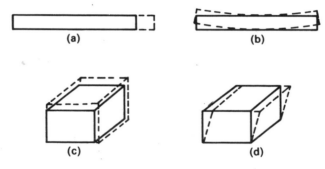

Fig. 1·11

On the other hand, if different parts of the body suffer different displacements the body will alter either its shape or its size, or both. It is said to be *deformed*. Deformation is thus a result of differential displacement of various points within the body.

The term *deformation* denotes a change in the size or shape of a body (Fig. 1.11). A bar of material might become longer or shorter, or it might become curved. A block can increase or decrease in volume, or it might alter its shape. Deformation is merely a change of geometry.

Linear deformation refers to change in length (Fig. 1.11 (a)). Any variation in size can be expressed in terms of linear deformations, since areas and volumes can be expressed in terms of linear dimensions. In Fig. 1.11 (c), for instance, the total volume change is the volumetric deformation. It can be calculated in terms of the original block dimensions and the linear deformations parallel to the edges. Deformations generally will be denoted by *e*.

Shear deformation is essentially a change of shape. It is measured by the relative displacement of two parallel planes in a direction tangential to the planes (Fig. 1.11 (d)).

Strain is the ratio of deformation to the original size of the body. Since it is the ratio of two similar quantities, strain is itself dimensionless.

Consider, first, *linear strain* or *direct strain*, which will be denoted by ϵ. In a body the strain may be uniform or it may vary from point to point. Suppose a bar of length L (Fig. 1.12) increases in length by an amount e. If the strain, ϵ, is the same for all elements along the bar, then

$$\epsilon = \frac{e}{L} \tag{1.6}$$

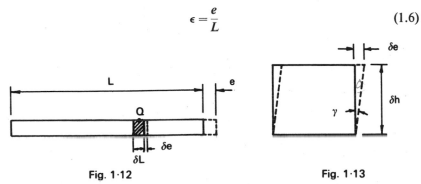

Fig. 1·12   Fig. 1·13

However, if the strain varies from point to point along the bar, the strain at any point Q is defined in terms of the small extension δe of an element of length δL adjacent to Q. The limit of the ratio of δe to δL is taken to be the strain. Thus, at Q the strain is

$$\dot{\epsilon} = \operatorname*{\mathcal{L}}_{\delta L \to 0} \frac{\delta e}{\delta L} = \frac{de}{dL} \tag{1.7}$$

Linear strains may be tensile or compressive and these are denoted as positive and negative respectively.

We now consider *shear strain*, which will be denoted by γ.

Fig. 1.13 shows a small element, of height δh, which suffers shear distortion. As a result of this the relative displacement of the top and bottom surfaces is δe. The shear strain is defined as

$$\text{shear strain} = \frac{\delta e}{\delta h}$$

or the limit of this as $\delta h \to 0$ if the strain is variable. Since the angle γ (Fig. 1.13) is small we can take $\tan \gamma \approx \gamma$ and since $\delta e/\delta h$ is $\tan \gamma$, we then have

$$\text{shear strain} = \gamma \approx \operatorname*{\mathcal{L}}_{\delta h \to 0} \frac{\delta e}{\delta h} = \frac{de}{dh} \tag{1.8}$$

where γ is the radian measure of the angle of distortion.

The above definitions, both for linear strain and shear strain, are useful only if the deformation is small compared with the original size of the element. A limit of 5% strain is reasonable (de/dl or $de/dh = 0.05$). Within this limit the extension can be divided either by the original length or the final length to obtain the linear strain without great error. Also, for shear strain, the angle γ can be taken as approximately equal to $\tan\gamma$. Beyond this limit other definitions are used.

Although strain is often caused by stress, it may also arise from other causes. For instance, it may be caused by a variation of temperature, in which case it is called temperature or *thermal* strain. In some materials a change of moisture content will cause *hygroscopic* strain. Concrete and timber exhibit this characteristic to a marked degree.

Problems

1.1   A straight bar of diameter 0.8 in sustains an axial tensile force of 12,000 lb. Find the stress on a normal cross-section of the bar. What type of stress acts on the cross-section?

1.2   A bar has a circular cross-section $\frac{3}{4}$ in diameter. If there is a uniform tensile stress of 12,000 psi over this section, what is the total force in the bar?

1.3   A bar has a rectangular cross-section 1.5 in wide and 2.5 in deep. The stress on the cross-section varies linearly from 12,000 psi at the top of the bar to zero at the bottom. The stress is constant across the width of the bar.

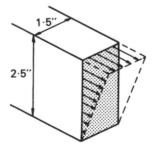

Fig. P1·3

(*a*) Find the resultant force on the cross-section and the position at which it acts.

(*b*) Replace this resultant force by a force N acting at the centre of the cross-section plus a couple M.

1.4   A bar has a cross-section in the shape of an isosceles triangle *ABC* of width 3 in and depth 6 in (Fig. P1.4). The cross-section sustains tensile stress which varies linearly from zero at *C* to 10,000 psi at the edge *AB*. The stress is uniform across any line parallel to *AB*.

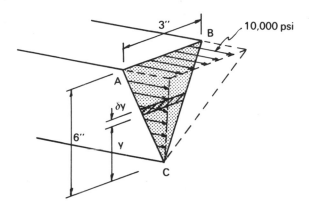

Fig. P1·4

What is the force on a strip of area parallel to *AB* whose width is δy and whose distance from *C* is *y*? What is the total force on *ABC*, and where does the resultant force act?

1.5   The cross-section of a bar is a rectangle 6 in wide and 9 in deep. The compressive stress on the section varies parabolically over the depth of the bar, being zero at the top and bottom and a maximum of 2000 psi at mid-depth (Fig. P1.5). The stress is uniform across the width. Find the total compressive force on the cross-section.

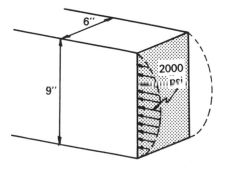

Fig. P1·5

1.6 An I-section beam has flanges 8 in wide and $1\frac{1}{2}$ in deep. The web is 15 in deep and 1 in wide. The overall depth of the beam is 18 in. The tensile stress on the cross-section varies linearly from 11 ksi at the top to 2 ksi at the bottom. Find the magnitude and position of the resultant of these stresses.

1.7 A rectangular plane $ABCD$ is 2 in × 6 in. There is a uniform shear stress of 4000 psi over the plane (Fig. P1.7). Find the total shear force on $ABCD$.

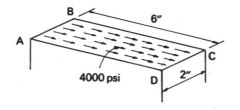

Fig. P1·7

1.8 The cross-section of a bar is in the shape of a semicircular arc of mean radius 10 in and thickness $\frac{1}{4}$ in (Fig. P1.8). The cross-section sustains a shear stress of uniform intensity 2000 psi, the direction of which is everywhere parallel to the circumference of the circle.

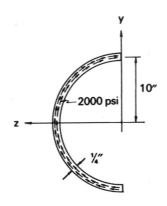

Fig. P1·8

(*a*) Find the magnitude and position of the resultant of these stresses.

(*b*) Find the force S acting through the centre of the circle, and the couple T which are together equivalent to this resultant.

1.9 A bar 80 in long decreases in length by 0.2 in. What is the strain in the bar, assuming that the strain is uniform along the length?

1.10 A bar of steel 200 in long has its temperature raised 60°F throughout its length. If the coefficient of linear expansion is 12×10^{-6} per °F, find the thermal strain in the bar and the total increase of length.

1.11 A bar AB, 160 in long, is originally at 40°F. It is heated in such a manner that the temperature varies linearly from 100°F at A to 40°F at B. Find (a) the thermal strain 40 in from A and (b) the total extension of the bar. Take $\alpha = 8 \times 10^{-6}$ per °F.

1.12 A cube of material is 3 in along each edge. If it suffers a tensile strain of 0.0005 in each direction, find the increase of volume of the block (volumetric deformation). Express this deformation as a fraction of the original volume (volumetric strain).

1.13 A plank of wood is 10 ft long, 15 in wide and 1 in thick (Fig. P1.13). During the drying process it suffers shrinkage strains of 0.0005 in the length (x direction), 0.04 in the width (y direction) and 0.07 in the thickness (z direction). Express the total decrease of volume as a fraction of the original volume. Also express the decrease of area of the top of the plank (the xy plane) as a fraction of its original area.

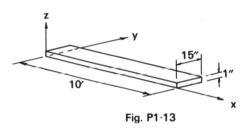

Fig. P1·13

1.14 A block of brass is 2 in cube. It is given a shearing strain of 0.0005 as shown in Fig. P1.14. What is the displacement of plane BC relative to plane AD? By how much does the diagonal AC extend?

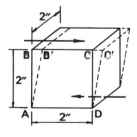

Fig. P1·14

CHAPTER 2

Analysis

2.1 Objective of Analysis

The general purpose of making an analytical study was discussed in Chapter 1. We may briefly restate the objective as the determination of stresses and strains throughout the structure. From these we can calculate its deformed shape and its margin of safety against collapse. As indicated previously, an accurate determination of strains and stresses at all points is usually impossible and simplifying assumptions are made. These will be discussed in more detail in later chapters.

When forces are applied to a body and the body deforms, the forces will do work. If the body behaves elastically, it will return to its original shape and size upon removal of the forces. The energy corresponding to the work done in deforming the body is thus recovered. Since the energy is potentially recoverable from the deformed body, we say that it has been stored in the form of *elastic strain energy*.

The calculation of this quantity, strain energy, is used as a basis for some methods of analysis. Strain energy is also an important consideration in the design of certain components and structures. It can be calculated once the stresses and strains are known, and is a derived quantity rather than a fundamental one. The question of stability is also important. This will be discussed in Chapter 10.

2.2 The Method of Analysis

The solutions of different problems vary in difficulty and complexity. Nevertheless, they all follow fundamentally the same process. Unless this procedure is clearly understood, the student will find the more complex problems very difficult to solve. For this reason, the emphasis in the present book is on the steps of the process, rather than on the answer to any particular problem. In the main, we shall study fairly simple problems so that the process may be easily appreciated.

For the purpose of analysis we imagine a structure to be composed of a number of elements. As a result of external loads (or other influences), these elements become deformed. The deformed elements will fit together to make a structure slightly different in shape to the unloaded structure.

In the simplest type of problem we can therefore proceed as follows:

1. Choose the elements into which we propose to divide the structure.

2. Find the force on each element from the given external forces.

3. From these element forces compute the element deformations.

4. Find the deformed shape of the complete structure from the deformations of the elements.

Step 1 will be considered further in the next section.

Step 2 calls for a knowledge of the principles of statics. It will be assumed that the student is already familiar with these. In the case of bars, the element forces are in fact the stress-resultants (axial force, bending moment, etc.) at

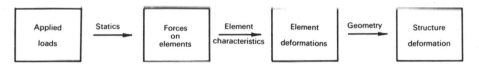

Fig. 2·1

the various cross-sections. The manner in which these are determined, by the use of statics, from the external forces is summarized in Appendix A.

Step 3 presupposes that the force-deformation relation for the chosen elements is already known. The determination of these relations will form the main subject of discussion in Chapters 3, 4, 5 and 6.

Step 4 calls for a knowledge of the principles of geometry. It will be assumed that the student is already familiar with these. (Perhaps this is one of the assumptions which is not justified.)

The steps in this simple analysis could then be illustrated diagrammatically as in Fig. 2.1.

Why is this simple scheme not always possible? We note first that it is necessary that the deformed elements must be capable of being fitted together to form a continuous, though deformed, structure. This requirement is known as the *Principle of Compatibility*. This requirement may or may not impose restrictions upon the deformations of the individual elements. Let us illustrate this by two examples.

In Fig. 2.2 we have a bar of homogeneous material and having a cross-section of 3 sq in. End forces of 75 kips are applied. The bar is divided into elements, each of which is a short length of bar. The forces N on a typical element are shown. It is obvious from statics that $N = 75$ kips. Under the action of the N forces each element stretches and the bar becomes longer. It does not matter how much each element stretches, they will still fit together end to end.

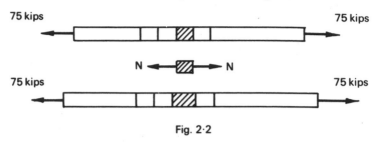

Fig. 2·2

Fig. 2.3 shows three bars joined together at the ends. The middle bar is made of material D and the outer bars are made of material C, which is much stiffer than D. The cross-section of each bar is 1 sq in and the whole assembly is subjected to end forces of 75 kips. The structure is divided into elements as shown. Forces on typical elements are N_1, N_2, and N_3, but statics does not tell us the value of these explicitly. If we were to assume that $N_1 = N_2 = N_3 = 25$ kips then the middle element would stretch more than the outer elements, since D is less stiff than C. So when we come to put the deformed elements together they do not fit. The upper and lower bars have small gaps between the elements.

Clearly the force of 75 kips must be apportioned between elements C and D so that *after the elements deform they still fit together*. It is also necessary that $N_1 + N_2 + N_3 = 75$. We do get some information from statics but not enough

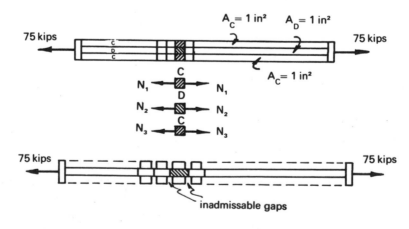

Fig. 2·3

to enable us to find the element forces explicitly. We must use the compatibility condition as well.

For this type of problem we can proceed as follows:

1. Choose the elements into which we propose to divide the structure.

2. Write down the static relations between the (unknown) element forces and the external forces.

3. Calculate the element deformations in terms of the element forces.

4. Write down the equations which express the fact that the element deformations are compatible.

5. Solve the equations from 2 and 4 (there will be just sufficient) and thus find the element forces and deformations explicitly.

6. Find the deformed shape of the complete structure from the deformations of the elements.

It can be seen that in this latter problem, where the elements are not free to deform independently of their neighbours, the determination of element forces and element deformations have become interrelated problems. An important consideration is the choice of elements which will lead to a successful solution of the problem.

2.3 Choice of Elements

For a particular problem, the correct choice of elements is perhaps the most important step in the analysis. In the most general type of problem it is necessary to break the structure down into infinitesimal elements. If we are using Cartesian co-ordinates each element will have dimension $dx \times dy \times dz$. Although the analytical process then follows the lines set out above, we often encounter mathematical difficulties with such problems and they will not be studied at this stage.

We shall be concerned with the behaviour of bars and these will be broken into elements contained between two cross-sections close together (Fig. 2.4). A typical element will be referred to in later chapters as an *elemental slice* of the bar. Each element has an infinitesimal dimension dx in the direction of the bar length, but its other two dimensions are finite although small. The introduction of these finite dimensions greatly simplifies the calculations, but at the same time it necessitates making assumptions about the distribution of strains across the finite faces of the element. Inevitably these assumptions will involve approximation. As long as they do not lead to gross inconsistencies, and as long as they are reasonably in accord with experimental observation, the assumptions can be maintained. When we cannot make a reasonable guess as to the strain distribution, or when we require a more accurate result, the elementary slice of Fig. 2.4 must be further subdivided, with consequent increase in the analytical work.

Where we can get by with elements having one infinitesimal dimension and two finite (small) dimensions, we refer to the structure as a bar structure or one-dimensional structure. When we use an element such as that in Fig. 2.5 with two infinitesimal dimensions and one finite (small) dimension, we say the structure is two-dimensional. These latter structures are also called plates (flat) or shells (curved). Finally, with elements $dx \times dy \times dz$, the structure is

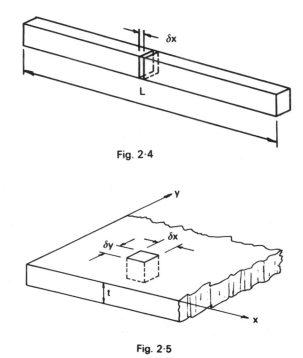

Fig. 2·4

Fig. 2·5

called three-dimensional. These terms depend not only on the appearance of the structure, as such, but also on the loading. Thus, in the neighbourhood of point loads, simplifying assumptions become invalid, and consequently any structure must be treated as three-dimensional, if the stresses and strains are required in the vicinity of the point loads. Special conditions, such as radial symmetry, sometimes permit a choice of elements which simplifies the particular problem.

When a bar or shell has been fully analysed by these means, it may be used as an element in a more extensive structure. For instance, a frame is a structure composed of bars, and in the analysis of the frame, the bars can be regarded as the elements. In this way the analysis of simple components can be used as a basis for the solution of more complex problems. This aspect of analysis is referred to briefly in Chapter 9.

2.4   Compatibility

The principle of compatibility imposes certain restrictions upon the deformation of the elements of a body, or upon the deformation of the body as a whole. Normally a structure is expected to remain continuous under load, temperature

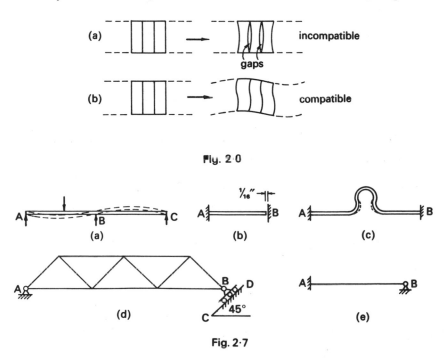

Fig. 2·0

Fig. 2·7

change and so on, and consequently the elements of the structure must deform in such a way that they still fit together without gaps. Naturally, these conditions will not apply when the body becomes cracked or fractured.

In Fig. 2.3 we have seen an example where, unless the increases in the lengths of adjacent elements are compatible, the elements cannot be fitted together. Fig. 2.6 illustrates the deformation of elements of a bar. These have finite dimensions in the plane of the bar cross-sections. Consequently, unless the finite faces of adjacent elements deform similarly, compatibility will be violated. The deformation of the complete structure is also governed by conditions which are given in any theoretical problem. Several external compatibility conditions are illustrated in Fig. 2.7. Fig. 2.7 (a) shows a two-span beam resting on supports at A, B and C. The implication here is that the beam cannot suffer any vertical movement, either upwards or downwards, at these three points. Hence, if a load W is placed on the beam it must deform into a curve which passes through the points A, B and C of the original undeformed shape.

Fig. 2.7 (b) shows a bar lying between two rigid supports A and B a fixed distance apart. Initially there is a gap of $\frac{1}{16}$ in between the end of the bar and the support B. If the bar is heated it will expand, and the total expansion must be less than or equal to $\frac{1}{16}$ in. Fig. 2.7 (c) shows an expansion bend of a type common in pipes which are to carry fluids at high temperatures. As the pipe is heated its shape will alter but if it is clamped at A and B, the distance between A and B must remain constant. This will impose a condition on the deformation at the loop. The gap at C will be reduced and the loop will suffer some bending.

Such restrictions upon movement at support points are indicated by conventional signs. The pin support at A in Fig. 2.7 (d) indicates that the frame, or truss, can rotate around this joint but cannot move vertically or horizontally. At B a roller joint is shown. This implies that the point B can move in the direction of the roller plane CD but not perpendicular to it. The rollers are intended to prevent motion away from the plane as well as motion towards the plane. In Fig. 2.7 (e), the fixed support at A indicates that rotation as well as translation of that end of the beam is inhibited. The pin support at B permits rotation but not translation. The beam is said to be *simply supported* at B and *direction-fixed* or *built-in* at A.

In practical problems these support conditions are not always known with certainty. However, some reasonable assumptions must be made before the analysis can proceed. It is then recognized that the results of the analysis are dependent upon the fulfilment of the assumptions.

2.5   Material Properties

When computing the deformation of an element we need to consider both the size of the element and the strain (which is the deformation of an element of unit size). The three main causes of strain are (*a*) stress (*b*) temperature change and (*c*) change of moisture content. The quantitative relationships between strain and each of these other quantities is expressed in the form of *material properties*. The nature of the relationships is discussed in books on materials science, and it may eventually be possible to derive these material properties from the physical structure of the material. At the present time, however, these material properties are determined by experiment.

The relation between stress and strain for a given material is usually quite complex. A detailed study of this subject is essential if a clear understanding is to be achieved. It is intended here to do little more than define certain terms which must be used in analysis and which appear in examples and problems.

The experiments for the determination of the stress–strain relation are usually carried out on long narrow bars in which the stress is uniaxial. Whether the results are valid for other conditions of stress is a question which needs careful consideration. This will be referred to again in Chapter 8.

Suppose that the results of such an experiment can be expressed as a simple graph (Fig. 2.8). This is the stress–strain graph for the material. For use in analysis it is more convenient if the graph can be replaced by an algebraic expression either of the form $\sigma = f_1(\epsilon)$ or of the form $\epsilon = f_2(\sigma)$. Frequently, by making small approximations to the shape of the graph, fairly simple algebraic expressions can be obtained.

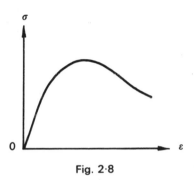

Fig. 2·8

For most materials, the portion of the graph for some distance from the origin is either straight or nearly straight. Provided the stresses in the structure remain within this *proportional range* the relationship will be in the form

$$\sigma = E \cdot \epsilon \tag{2.1}$$

for direct stresses and strains, or

$$\tau = G \cdot \gamma \tag{2.2}$$

for shear stresses and strains. The constant E is known as *Young's modulus of elasticity*, and the constant G is called the *shear modulus of elasticity*. These elastic moduli are a measure of the material's resistance to deformation. Perhaps the term "stiffness constant" would be better than "modulus of elasticity".

For some materials the graph can reasonably be approximated by two straight lines as in Fig. 2.9. The region AB is then the proportional range and A is the *yield point*. The stress in the range AB is the *yield stress* (σ_y) and the strain at A is the yield strain (ϵ_y).

For strain-hardening materials, the proportional range is followed by a range in which the graph is curved (Fig. 2.10 (a)). If the curve is not too pronounced it may be possible to represent the complete behaviour by two straight lines OA and AC. Sometimes the strain-hardening range takes place after a yield range (Fig. 2.10 (b)).

Some of the examples and problems of later chapters are based on materials for which the complete stress–strain graph is approximately parabolic, with no proportional range.

The slope of the stress–strain graph at any point K (Fig. 2.11) is called the *tangent modulus* at K and is denoted by E_t. Its value is $d\sigma/d\epsilon$ and it represents the instantaneous stiffness, so to speak, of the material at this particular stress. It is of great importance if we wish to discuss incremental deformation of a

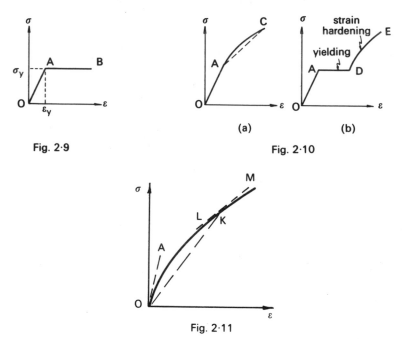

Fig. 2·9   Fig. 2·10

Fig. 2·11

structure due to incremental loading. The slope of the tangent OA at the origin is sometimes called the *initial tangent modulus*, E_0. Clearly this will coincide with Young's modulus or the shear modulus if the material has a proportional range. The slope of the line OK (Fig. 2.11) is called the *secant modulus* at K. This gives a measure of the average stiffness of the material up to point K.

When the stress is within the proportional range (if any), it is often said that the material is behaving *linearly*. It is also said to be obeying Hooke's Law. Conversely, where stress and strain are not proportional the term *non-linear behaviour* is used. These terms are convenient although not strictly accurate.

The foregoing remarks are based on the supposition that the experimental stress–strain curve is a simple line as in Fig. 2.8. This will be the case when the test specimen is subjected to monotonically increasing load up to failure over a fairly short period of time. However, if at some stage after the proportional range has been exceeded the stress is reduced or removed, the test specimen may not retrace the graph back to the origin. Instead, the graph may pursue a path *GH* (Fig. 2.12) parallel to the initial proportional range *OA*. The part *GH* is called an *unloading curve* and its slope will be *E* in the case of direct stress or *G* in the case of shear stress.

If the stress is of long duration, the corresponding strain may not have a specific value but may increase with time. Such an increase of strain over a time period is called *creep*. If the stress is within the proportional range the creep will usually be negligible. For higher stresses the creep strain can represent a substantial increase over the instantaneous strain. For moderate

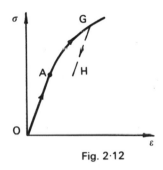

Fig. 2·12

stresses the creep strain may develop over months or even years, while for very high stresses the creep may take place in a matter of minutes.

It is re-emphasized that the above descriptions do not necessarily correspond in detail to the actual behaviour of the test specimen, but represent a close approximation which is more convenient for use in analytical work.

The question of strains resulting from temperature change is dealt with in physics. The relation is expressed by a linear coefficient of thermal expansion (α) such that

$$\epsilon = \alpha T \tag{2.3}$$

where T is the temperature rise. This is equivalent to assuming that the strain–temperature relation is a linear one with a constant of proportion equal to α. This linearity may not be valid if the range of temperature is very large but it is sufficiently accurate for most problems.

The relation of strains to moisture variation in such materials as concrete and timber is extremely complex. Where such strains are introduced in problems, the necessary numerical data will simply be stated.

2.6 Principle of Superposition

The deformation of a structure caused by several loads acting together is often conveniently obtained by computing the deformation due to each load separately, and then adding the results. This method is known as the *principle of superposition*. It is most important to bear in mind that this principle is valid only if every component of the structure is behaving in a proportional manner, i.e., if the imposed force is proportional to the deformation produced.

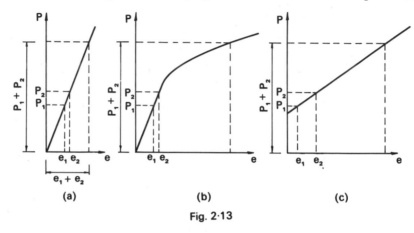

Fig. 2·13

Fig. 2.13 (a) shows a force–deformation graph for a particular component. This component has a linear response. Individual loads P_1 and P_2 cause deformations e_1 and e_2 respectively. From similar triangles it is clear that load $(P_1 + P_2)$ will cause a deformation $(e_1 + e_2)$. Fig. 2.13 (b) illustrates a non-linear relationship and it is obvious that the above proposition is not valid. Fig. 2.13 (c) illustrates a force–deformation relation which is linear, though not proportional. Here, again, the deformation due to the combined load is not the sum of the deformations due to the individual loads.

Clearly, *linearity* of the force–deformation relation is not sufficient. The relation must be *proportional* if the principle of superposition is to be valid. It might be noted that if we are interested only in the result for the combined loads, then it is only required that the *resultant* force and deformation shall be in the proportional range. Whether individual loads and deformations fall within this range is of no consequence.

Problems

2.1 A bar is made of mild steel having a yield stress of 38,000 psi. If the area of cross-section of the bar is 1.6 in² what tensile force will just cause the bar to yield?

2.2   A bar 50 in long and having a cross-section 0.04 in² is made of hard-drawn steel for which the stress and strain are related by the equation

$$\epsilon = \left(\frac{\sigma}{31 \times 10^6}\right) + 80\left(\frac{\sigma}{31 \times 10^6}\right)^2 \quad (0 < \sigma < 190{,}000)$$

where σ is measured in psi. If a tensile force of 3.6 kips is applied to the bar find
   (a) the stress on a cross-section;
   (b) the strain in the bar;
   (c) the total elongation of the bar.

2.3   What force would be required to produce a total elongation of 0.35 in in the bar of problem 2.2?

2.4   For the material of problem 2.2 find
   (a) the initial tangent modulus;
   (b) the tangent modulus when the stress is 10^5 psi;
   (c) the secant modulus when the stress is 10^5 psi.

2.5   A rigid girder ABC (Fig. P2.5) is supported as shown by three bars of equal length. If the lengths of AD and CF increase by 0.03 in and 0.20 in respectively, find the extension of bar BE if the girder ABC remains straight.

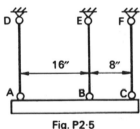

Fig. P2·5

2.6   In Fig. P2.6, the bar AC is twice as long as the bar BD. The bars AC and BD extend in such a way that AB remains horizontal. Find the ratio of the strains in AC and BD.

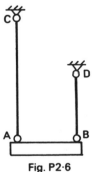

Fig. P2·6

2.7 In Fig. P2.7, the bars *AC* and *BD* are each 100 in long. The beam *AB* is constrained by a guide to move vertically. If *AC* contracts by 0.1 in, what deformation must *BD* suffer so that *AB* remains horizontal?

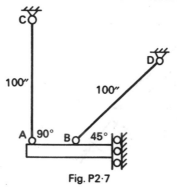

Fig. P2·7

2.8 In Fig. P2.8, *AB* and *BC* are each 100 in long. If *AB* suffers a tensile strain of 0.002 and *BC* suffers a compressive strain of 0.001, find (*a*) the change in length of each bar and (*b*) the horizontal and vertical movement of pin *B*. In addition to an extension, bar *AB* will undergo a very small rotation. During this rotation it can be assumed that *B* moves perpendicular to *AB*. Similar remarks apply to bar *BC*.

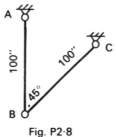

Fig. P2·8

2.9 In Fig. P2.9, the slender bar *AB* is pinned at *A* and welded at *B* to a cross-piece *CBD*. The cross-piece is pivoted at *B* and is also held by the two bars *CE* and *DF*. If the bar *CE* suffers a tensile strain of 0.002 while the length of *AB* remains constant, find the strain in *DF*. As a result of these strains the bar *AB* will bend. Find the angular rotation of the end *B* in radians.

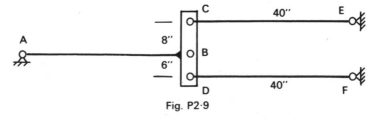

Fig. P2·9

CHAPTER 3

Axial Force and Deformation

3.1 Basic Ideas of Simple Bar Theory

In this book we are concerned with the stresses and deformations in bars. In section 2.3 it was stated that such components can be regarded as being made up of *elemental slices* each of which is contained between two cross-sections *C* and *D* (Fig. 3.1) a distance dx apart in the direction of the axis of

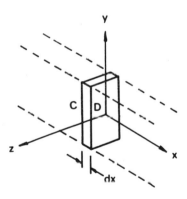

Fig. 3·1

the bar. The y and z axes are taken as the principal axes of the cross-section (see Appendix B). The finite dimensions of such an element in the yz plane make it necessary to introduce assumptions as to the distribution of stress or strain over this plane.

It will be assumed that as the bar as a whole deforms, *the two cross-sections C and D (Fig. 3.1) remain plane and undistorted*. This assumption forms the basis of the Simple Theory of Bars.

If the two faces *C* and *D* remain plane and undistorted, then the element deformation can be conveniently described in terms of six component displacements of face *D* relative to face *C*. These component deformations are indicated in Fig. 3.2. The reader is advised to study these six types of deformation carefully. The first (Fig. 3.2 (a)) is a movement of D (relative to C) in the x direction; the second is a movement of D in the y direction; the third is a movement of D in the z direction; the fourth is a rotation of D around the x axis; the fifth is a rotation of D around the y axis; the sixth is a rotation of D around the z axis. The rotations are shown as positive according to the usual right-hand screw convention.

We note that the total force transmitted across face D (in the complete bar) can also be broken into six components (see Appendix A). It would be very convenient if the displacement of face D in the x direction (axial deformation, Fig. 3.2 (a)) could be associated simply with the stress-resultant in the x direction (axial force), and so on through the six components. Finally, we should have the rotation of D around the z axis (bending deformation, Fig. 3.2 (f)) associated with the bending moment M_z.

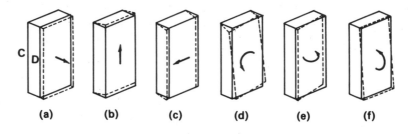

Fig. 3·2

The idea that each component deformation might be associated only with the corresponding stress-resultant and with no other is very attractive. The relation of axial deformation (displacement of face D in the x direction) to axial force is examined in the present chapter. The relation of the shear deformations (displacement of D in the y and z directions) to the respective shear forces is discussed in Chapter 5. Chapter 6 deals with the relation between the torsional deformation and the twisting moment. In Chapter 4, the relation between the bending deformations (rotation of D about the y and z axes) and the respective bending moments is examined.

The basic hypothesis is that the faces C and D of the elemental slice remain undistorted and that each of the six components of deformation is associated with the corresponding component stress-resultant and with no other. In the case of axial deformation and bending deformations, it is found that this assumption is true in special cases and approximately true in a large number of practical cases. For shearing deformations (Chapter 5) it will be found that the hypothesis is quite untenable. There is a way of getting around this difficulty and of computing stresses associated with shear force in many practical problems. As regards torsional deformation it will be found (Chapter 6) that the hypothesis is applicable to bars with certain types of cross-section but not to others.

3.2   Stresses due to Axial Force

Consider a straight bar of homogeneous material but not necessarily of uniform section (Fig. 3.3). The cross-sections CC and DD which define a typical elemental slice are both normal to the axis of the bar and are therefore parallel.* All fibres (elements of material parallel to the bar axis) such as GH in the slice CD are of equal length.

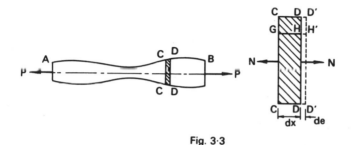

Fig. 3·3

If we now allow the face DD to move in the x direction by an amount de to position $D'D'$, every fibre will extend by de and since they were all originally of length dx the strain will be the same for all fibres namely

$$\epsilon = de/dx$$

Since the material is homogeneous the stress will thus be uniform over the cross-section. Note that this is true whether the stress is within the proportional range or not. The resultant of these uniform stresses is σA (see Chapter 1) and acts at the centroid of the cross-section. Provided we ensure that the axial force, N, acts through the centroid of the cross-section we can equate N to the resultant of the stresses. Then

$$\sigma = \frac{N}{A} \tag{3.1}$$

Here, A is the area of the section at which we are finding the stresses. This area may vary from point to point along the bar as in Fig. 3.3 (a).

---

* This involves an assumption that the bar axis in the vicinity of the element is straight. Provided it is nearly straight the errors involved are negligible.

3.3   Deformation due to Axial Force

The deformation of a bar due to axial force will, of course, depend upon the stress–strain relation for the particular material. We consider, first, deformation within the proportional range. Then the deformation of an elemental length, dx, of the bar is given by

$$de = \epsilon\, dx = \frac{\sigma}{E} dx = \frac{N\, dx}{EA} \tag{3.2}$$

The total elongation of the bar is thus

$$e = \int_0^L \frac{N\, dx}{EA} \tag{3.3}$$

where L is the bar length.

In the case of a straight bar of constant section, A, and subjected to end forces P, the axial force N is everywhere equal to P so that equation (3.3) yields

$$e = \frac{PL}{EA} \tag{3.4}$$

Now, essentially our problem is to relate the bar extension to the applied end forces. Hence, it is convenient to write equation (3.4) as

$$e = fP \tag{3.5}$$

where

$$f = \frac{L}{EA} \tag{3.6}$$

The factor f is called a *flexibility coefficient* and is constant while the material is behaving linearly.

The same relation can alternatively be expressed in the form

$$P = ke \tag{3.7}$$

where

$$k = \frac{EA}{L} \tag{3.8}$$

The factor k is called a *stiffness coefficient*. Evidently it is the reciprocal of the flexibility coefficient in this example.

The terms stiffness and flexibility have more or less their usual everyday significance. Thus, if we take a second bar geometrically similar to the first but made of material with larger E, then we can say that the second bar is more stiff or less flexible than the first. The coefficients k and f will indicate this numerically (equations 3.6 and 3.8).

Also from equation (3.5) we see that the flexibility coefficient, f, could be defined as "the extension produced by unit end force". Similarly, from equation (3.7), the stiffness coefficient could be defined as "the end force required to produce unit extension". However, this last definition must not be taken too literally. For example, if we are working in feet units it is unlikely that the bar can actually be stretched by 1 ft.

The stiffness coefficient may also be defined as the slope of the force–deformation graph of the bar (the graph of P against e). This concept emphasizes the relation of the stiffness coefficient to Young's modulus, which is, in effect, the stiffness coefficient of a cube of the material of unit size.

The flexibility and stiffness coefficients will always depend partly on the material properties (represented here by E) and partly on the geometrical properties of the particular bar (A/L in the case of a uniform bar).

We can express equation (3.2) in the form

$$de = \left(\frac{dx}{EA}\right) N$$

which shows that (dx/EA) is simply the flexibility of the elemental length dx.

For bars of variable cross-section the area, A, in equation (3.3) must be expressed as a function of x.

Example 3.1. Fig. 3.4 (a) shows a bar AB 120 in long. The cross-section is $\frac{1}{2}$ in square at A and $1\frac{1}{2}$ in square at B, and the bar is uniformly tapered. Find the extension of the bar due to tensile end forces of 2000 lb. Also find the flexibility and stiffness coefficients for this bar. The material is homogeneous and for the stresses of this problem it obeys Hooke's Law with $E - 20 \times 10^6$ psi.

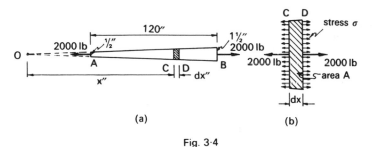

(a)   (b)

Fig. 3·4

Solution. Consider, first, the extension of a typical element (Fig. 3.4(b)). In the case of tapered bars it greatly simplifies the algebra if the distance x is measured from the origin of taper (O in Fig. 3.4 (a)). By similar triangles, O is found to be 60 in beyond A.

Then at C, x inches from O, the width of the bar is $x/120$ in and the area of cross-section is

$$A = \frac{x^2}{14,400} \text{ in}^2$$

The elemental extension is then

$$de = \frac{N\,dx}{EA} = \frac{2000\,dx}{20 \times 10^6}\,\frac{14{,}400}{x^2}$$

$$= \frac{1.44\,dx}{x^2}\,\text{in}$$

The total extension is

$$e = \int_A^B \frac{1.44\,dx}{x^2} = \int_{60}^{180} \frac{1.44\,dx}{x^2}$$

$$= 0.016\,\text{in}$$

To find the flexibility coefficient of this bar directly we could carry out the above with end forces of 1 lb. Clearly this would give

$$f = \frac{0.016}{2000} = 8 \times 10^{-6}\,\text{in/lb}$$

The stiffness coefficient of the bar is

$$k = \frac{1}{f} = 0.125 \times 10^6\,\text{lb/in}$$

Thus, the force required to extend the bar 1 in is 0.125×10^6 lb (provided the material still obeyed Hooke's Law).

If the material does not behave linearly throughout the range of stress, then, instead of replacing ϵ by σ/E in equation (3.2), the appropriate stress–strain function is used.

Example 3.2. The bar of Fig. 3.5 (a) is made of high tensile steel for which the stress–strain diagram is linear up to a stress of 100 ksi ($E = 30 \times 10^3$ ksi) and thereafter obeys the relationship (σ in ksi)

$$\epsilon = \frac{1}{300}(10^{-4}\sigma^2 - 10^{-2}\sigma + 1) \tag{3.9}$$

The bar is circular in section, being 2 in diameter at the ends and 1 in diameter in the central portion.

Find the extension of the bar under tensile end forces of 120 kips.

Solution. Throughout the bar $N = P = 120$ kips.

In the end portions:

$$A = 3.142\,\text{in}^2$$

$$\sigma = \frac{120}{3.142} = 38.2\,\text{ksi}$$

Since this value is within the proportional range, then for each end part

$$e = \frac{NL}{EA} = \frac{\sigma L}{E} = \frac{38.2 \times 36}{30 \times 10^3} = 4.59 \times 10^{-2}\,\text{in}$$

In the central portion:

$$A = 0.785 \text{ in}^2$$

$$\sigma = \frac{120}{0.785} = 152.8 \text{ ksi}$$

This is above the proportional limit, hence the strain will be given by equation (3.9)

$$\epsilon = \frac{1}{300} [10^{-4}(152.8)^2 - 10^{-2}(152.8) + 1]$$

$$= 6 \times 10^{-3}$$

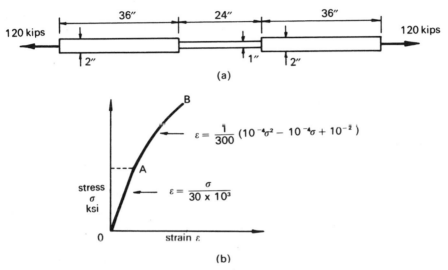

(a)

$$\varepsilon = \frac{1}{300} (10^{-4}\sigma^2 - 10^{-4}\sigma + 10^{-2})$$

$$\varepsilon = \frac{\sigma}{30 \times 10^3}$$

stress
σ
ksi

strain ε

(b)

Fig. 3·5

The extension of the central part is then

$$e = \epsilon L = 6 \times 10^{-3} \times 24 = 14.4 \times 10^{-2} \text{ in}$$

The total extension is

$$e = (4.59 \times 2 + 14.4) 10^{-2}$$

$$= 0.236 \text{ in}$$

It might be noted that at the abrupt changes of section, the basic assumption that the cross-sections remain plane is invalid. The transfer of stress must be confined to the central region of the cross-section. A fair idea of the way the stress is distributed in the bar can be obtained by imagining the bar as a hollow tube and the tensile force as a fluid entering by a point at one end and flowing out by a point at the other. At a short distance from the entry point the flow will spread out over the cross-section of the tube. When the flow approaches the constriction, the lines of flow must converge in order to enter the narrower part. This convergence of flow lines corresponds to a concentration of stress.

In each of the above problems the axial force has been uniform throughout the bar. This will not always be the case. A common cause of variation of N is the self-weight of a vertical bar. In such problems N must be regarded as a function of x in equation (3.3).

The flexibility and stiffness coefficients as defined by equations (3.6) and (3.8) provide a relation between end forces P applied to a bar and the corresponding bar deformation. There are many ways in which a bar can be loaded and many types of deformation which might be of practical interest. Some of these will be investigated in subsequent chapters. Provided the bar behaves in a linear manner, we can select any system of loading and any deformation which is related thereto and express the relationship in the forms

$$e = fP \quad \text{or} \quad P = ke$$

In this chapter we have used the relationship only for end axial loads and the related bar extension.

The concept need not be restricted to the proportional range. If the stresses in any region exceed the proportional limit then the force–deformation graph for the bar will be non-linear. In the proportional range we speak of the (constant) elastic modulus of the material and the (constant) stiffness coefficient of the bar. Beyond the proportional range we use the term tangent modulus for the slope of the stress–strain curve at any point. In the same way we can speak of the *tangent stiffness* of the bar, signifying the slope of the force–deformation graph. This value will, like the tangent modulus, vary with the load. For a uniform bar with end loads we shall have

$$k_t = \frac{E_t A}{L} \quad \text{and} \quad f_t = \frac{L}{E_t A}$$

where E_t, k_t and f_t are the tangent modulus, tangent stiffness and tangent flexibility respectively.

Example 3.3. Find the tangent stiffness and tangent flexibility for the bar of example 3.2 at the given load of 120 kips.

Solution. For $0 < \sigma < 100$ ksi, $E = 30 \times 10^3$ ksi.
∴ for each end portion

$$f_t = \frac{L}{EA} = \frac{36}{30 \times 10^3 \times 3.142} = 3.82 \times 10^{-4} \text{ in/kip}$$

For $\sigma > 100$ ksi the stress–strain law is

$$\epsilon = \frac{1}{300} (10^{-4} \sigma^2 - 10^{-2} \sigma + 1)$$

$$\frac{1}{E_t} = \frac{d\epsilon}{d\sigma} = \frac{1}{300} (2 \times 10^{-4} \sigma - 10^{-2})$$

For $\sigma = 152.8$ ksi

$$\frac{1}{E_t} = \frac{1}{300} \left(\frac{2 \times 152.8}{10^4} - \frac{1}{10^2} \right) = 6.84 \times 10^{-5}$$

∴ for the central portion

$$f_t = \frac{L}{E_t A} = \frac{6.84 \times 10^{-5} \times 24}{0.785} = 2.05 \times 10^{-3} \text{ in/kip}$$

The flexibility of the whole bar is thus

$$f_t = (2 \times 3.82 \times 10^{-4}) + (2.05 \times 10^{-3})$$
$$= 2.81 \times 10^{-3} \text{ in/kip}$$

The bar stiffness is

$$k_t = \frac{1}{f_t} = 355 \text{ kips/in}$$

Note that the flexibility of each part is the extension under unit load. When the several parts are joined together end to end (in series) the extension of the whole is the sum of the extensions of the parts, so the flexibility of the whole is the sum of the flexibility of the parts. The stiffness coefficients are the reciprocal of the flexibility coefficients. Hence, the stiffness of the whole bar is less than that of any component.

On the other hand, if several bars are joined together side by side (in parallel) the reverse situation applies. The stiffness of the combined bar is now the sum of the individual bar stiffnesses. The reciprocals of individual flexibilities are added to obtain the reciprocal of the flexibility of the compound bar.

3.4 Thermal Deformation

Temperature change will induce strain and, therefore, axial deformation, in a bar. The thermal strain is

$$\epsilon = \alpha T$$

and hence the total deformation is

$$e = L\alpha T \tag{3.10}$$

where T is the temperature rise and α is the coefficient of linear expansion.

If axial force and temperature variation occur simultaneously the total deformation will be

$$e = \frac{NL}{EA} + L\alpha T \qquad (3.11)$$

provided the stress is within the proportional range.

Sometimes the free expansion or contraction of a bar is restricted by end restraints. A change in temperature will then be accompanied by induced stresses which can be calculated directly from equation (3.11).

Example 3.4. A bar of length 120 in is prevented from expansion or contraction by rigid end restraints. Find the stress induced by a temperature rise of 60°F. $\alpha = 8 \times 10^{-6}$ per °F and $E = 20 \times 10^{6}$ psi.

Solution. Equation (3.11) may be written

$$e = \frac{\sigma L}{E} + L\alpha T$$

In the present problem the total expansion is zero.

$$0 = \frac{\sigma L}{20 \times 10^{6}} + 8 \times 10^{-6} \times 60L$$

Thus $\sigma = -9600$ psi

3.5   Strain Energy

A pair of forces applied to the ends of a bar cause the bar to extend or contract. During this deformation work will be done. If we plot a graph of load against bar extension (or contraction), the area under this curve will be equal to the work done by the applied forces. This work is called the *work of deformation*. In Fig. 3.6 the area OQN gives the work of deformation up to the point Q.

Some of the energy corresponding to this work can be recovered if the bar is unloaded. The curve of unloading for many materials would be the straight line QM, which is parallel to the initial straight portion OA (or parallel to the tangent at the origin). In such a case the recovered energy would be represented by the area MQN. This recoverable energy is called *elastic strain energy*.

Of the remainder of the work done, some has been used to cause inelastic deformation in the form of crystal dislocations or other physical changes in the structure of the material. This energy is called *inelastic strain energy*. The remainder of the work of deformation has been converted into heat energy, and may be considered as lost to the system.

While the bar is loaded within the proportional range, all the work of deformation is stored as elastic strain energy, since it is all recoverable. If we assume that Hooke's Law applies, then the end force and the extension are

proportional to one another. Thus, if the bar is loaded up to a point such as
K (Fig. 3.6) the elastic strain energy is given by

$$U = \tfrac{1}{2}P \cdot e \qquad (3.12)$$

where e is the extension caused by the force P and U is the strain energy stored
in the bar.

It might be thought that if a force P causes an extension e, the work done
would be Pe. We note, however, that the act of extending (or contracting) must
occupy a certain amount of time and that in the present instance we are
considering equilibrium conditions at every instant during that period of time.

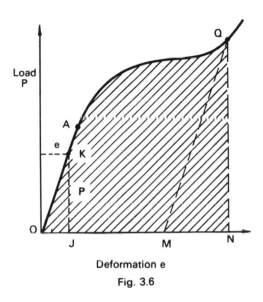

Fig. 3.6

Now the internal stresses build up only as the strain increases, so that initially
when the strain is zero the stresses are also zero. Thus, if the full force P were
to be applied from the beginning when $e = 0$, equilibrium would not prevail.
To maintain equilibrium throughout, the application of the force must *keep
pace* with the extension (see Fig. 3.6), which means that the bar is subjected
not to a constant force P but to a force which increases gradually from 0 to P.
It is easier to appreciate the necessity for applying the force gradually when
the extension is large enough to be clearly visible. We may suppose that the
bar is made of rubber (the elastic modulus of which is small) or that it is
replaced by a spring (Fig. 3.7). If the force P is applied all at once, an oscillation
will be set up, showing that equilibrium has been destroyed. To obtain a static
deflection of the spring the force must be applied gradually. Exactly the same

conditions exist when a force is applied to a short metal bar, but since the amplitude of the oscillation is much smaller it is less readily appreciated.

For a bar of given dimensions and known material the relation between P and e can be found. Hence, from equation (3.12) the strain energy can be

Fig. 3·7

re-expressed in terms of either P or e. For a bar of length L and constant section A,

$$e = fP \quad \text{or} \quad P = ke$$

The strain energy in such a bar would then be

$$U = \tfrac{1}{2}Pe$$
$$= \tfrac{1}{2}P \times fP$$
$$= \tfrac{1}{2}fP^2 \tag{3.13}$$

or

$$U = \tfrac{1}{2}Pe$$
$$= \tfrac{1}{2}(ke)e$$
$$= \tfrac{1}{2}ke^2 \tag{3.14}$$

For the particular tapered bar of example 3.1, it was found that the force of 2000 lb caused an extension of 0.016 in. The strain energy in the bar would be

$$U = \tfrac{1}{2}Pe$$
$$= \tfrac{1}{2} \times 2000 \times 0.016$$
$$= 16 \text{ in-lb}$$

In example 3.2 a bar was considered in which the centre portion was strained beyond the elastic limit, while the end portions were not. In each end part a force of 120,000 lb caused an elastic extension of 0.0153 in, giving rise to a strain energy

$$U = \tfrac{1}{2} \times 120,000 \times 0.0153$$
$$= 918 \text{ in-lb}$$

For the central part, the force–extension graph would take the same shape as the stress–strain graph OB (Fig. 3.5 (b)) and the work done would be equal to the area under this force–extension graph. Not all of this work would be recoverable, as explained above. The work of deformation per unit volume is equal to the area under the stress–strain curve up to the actual stress in the bar (152.8 ksi) and this is found to be 406 in-lb/in^3. The volume of this part of the bar is 18.79 in^3, so that the total work of deformation is 7623 in-lb.

The recoverable part of this energy can be calculated if we assume that upon unloading, the stress–strain curve has a constant slope of 30×10^3 ksi (Fig. 3.5 (b)). The recovery of energy per unit volume will then be

$$\tfrac{1}{2}\sigma\left(\frac{\sigma}{E}\right) = \tfrac{1}{2} \times \frac{152.8^2}{30 \times 10^3} = 0.389 \text{ in-kips/in}^3 = 389 \text{ in-lb/in}^3$$

The total energy recovered will be 7309 in-lb.

It will be observed that the work of deformation is far greater in the central portion of the bar than in the end portions. This is mainly because of the smaller cross-section (hence greater deformation) and to a small extent due to the inelastic deformation. If the stress–strain curve had departed more from the initial straight line, the latter effect would have been much more pronounced.

When the axial force varies along the bar, we must find the energy within a typical element and then find the total energy by integration. If the axial force at a typical section is N the deformation of the element is $(N dx)/(EA)$. The energy within the slice is then

$$dU = \tfrac{1}{2}N\frac{N\,dx}{EA}$$

$$= \frac{N^2\,dx}{2EA}$$

Hence

$$U = \int_0^L \frac{N^2\,dx}{2EA} \tag{3.15}$$

The energy per unit volume of material is given by

$$U_1 = \tfrac{1}{2}\sigma\epsilon = \tfrac{1}{2}\sigma^2/E = \tfrac{1}{2}\epsilon^2 E \tag{3.16}$$

Or, if the stress is beyond the limit of proportionality, we may say that the area under the stress–strain curve represents the "work of deformation" per unit volume. When the stress is uniform throughout a bar, the total energy can readily be found by multiplying the area under the stress–strain curve by the volume of the bar.

It is important to note that the strain energy in an element of volume depends only upon the stress and strain in the element and not upon the external forces. The strain energy in a body is only equal to the work done by the external forces provided that equilibrium prevails. Equation (3.15) is valid irrespective of whether N is equal to the external load P or not.

3.6   Non-homogeneous Bars

The basic assumption of axial deformation is that the strain is uniform over the cross-section of the bar. For a homogeneous material this leads to a uniform stress distribution. When the material is not homogeneous the stress will vary in proportion to the stiffness of the material.

It frequently happens that material properties vary across the section but not along the length of the bar. For such bars the "stress block" concept (see Chapter 1) is useful.

Example 3.5. A bar has the cross-section shown in Fig. 3.8 (a). Materials C and B both obey Hooke's Law and E_B is six times E_C. If the bar undergoes axial deformation and the stress in C is 2000 psi find the total force on the cross-section and the point at which it acts. It is assumed that no slip occurs between materials B and C so that the strains in the two materials will be equal.

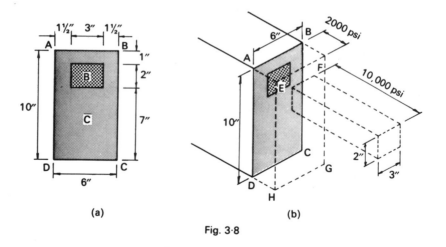

Fig. 3·8

Solution. The stress block, or diagram of normal stresses, will be as shown in Fig. 3.8 (b). The stress ordinates over the region B will be six times as large as those in region C (i.e., $\sigma_B = 12{,}000$ psi) and will therefore exceed σ_C by 10,000 psi.

The simplest division of the stress block is into volume $ABCDEFGH$ with resultant N_1 and the projecting volume with resultant N_2.

$N_1 = 2000 \times 10 \times 6 = 120{,}000$ lb acting at the centre of $ABCD$.

$N_2 = 10{,}000 \times 3 \times 2 = 60{,}000$ lb acting at the centre of B, i.e., 3 in above the centre of $ABCD$.

The resultant force is therefore

$$N = 180,000 \text{ lb acting 1 in above the centre of } ABCD.$$

To accord with the basic hypothesis of uniform strain it would be necessary in such a bar to locate the axial force 1 in above the centre of the cross-section. This is possible since the location of N is arbitrary (see Appendix A).

The concept of a "transformed area" provides a convenient stratagem for problems of this type. One of the materials is replaced by an equivalent area of the other. "Equivalent" means that it has the same stiffness as the original and the same centroid. In the above problem area B might be replaced by an area six times as large but having a modulus equal to that of C, thus leaving the product EA unchanged for part B. The area into which B is transformed must have its centroid at the centre of B. One possible arrangement is shown in Fig. 3.9 in which the shaded area is six times that of B and has the same centroid.

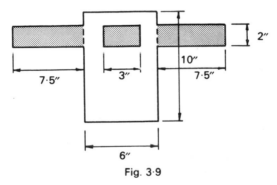

Fig. 3·9

The transformed section has the advantage that it has a constant modulus of elasticity. It is the equivalent homogeneous bar made of material C. The advantage is seen if we try to find the stresses produced by an axial force of, say, 200,000 lb.

The total area of the transformed section (Fig. 3.9) is 90 in². Hence the stress (in material C) is

$$\sigma_C = \frac{200,000}{90} = 2222 \text{ psi}$$

In material B the stress will be six times as great. The axial force must act at the centroid of the transformed section which is 6 in above the bottom.

The original section (Fig. 3.8 (a)) can be transformed to an equivalent section of material B. For this, area C must be replaced by an area one-sixth of the size and having the same centroid as C.

Suppose a bar has a modulus of elasticity which varies linearly from E_1 to E_2 (Fig. 3.10 (a)). We can replace the cross-section by the equivalent section having a modulus E_1 (Fig. 3.10 (b)). The bar with this section can then be treated as a homogeneous bar with modulus E_1.

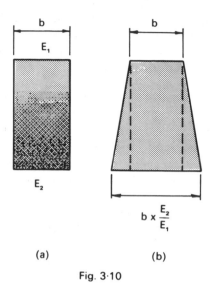

Fig. 3·10

3.7   The Principle of St Venant

In section 3.3 we considered a bar which had abrupt changes of section (example 3.2). It was noted that the Simple Theory of Bars was invalid at these abrupt section changes and also at the ends of the bar where the external force was applied at a point instead of being distributed uniformly over the end cross-section. Does this mean that our solution to this problem is slightly inaccurate, or is it hopelessly wrong?

Let us consider first the end face of the bar. If the external force P were uniformly distributed, all would be well. We can consider the actual load, which is applied at a point (or at least over a small area) as being the combination of two systems (Fig. 3.11). The first is a distributed load of P/A which agrees with our hypothesis. The second is a self-equilibrating system, or a system whose resultant is zero. The question now is to what extent this second system invalidates our simple theory.

We make use of the Principle of St Venant which tells us that if a self-equilibrating force system such as that of Fig. 3.11 (c) is applied to a body over an area characterized by a dimension d, it will give rise to stresses which are usually negligible beyond a distance d from where the system is applied.

This is an extremely useful principle. In this instance it tells us that within a distance of one bar diameter from the end of the bar, the effects of system

3.11 (c) will be negligible, so that the stresses will be uniformly distributed across the section irrespective of the manner in which the force P is applied at the end. At the abrupt change of section, the same argument indicates that the stress disturbance will die out within a distance d on either side of the discontinuity. The percentage error involved in making use of the simple theory is thus seen to depend on the ratio of the length to diameter of the sections of the bar. The theory may not be valid within a length d (= diameter) at each end. If the bar is long compared with d this will not matter except in so far as we are interested in the stresses in the disturbed region. These we cannot calculate by this theory.

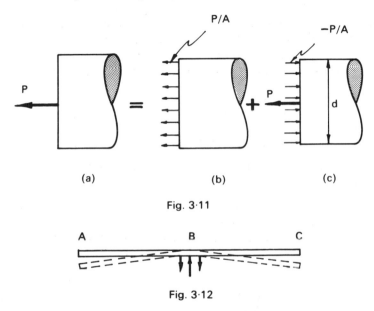

Fig. 3·11

Fig. 3·12

Another way of thinking of the principle is to consider a force system applied to a body within a small region of diameter d. As regards stresses and strains at points beyond a distance d from this region, we need only to know the magnitude and position of the resultant applied force. Within a distance d the stresses and strains will depend upon the actual distribution of the applied forces. At some points this may make an enormous difference.

If the stresses due to a self-equilibrating system such as that of Fig. 3.11 (c) die out within a distance d, so also must the strains. It does not necessarily follow that *displacements* at remote points are negligible. This depends upon the nature of the strains and of the body. The bar of Fig. 3.12 is very long and slender. A self-equilibrating force system applied in the vicinity of B causes a bend which can produce quite large displacements at A and C if these points are far enough from B. The stresses and strains are confined to the neighbour-hood of B, BA and BC rotating as rigid bodies.

Is the principle universally valid? It can always be applied to what we might call *solid* bodies. However, if we regard a pin-jointed truss as a body, then the principle will not, in general, apply to such a body. It is not pertinent to discuss the application of the principle in detail at this stage, but it should be noted that there are instances where it is invalid.

Problems

3.1　A round tie-bar of mild steel is 18 ft long and $1\frac{1}{2}$ in in diameter. It extends $\frac{1}{16}$ in under a tensile force of 15,000 lb. Find the tensile stress in the bar, the value of Young's modulus of elasticity, the strain energy stored in the bar, and the flexibility and stiffness coefficients for the bar (use inch and pound units).

3.2　A steel cable 1500 ft long and of uniform section hangs vertically down a mine shaft. What is the maximum stress in the cable and the extension due to its own weight? Steel weighs 490 lb/ft³. Take $E = 30 \times 10^6$ psi.

3.3　Find the strain energy in the cable of problem 3.2 if the cable has a cross-section of 4 sq in.

3.4　A certain specimen of monel metal is found to have a stress–strain relationship as shown in Fig. P3.4. It is linear up to a stress of 65,000 psi ($E = 23.7 \times 10^6$). Above this stress the relationship is given by the following values.

ϵ	σ (psi)
0.003	70,250
0.004	88,000
0.005	102,000
0.006	112,500
0.007	119,500

During unloading, the stress–strain graph is as shown by the dotted line, which is parallel to the initial straight portion.

A straight bar of this material is 80 in long and has a cross-section of 0.5 in². A tensile force of 50,000 lb is applied gradually to the bar. Find, approximately, the work done by this force in extending the bar. How much of this work is recovered when the force is removed? What has happened to the remainder of the work? Would you describe the force system as conservative or non-conservative?

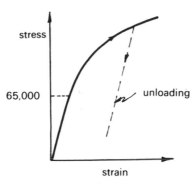

Fig. P3·4

3.5   For the bar of problem 3.4 find the flexibility and stiffness coefficients when the stress is within the proportional range. Also find, approximately, the tangent stiffness and tangent flexibility when the stress is 10^5 psi.

3.6   A steel piston-rod $1\frac{1}{2}$ in diameter is subjected alternately to compressive and tensile forces. The tensile and compressive stresses are each 8000 psi. Two points A and B are marked on the rod 4 ft apart when the rod is unloaded. If $E = 30 \times 10^6$ psi, find (a) the effective load on the piston and (b) the difference between the greatest and least distances between A and B.

3.7   A bar AB is made of a titanium alloy ($E = 17 \times 10^6$ psi). It is 120 in long and is circular in cross-section. It is tapered so that its diameter, d, is given by $d = kx^2$, where x is the distance (inches) from a point 80 in beyond A (see Fig. P3.7). The diameter at B is 1 in. The bar is placed in tension by a force of 3000 lb at each end. Find (a) the maximum stress in the bar and (b) the elongation. Also, find (c) the stiffness coefficient for the bar.

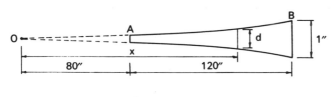

Fig. P3·7

3.8   A uniformly tapered bar of steel 1000 ft long is suspended vertically in a shaft and supports a load of 2000 lb attached to the lower end. The diameter of the bar is $\frac{1}{2}$ in at the lower end and 1 in at the upper end. Find the total extension of the bar due to its own weight and the suspended load. Take the specific weight of steel as 490 lb/ft^3 and E as 30×10^6 psi.

3.9 A bar of square cross-section 5 ft long tapers uniformly from $\frac{1}{2}$ in square at one end to 1 in square at the other. It is made of a material whose stress–strain relationship can be expressed approximately by the equation

$$\epsilon = 8 \times 10^{-8}(10^{-3}\sigma^2 + \sigma)$$

where σ is in psi.

(a) What extension will be caused by end forces of 1500 lb?

(b) Would forces of 3000 lb produce an extension of twice as much, more than twice, or less than twice?

3.10 A straight bar ABC (Fig. P3.10) is cut from a sheet of aluminium $\frac{1}{8}$ in thick. The part AB is 40 in long and $\frac{1}{2}$ in wide, and the part BC is 40 in long and 1 in wide. It is subjected to end forces of 250 lb and the material obeys Hooke's Law ($E = 10 \times 10^6$ psi).

(a) Find the strain energy stored in each of the parts.

(b) If the part AB were made narrower, the axial force remaining the same, would you expect the strain energy to increase or decrease?

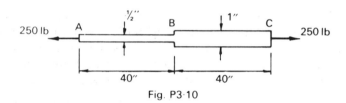

250 lb   A $\frac{1}{2}''$ B   1''   C   250 lb

40″   40″

Fig. P3·10

3.11 Three bars of different materials A, B and C each 10 ft long are joined end to end to form a composite bar 30 ft long. The stiffness coefficients for the bars are $k_A = 10$ kips/in, $k_B = 20$ kips/in and $k_C = 40$ kips/in. Find the stiffness and flexibility of the composite bar.

3.12 A reinforced concrete column is 12 in square and 10 ft long. In each corner is embedded a 1 in diameter bar of mild steel (Fig. P3.12). For concrete $E = 5 \times 10^6$ psi and for steel $E = 30 \times 10^6$ psi. The column carries a load of 350 kips.

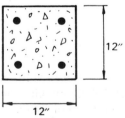

12″

12″

Fig. P3·12

(a) Use the transformed area method to find the stresses in the steel and concrete just after the load is applied. Also, find the strain energy in the system at this time.

(b) The creep characteristic of the concrete is such that under constant stress, the strain at the end of 1 year is 3 times the elastic strain at the instant of loading. In the present problem, if the variation of concrete stress during the year is ignored, the creep will be equivalent to a reduction of E to one-third of its original value. On this assumption, find the steel and concrete stresses one year after the load is applied.

3.13 A concrete beam is 20 ft long, 16 in high and 8 in wide. Due to water movement before the concrete had set, the elastic modulus of the hardened concrete varies from 5×10^6 psi at the bottom to 4×10^6 psi at the top of the beam. On the assumption that axial force produces uniform strain over the cross-section, sketch the corresponding stress distribution. (a) Where must a compressive end force P be located to produce pure axial compression? (b) By how much would the beam contract due to an axial compression of 250,000 lb? and(c) what would be the maximum compressive stress?

3.14 A bar of material A ($E = 10^4$ ksi) is 10 ft long and has a circular cross-section with an outside diameter of 2 in. Embedded in the centre of this material is a rod of material B ($E = 2.4 \times 10^4$ ksi) with a diameter of 1 in.

(a) Use the transformed area method to find the stress in each material when the axial force is 60 kips.

(b) Find the stiffness coefficient for the composite bar.

3.15 A brass bar 1 in wide and $\frac{1}{2}$ in thick is formed into a ring of outside diameter 30 in (Fig. P3.15). The ring is put into compression by a uniform radial force of 160 lb per inch of circumference. Find the axial force in the bar, the stress, and the change in the diameter of the ring. (E for brass = 15×10^6 psi.)

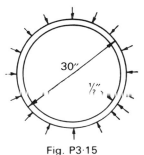

Fig. P3·15

3.16 A bar of nickel is 20 in long and is held between two supports which are assumed to be immovable. If the bar is heated 80°C find the stress induced in the bar. (E is 22×10^6 psi and the coefficient of linear expansion, α, is 10×10^{-6} per °C.)

3.17   A bar is made of material whose stress–strain graph comprises two straight lines. For $0 < \sigma < 7$ ksi the behaviour is proportional with $E = 20 \times 10^3$ ksi. For stresses above 7 ksi

$$\sigma = 10^4 \, \epsilon + 3.5 \quad \text{(where } \sigma \text{ is in ksi)}$$

If the bar is held between immovable supports, find the stress induced in the material by a fall in temperature of 80°F. Take $\alpha = 8 \times 10^{-6}$ per °F.

3.18   A steel rod AB is 48 in long and 1.5 in² in area. It rotates about the end A at 500 rpm.

(a) Calculate the axial force in the rod at x in from end B.

(b) Find the total extension of the rod.

(c) At what speed of rotation will the material yield, and where will yielding begin?

($E = 29 \times 10^6$ psi and the yield stress is 40,000 psi. Steel weighs 0.28 lb/in³.)

Note: Attention should be paid to the units in this question.

CHAPTER 4

Bending

4.1 Stresses due to Bending

In this chapter we discuss the type of deformation referred to previously in Figs 3.2 (e) and (f), and shown again in Fig. 4.1 (b). This is called a *bending deformation* and it results in the bending of the bar as a whole. In many structures this is the major cause of the distortion of the complete structure.

We refer once more to an element of the bar contained between cross-sections C and D a distance dx apart (Fig. 4.1 (a)). Following a procedure analogous to that of the previous chapter, we now consider the effect of rotating plane D relative to plane C about some axis z (Fig. 4.1 (b) and (c)). If z is within the cross-section, this deformation will produce compressive stresses above z and tensile stresses below. Can z be chosen so that these stresses constitute a couple equal to the bending moment M_z caused by the external loading?

For the present we consider a material which is behaving linearly. *In such a case* it can be shown that if O (Fig. 4.1 (c)) is the centroid of the cross-section and y and z are principal axes (Appendix B), then a rotation of plane D (relative

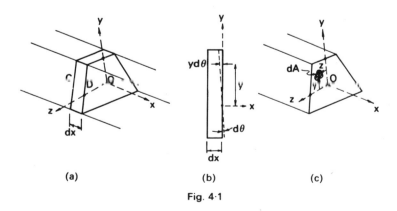

(a) (b) (c)

Fig. 4·1

to C) about the z axis will give rise to stresses whose resultant is a couple around the z axis. Similarly, a rotation about the y axis will induce stresses which form a couple about the y axis.

Irrespective of the axis of the bending moment produced by the external loads, we can resolve this bending moment into a couple M_z around the z axis and a couple M_y around the y axis. We then know that the M_y bending moment will be associated with bending about the y axis, while M_z will be associated with bending about the z axis. In what follows, only the bending moment M_z will be considered since the results for M_y can be obtained merely by interchanging symbols y and z.

The proof of the foregoing is as follows. The rotation of plane D (relative to C) about axis z by an amount $d\theta$ (Fig. 4.1 (b)) causes strains which vary linearly with y.* (This is true irrespective of the material properties.) In fact

$$\epsilon = -\frac{y\,d\theta}{dx} = -\left(\frac{d\theta}{dx}\right)y \tag{4.1}$$

Provided σ is proportional to ϵ,

$$\sigma = -E\left(\frac{d\theta}{dx}\right)y \tag{4.2}$$

or for convenience,

$$\sigma = By.$$

The force acting on an element of area dA at (y,z) (Fig. 4.1 (c)) is

$$dF = By\,dA$$

The resultant force on the section is

$$F = B \int y\,dA \tag{4.3}$$

Since axis z passes through the centroid $\int y\,dA$ is zero, being the first moment of A about z (see Appendix A). Hence the stresses form a couple.

The total moment around the y axis is

$$M_y = \int dF\cdot z = B \int yz\,dA \tag{4.4}$$

Since y and z are principal axes $\int yz\,dA$ is zero, being the product of inertia of A about axes y and z (see Appendix A). Hence the moment of the stresses about the y axis is zero.

The total moment around the z axis is

$$M_z = \int dFy = B \int y^2\,dA \tag{4.5}$$

---

* This assumes that the bar axis is straight, so that sections C and D are parallel and dx is the same for all fibres.

The term $\int y^2 dA$ is the second moment of area (or moment of inertia) of A about axis z, and may be denoted by I_{zz}. Hence

$$M_z = BI_{zz} = \frac{\sigma I_{zz}}{y} \qquad (4.6)$$

While we are dealing with moments about the z axis only, I_{zz} will be called simply I. When M_z is positive, the stresses above the z axis are compressive. So, strictly speaking, equation (4.6) should be written

$$\sigma = -\frac{My}{I} \qquad (4.7)$$

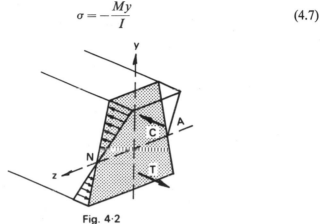

Fig. 4·2

However, it is usually obvious from practical considerations which side of the bar is in tension, and the minus sign is often omitted. Equation (4.7) gives the value of the normal stress at any point, with ordinate y, due to the bending moment M_z. The stresses and strains at all points on the z axis are zero, and this axis is referred to as the neutral axis (N.A.) of bending.

The stresses due to bending are distributed as shown in Fig. 4.2. For positive bending, the stresses above the N.A. are compressive with a resultant C, while the stresses below the N.A. are tensile with a resultant T. They are equal and constitute a couple equal to the bending moment, M_z.

It should be noted that the principal axes of the section are associated with the bending stress problem only while stresses are within the proportional range. Non-linear distributions of stress are examined in section 4.4.

Example 4.1. A steel beam has the cross-section shown in Fig. 4.3. If the bending moment is +750,000 lb-in find the stress at the top and bottom surface and at a point 3 in above the neutral axis.

Solution. By symmetry the neutral axis z–z is at the mid-depth. The moment of inertia is (see Appendix B)

$$I = 405.67 \text{ in}^4$$

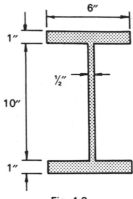

Fig. 4·3

At the top of the beam, $y = +6$ in and

$$\sigma = \frac{-750,000(+6)}{405.67} = -11,100 \text{ psi}$$

At the bottom, $y = -6$ and

$$\sigma = \frac{-750,000(-6)}{405.67} = +11,100 \text{ psi}$$

When $y = +3$

$$\sigma = \frac{-750,000(+3)}{405.67} = -5500 \text{ psi}$$

While the stress distribution is linear, the maximum stresses occur at the points farthest from the neutral axis, namely the extreme fibres. If the ordinates of the lower and upper extremities of the section are y_1 and y_2 respectively,

$$\sigma_{max} = \frac{My_1}{I} \quad \text{or} \quad \frac{My_2}{I}$$

We may put $I/y_1 = Z_1$ and $I/y_2 = Z_2$, where Z_1 and Z_2 are called *section moduli*. Then the extreme fibre stresses are M/Z_1 and M/Z_2.

Example 4.2. A beam has the section shown in Fig. 4.4. Find the section moduli. What are the extreme fibre stresses caused by a bending moment of $-30,000$ lb-ft?

Solution. The centroid is found to be 9.375 in above the bottom. The moment of inertia about the neutral axis is 419 in⁴.

$$y_1 = -9.375 \text{ in} \quad \text{and} \quad y_2 = 4.125 \text{ in}$$

$$Z_1 = \frac{I}{y_1} = 43.7 \text{ in}^3 \quad \text{and} \quad Z_2 = \frac{I}{y_2} = 101.5 \text{ in}^3$$

$$\sigma_1 = \frac{M}{Z_1} = \frac{30,000 \times 12}{43.7} = 8240 \text{ psi}$$

$$\sigma_2 = \frac{M}{Z_2} = \frac{30,000 \times 12}{101.5} = 3550 \text{ psi}$$

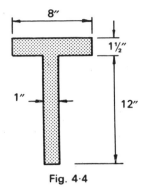

Fig. 4·4

Since the bending moment is negative, we know that the tensile stresses occur at the top. Hence we can say

$$\sigma_1 = -8240 \text{ psi} \quad \text{and} \quad \sigma_2 = +3550 \text{ psi}$$

4.2   Deformation due to Bending

Each elemental slice, which before deformation has parallel faces, after deformation has its faces inclined to one another at an angle $d\theta$. Thus when all the deformed elements are re-assembled they will form a bent bar. The initial centroidal axis LM (Fig. 4.5) *deflects* to the shape $L'M'$. It should be

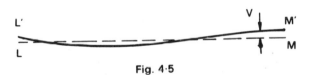

Fig. 4·5

noted that in practical problems the deflection v is usually very small compared with the length of the bar. The deformation is defined when v is expressed as a function of x. This expression will be the equation of the deflected curve, the original centroidal axis being the x axis. The positive direction of v is taken the same as the positive direction of y, the principal axis of the cross-section.

The function v will depend partly on the *shape* of the deformed bar, which is governed by the deformation of the elements, and partly on the *position* of the curve, which is determined by the support conditions or boundary conditions. Fig. 4.6 illustrates three identical deformed shapes differently oriented with respect to the original axis on account of different support conditions.

It should be mentioned that in this chapter we are considering only the deformation caused by the *bending* of the elements. They may suffer other types of deformation which will in turn cause additional deformation of the bar as a whole.

The element deformation determines the *curvature* of the deformed bar at the given point. Fig. 4.7 shows that the radius of curvature, R, is related to the bending of the element by $dx = R\, d\theta$.

The curvature is ρ, where

$$\rho = \frac{1}{R} = \frac{d\theta}{dx}$$

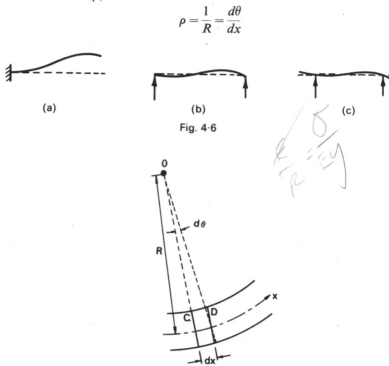

(a)   (b)   (c)

Fig. 4·6

Fig. 4·7

From equation (4.2)

$$\frac{d\theta}{dx} = \frac{\sigma}{Ey}$$

and from equation (4.7)

$$\frac{\sigma}{y} = \frac{M}{I}$$

$$\therefore \quad \rho = \frac{d\theta}{dx} = \frac{M}{EI} \tag{4.8}$$

We might say that the *bend* in an element is $d\theta$ where

$$d\theta = \frac{M\,dx}{EI} = \left(\frac{dx}{EI}\right)M \tag{4.9}$$

Note the similarity of this expression to that giving the axial deformation of an element namely $de = (dx/EA)N$ (p. 33). Clearly the factor dx/EI defines the element flexibility. The flexural rigidity EI is the product of the material stiffness, E, and a section property I.

To find the equation of the deflected shape, we note that

$$\rho = \frac{d^2v/dx^2}{[1 + (dv/dx)^2]^{3/2}}$$

and provided dv/dx is negligible, as it usually is, we can write

$$\rho = \frac{d^2v}{dx^2}$$

Consequently

$$\frac{d^2v}{dx^2} = \frac{d\theta}{dx} = \frac{M}{EI} \qquad (4.10)$$

Evidently θ is dv/dx, the slope of the deformed bar.

When M/EI is known at each point, v may be obtained by integration.

Example 4.3. Find the slope and deflection at the free end of the cantilever AB in Fig. 4.8. Take $E = 20 \times 10^6$ psi and $I = 25$ in^4.

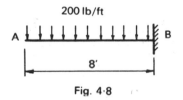

200 lb/ft

A ↓↓↓↓↓↓↓↓↓↓↓↓↓↓↓ B

8'

Fig. 4·8

Solution. Note that consistency of units is important. Suppose we choose pound and inch units. Take A as origin.

Then at x (inches) from A, the bending moment is

$$M = \frac{-200}{12} \frac{x^2}{2} \text{ lb-in}$$

The curvature is

$$\frac{1}{R} = \frac{d^2v}{dx^2} = \frac{M}{EI} = \frac{-200x^2/24}{20 \times 10^6 \times 25} = \frac{-x^2}{6 \times 10^7}$$

$$\frac{d^2v}{dx^2} = \frac{-1}{6 \times 10^7} x^2$$

$$\frac{dv}{dx} = \frac{-1}{6 \times 10^7} \frac{x^3}{3} + c_1 \qquad (4.11)$$

$$v = \frac{-1}{6 \times 10^7} \frac{x^4}{12} + c_1 x + c_2 \qquad (4.12)$$

The first term expresses the *shape* of the curve oriented so that it has zero slope and deflection at the origin. The terms $(c_1 x + c_2)$ permit a rigid body displacement, according to the choice of c_1 and c_2, so that the curve can be re-oriented to fit the given support conditions. In this example the support conditions are that the slope and deflection are both zero at B. That is to say $dv/dx = 0$ when $x = 96$, and $v = 0$ when $x = 96$.

When the first condition is substituted in equation (4.11) we obtain

$$c_1 = \frac{1}{6 \times 10^7} \frac{96^3}{3}$$

The second condition, substituted in equation (4.12) gives

$$c_2 = \frac{-1}{6 \times 10^7} \frac{96^4}{4}$$

With these values, equation (4.11) gives the slope at any point and equation (4.12) gives the deflection at any point. At point A $(x = 0)$ we have

$$\frac{dv}{dx} = c_1 = \frac{1}{6 \times 10^7} \frac{96^3}{3} = 0.0049 \text{ radian}$$

$$v = c_2 = \frac{-1}{6 \times 10^7} \frac{96^4}{4} = -0.354 \text{ in}$$

The negative sign indicates downward deflection, which is to be expected.

Example 4.4. Find the end slope and deflection of the cantilever of Fig. 4.9. The rigidity EI is constant.

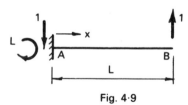

Fig. 4·9

Solution. With A as origin the bending moment at x is

$$M' = L - x$$

Since EI is constant it is convenient to leave it on the left-hand side and write

$$EI \frac{d^2 v}{dx^2} = M = L - x$$

$$EI \frac{dv}{dx} = Lx - \frac{x^2}{2} + c_1$$

$$EIv = L \frac{x^2}{2} - \frac{x^3}{6} + c_1 x + c_2$$

In this problem $v = dv/dx = 0$ at $x = 0$. Hence $c_1 = c_2 = 0$.
At $x = L$

$$\frac{dv}{dx} = \left(L^2 - \frac{L^2}{2}\right)\frac{1}{EI} = \frac{L^2}{2EI}$$

and

$$v = \left(\frac{L^3}{2} - \frac{L^3}{6}\right)\frac{1}{EI} = \frac{L^3}{3EI}$$

In this last example $L^2/2EI$ represents the rotation of B caused by a unit force. The term can therefore be regarded as a flexibility coefficient. Similarly, the deflection due to unit force is $L^3/3EI$ and this is another flexibility coefficient. The question of flexibility and stiffness coefficients due to bending is discussed in more detail in section 4.8.

In the two previous examples, the bending moment was a continuous function of x over the interval $O - L$. When M/EI is a discontinuous function, so also is $\int (M/EI)\,dx$ which gives the slope and $\int\int (M/EI)\,dx\,dx$ which gives the deflection. The bar itself bends into a smooth curve, of course, but this curve is not represented by the same mathematical expression throughout.

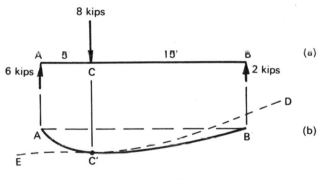

Fig. 4·10

For instance, the beam of Fig. 4.10 (a), when loaded as shown, will bend into a curve such as $AC'B$ (Fig. 4.10 (b)). The expression for the curve AC', if evaluated for all values of x may yield a graph such as $AC'D$. Similarly, the curve $C'B$ is part of a mathematical curve such as $EC'B$. These two curves will meet tangentially at C' so that the bar $AC'B$ appears as a smooth curve. It is important to realize that one cannot substitute a boundary condition at B, for example, into the expression for segment AC since there is no requirement that *this* curve should pass through B.

Boundary condition A applies to curve AC' while boundary condition B applies to curve $C'B$. There is the further requirement that the two different expressions AC' and $C'B$ should yield the same slope and the same deflection at C'.

Example 4.5. Find the slope and deflection at the centre of the beam of Fig. 4.10 (a) in terms of EI, which is constant.

Solution. For the segment AC, with origin at A, we have (using kip and feet units)

$$M = 6x \text{ kip-ft}$$

Thus

$$EI \frac{d^2 v}{dx^2} = 6x \tag{4.13}$$

$$EI \frac{dv}{dx} = 3x^2 + c_1 \tag{4.14}$$

$$EIv = x^3 + c_1 x + c_2 \tag{4.15}$$

The support condition at A, which applies to this segment, indicates that $v = 0$ when $x = 0$. Hence, from equation (4.15) $c_2 = 0$. The constant c_1 is at present undetermined.

For the segment CB, with origin at A, we have

$$EI \frac{d^2 v}{dx^2} = M = 6x - 8(x - 5)$$
$$= -2x + 40 \tag{4.16}$$

$$EI \frac{dv}{dx} = -x^2 + 40x + c_3 \tag{4.17}$$

$$EIv = \frac{-x^3}{3} + 20x^2 + c_3 x + c_4 \tag{4.18}$$

From the support at B, $v = 0$ when $x = 20$ ft. Hence from equation (4.18)

$$c_4 = -5333 - 20c_3 \tag{4.19}$$

It remains to equate the slopes of the two segments at C, and also their deflections. The slopes are given by equations (4.14) and (4.17) respectively. Evaluating these for $x = 5$ ft and equating, we obtain

$$75 + c_1 = 175 + c_3 \tag{4.20}$$

Similarly, evaluating the deflections from equations (4.15) and (4.18) and equating, we obtain

$$125 + 5c_1 = -4875 - 15c_3 \tag{4.21}$$

From equations (4.20) and (4.21) we find that

$$c_1 = -175 \text{ and } c_3 = -275$$

The complete expressions for the deformation are now:
Segment AC:

$$\frac{dv}{dx} = \frac{1}{EI} (3x^2 - 175)$$

$$v = \frac{1}{EI} (x^3 - 175x)$$

Segment CB:

$$\frac{dv}{dx} = \frac{1}{EI} (-x^2 + 40x - 275)$$

$$v = \frac{1}{EI} \left(\frac{-x^3}{3} + 20x^2 - 275x + 167 \right)$$

From the last two expressions we can evaluate the slope and deflection at the mid-point of the beam ($x = 10$)

$$\theta = \frac{dv}{dx} = \frac{25}{EI}\text{ radians}$$

$$v = \frac{-917}{EI}\text{ ft}$$

The slope is positive, i.e., sloping up from left to right. The deflection is negative, i.e., downward. Calculation throughout has been carried out in kip and feet units. It is therefore important to use these units when numerical values are substituted for E and I. The coefficients in the slope and deflection expressions are not non-dimensional.

It will be apparent from this example that when several discontinuities occur, the analytical work involved in the calculation of deformations becomes very laborious, although elementary. Some procedures for overcoming this difficulty are discussed in sections 4.6, 4.7 and 4.8.

In Appendix A it is shown that the statical relationships between the intensity of loading on a beam, the shear force and the bending moment can be expressed in terms of the operations of integration or differentiation. Since the same operations link bending moment with slope and deflection, all these quantities can be related in an orderly sequence.

Considering as before the shear force S_y and the bending moment M_z we have, from Appendix A,

$$w = \text{load intensity expressed as a function of } x$$

$$S = -\int w\,dx \tag{4.22}$$

$$M = -\int S\,dx \tag{4.23}$$

Now we also have

$$\theta = \frac{dv}{dx} = \int \frac{M}{EI}dx \tag{4.24}$$

$$v = \int \theta\,dx \tag{4.25}$$

We could, of course, start with the deflection, v, and write

$$\theta = \frac{dv}{dx} \tag{4.26}$$

$$M = EI\frac{d^2v}{dx^2} \tag{4.27}$$

$$S = -\frac{d}{dx}\left(EI\frac{d^2v}{dx^2}\right) \tag{4.28}$$

$$w = \frac{d^2}{dx^2}\left(EI\frac{d^2v}{dx^2}\right) \tag{4.29}$$

When EI is constant, the last equation can be written

$$w = EI\frac{d^4 v}{dx^4} \tag{4.30}$$

This is sometimes referred to as the differential equation of deflection of an elastic beam.

The signs of the terms in equations (4.22) to (4.30) depend on the sign conventions adopted for the various quantities. Here, both load and deflection are measured positive upward (i.e., in the same direction as y).

Note that equations (4.22) and (4.23) express relations of statics while (4.24) and (4.25) express geometrical relations.

It is possible to obtain deflections from loads by integration.

Example 4.6. The beam of Fig. 4.11 is direction-fixed at A and simply supported at B. Find the equation of the deflection curve. Take EI as constant.

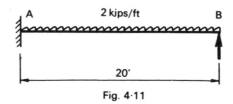

Fig. 4·11

Solution. Take A as origin. Use kip and feet units.

$$w = -2 \text{ (minus because the load is downward)}$$

$$S = -\int w\,dx = +2x + c_1$$

$$M = -\int S\,dx = -x^2 - c_1 x + c_2$$

$$EI\frac{dv}{dx} = \int M\,dx = \frac{-x^3}{3} - \frac{c_1 x^2}{2} + c_2 x + c_3$$

$$EIv = \int EI\frac{dv}{dx}\,dx = \frac{-x^4}{12} - \frac{c_1 x^3}{6} + \frac{c_2 x^2}{2} + c_3 x + c_4$$

The boundary conditions are $dv/dx = 0$ and $v = 0$ at $x = 0$, while $M = 0$ and $v = 0$ at $x = 20$. Substitution of these in the appropriate equations leads to values for the constants

$$c_1 = -25; \quad c_2 = -100; \quad c_3 = 0; \quad c_4 = 0$$

The equation of the deflection curve is thus

$$v = \frac{1}{EI}\left(-\frac{1}{12}x^4 + \frac{25}{6}x^3 - 50x^2\right) \text{ ft}$$

If the slope or deflection at isolated points are required, these may be calculated more readily by the methods described in section 4.7.

4.3   Strain Energy due to Bending

Provided the bending deformation of a bar is elastic (i.e., it disappears upon removal of the deforming forces), the work of deformation is stored as elastic strain energy. The deformation of a small element of the bar is pictured in Fig. 4.12 (a). If the strains are not only elastic but also proportional to stresses, the deformation $d\theta$ will be proportional to M. The force–deformation relation will be linear as in Fig. 4.12 (b).

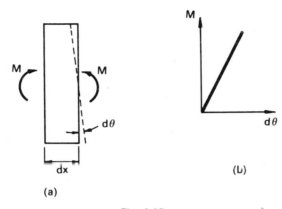

(a)

(b)

Fig. 4·12

The work done by a bending moment M will evidently be

$$dU = \tfrac{1}{2}M \, d\theta \tag{4.31}$$

Since M and $d\theta$ are related (equation 4.9), the strain energy can be re-expressed in alternative forms.

$$d\theta = \frac{M \, dx}{EI}$$

$$\therefore \quad dU = \frac{M^2 \, dx}{2EI} \tag{4.32}$$

Or we can put $M = EI\dfrac{d\theta}{dx}$, giving

$$dU = \tfrac{1}{2}EI\frac{d\theta}{dx} \cdot d\theta$$

$$= \tfrac{1}{2}EI\left(\frac{d\theta}{dx}\right)^2 dx$$

$$dU = \tfrac{1}{2}EI\left(\frac{d^2 v}{dx^2}\right)^2 dx \tag{4.33}$$

Then for the complete bar,

$$U = \int \frac{M^2\,dx}{2EI} \tag{4.34}$$

or

$$U = \int \frac{EI}{2}\left(\frac{d^2 v}{dx^2}\right)^2 dx \tag{4.35}$$

Note the similarity of equation (4.34) to that for strain energy due to axial deformation (equation 3.15, p. 41).

Provided equilibrium is preserved, the strain energy will always be equal to the work done by the external loads during the deformation.

Example 4.7. Find the bending strain energy stored in the beam of Fig. 4.13, in which EI is constant. Find the deflection under the load.

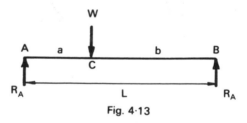

Fig. 4·13

Solution. Since M is discontinuous at C, we consider the two parts AC and CB separately. For segment AC:

$$R_1 = Wb/L$$

$$M = \frac{Wb}{L}x$$

$$M^2 = \frac{W^2 b^2 x^2}{L^2}$$

From equation (4.34)

$$U_{AC} = \int_0^a \frac{W^2 b^2 x^2}{2EIL^2}\,dx$$

$$= \frac{W^2 b^2 a^3}{6EIL^2}$$

For segment CB:

By taking B as origin and measuring x to the left we should obtain a value for U_{CB} similar to that for U_{AC} with a and b interchanged.

$$U_{CB} = \frac{W^2 a^2 b^3}{6EIL^2}$$

$$\text{Total } U = \frac{W^2 a^2 b^2}{6EIL}(a+b) = \frac{W^2 a^2 b^2}{6EIL}$$

If the deflection at C is v_C measured in the same direction as the load, the external work done is $\frac{1}{2}Wv_C$.

Hence

$$\tfrac{1}{2}Wv_C = \frac{W^2\,a^2\,b^2}{6EIL}$$

and

$$v_C = \frac{Wa^2\,b^2}{3EIL}$$

4.4 Non-linear Distribution of Stress

We have now to consider bars made of non-linear material or bars in which the stresses have exceeded the proportional limit. We revert to the basic idea of an element between cross-sections C and D. When section D rotates about some axis z, strains are induced which are proportional to the distance from z.

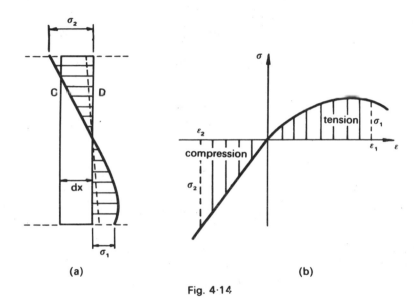

(a)   (b)

Fig. 4·14

That is, the *strain* distribution is linear irrespective of the material characteristics (on our hypothesis that sections C and D remain plane).

If the stresses are not proportional to strains, then the stress distribution will no longer be linear. Instead, it will follow the same pattern as the stress-strain curve of the material. Suppose that we deform the element shown in Fig. 4.14 (a) in such a way that the strains vary (linearly of course) from a tensile strain ϵ_1 at the bottom to a compressive strain ϵ_2 at the top. Suppose further that the stress–strain graph for the material is as shown in Fig. 4.14 (b). The stresses on the cross-section will vary from σ_1 to σ_2 in the same way as they do in Fig. 4.14 (b).

Thus for any assumed deformation we are able to plot the "stress-block". From this we can find the resultant force on the section as well as the total moment about any chosen axis. In simple problems, this presents no difficulty —some problems are easier than those involving stresses within the proportional range. On the other hand, depending on the stress–strain properties and on the section shape, some problems require numerical methods for their solution. Even these problems can hardly be called difficult since the assistance of computers renders numerical methods reasonably simple.

Consider first a material having the stress–strain graph shown in Fig. 4.15 (a).

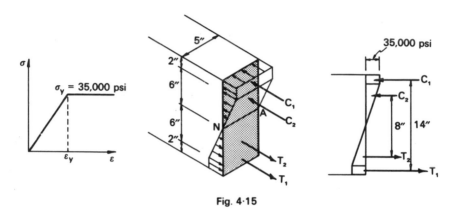

Fig. 4·15

Example 4.8. A beam of rectangular cross-section, 5 in wide and 16 in deep is made of mild steel with a yield stress of 35 ksi. The stress–strain diagram can be taken as in Fig. 4.15 (a), both in tension and compression. Calculate the resultant of the stresses when two inches of the section has yielded both at the top and bottom.

Solution. The stresses on the cross-section are shown in Fig. 4.15 (b). The stress block is divided into convenient sub-sections and the resultant of each is evaluated.

$$C_1 = 35 \times 2 \times 5 = 350 \text{ kips acting 7 in above N.A.}$$

$$C_2 = 35 \times \frac{6}{2} \times 5 = 525 \text{ kips acting 4 in above N.A.}$$

$$T_1 = C_1 \quad \text{and} \quad T_2 = C_2$$

The resultant force is zero and the resultant couple is

$$M = (35 \times 14) + (525 \times 8) = 9100 \text{ kip-in}$$

The bending deformation, $d\theta$, of an element cannot be calculated as $M\,dx/EI$. In this expression, I stands for $\int y^2\,dA$, but this integral does not occur unless the stress distribution is linear. Similarly, E is the (constant) slope of the stress–strain curve and this also is inapplicable to the present problem.

However, the deformation is easily calculated from strain considerations. Let ϵ_P and ϵ_Q be the strains at any two points P and Q on the cross-section (Fig. 4.16) and let c be the distance between P and Q measured normal to the neutral axis. (It is important to note that the strains vary linearly for the full depth of the beam even where the stress is constant as in the previous example.)

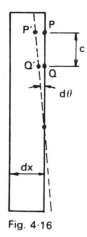

Fig. 4·16

The deformations at P and Q are

$$PP' = \epsilon_P\, dx \quad \text{and} \quad QQ' = \epsilon_Q\, dx$$

Hence

$$d\theta = \frac{\epsilon_P\, dx - \epsilon_Q\, dx}{c} = \frac{\epsilon_P - \epsilon_Q}{y_P - y_Q}\, dx \qquad (4.36)$$

Due account must be taken of the signs of the strain. Thus if P and Q are on opposite sides of the neutral axis the strains will in fact be added.

In example 4.8 the deformation at the section considered can best be found by taking P as the point 6 in above the neutral axis and Q on the neutral axis. Since $E = 30 \times 10^3$ ksi (Fig. 4.15 (a)) then $\epsilon_P = 35/(30 \times 10^3) = 1.17 \times 10^{-3}$.

Hence

$$d\theta = \frac{\epsilon_P - \epsilon_Q}{c}\, dx$$

$$= \frac{1.17 \times 10^{-3} - 0}{6}\, dx$$

$$= 1.94 \times 10^{-4}\, dx \text{ radian}$$

Before the shape of the deflected beam can be found, it is necessary to evaluate the curvature at a number of points along its length and so to obtain a graph of $\rho\ (= d\theta/dx)$. It is helpful first to construct a graph of M against ρ, the moment-curvature relationship for the particular beam cross-section. This is done simply by considering a series of values for M and calculating, by the above methods, the corresponding ρ values, or vice versa.

Once the $M-\rho$ graph is obtained the variation of ρ along the beam is easily obtained from the bending moment graph. The deflected shape of the beam can then be found according to the principles of section 4.2, although numerical methods will probably be required. This question is referred to again briefly in section 4.8 (p. 86).

In example 4.8 the neutral axis happens to pass through the centroid of the cross-section. This occurs because the section is symmetrical about this axis and the material properties are the same in tension and compression. Where this is not so, no general rule can be given and the position of the neutral axis will depend on the nature of the stress–strain curve. It will also, in general, vary with the bending moment.

For yielding materials such as mild steel (Fig. 4.15 (a)) the region above (or below) the neutral axis approaches a state of constant stress as the bending moment increases. Although there will always be a small region near the neutral axis where the stress is less than σ_y this has little effect on the moment. The plastic moment, M_P, is defined as the limiting bending moment that a section can withstand, and is calculated as if the whole section sustains the yield stress, σ_y, either compressive or tensile.

Example 4.9. Compute the plastic moment of the T-section shown in Fig. 4.17 if the yield stress is the same in tension and compression.

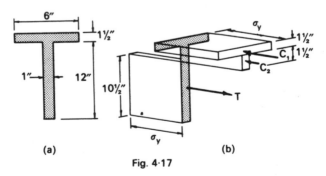

Fig. 4·17

Solution. Since both the tensile stresses and the compressive stresses are uniform, the resultant force will be zero if the areas above and below the neutral axis are equal.

The total area of the section is 21 in². The neutral axis is therefore located at 10.5 in above the bottom of the section. The stress blocks are shown in Fig. 4.17 (b).

$$C_1 = 6 \times 1\tfrac{1}{2} \times \sigma_y = 9\sigma_y \text{ at 2.25 in above N.A.}$$

$$C_2 = 1 \times 1\tfrac{1}{2} \times \sigma_y = 1.5\sigma_y \text{ at 0.75 in above N.A.}$$

$$T = 10.5\sigma_y \text{ at 5.25 in below N.A.}$$

Taking moments about N.A. we have

$$M_P = (9\sigma_y \times 2.25) + (1.5\sigma_y \times 0.75) + (10.5\sigma_y \times 5.25)$$

$$= 76.5\sigma_y$$

It will be seen that in this problem the neutral axis does not pass through the centroid.

If the section possesses no axis of symmetry, then the principal axes of the section have no special significance when the stress distribution is non-linear. That is to say, if z is a principal axis and we bend an element of the beam about z, then in general the stresses induced will not form a couple around the z axis. For such a section and for a given bending moment it may be necessary to seek the position and direction of the neutral axis by trial and error.

4.5   Non-homogeneous Beams

The stress-block concept can also be employed for the analysis of beams whose cross-section contains different materials, each stressed within its proportional range.

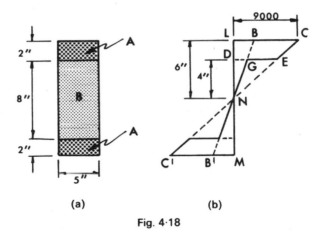

(a)   (b)

Fig. 4·18

For instance, Fig. 4.18 (a) shows a beam whose outer layers are formed of material A with an elastic modulus of 20×10^3 ksi. The interior is composed of material B having an elastic modulus of 5×10^3 ksi. When the extreme fibre stress is 9 ksi the stress distribution will be as shown in Fig. 4.18 (b). If the section were of material A throughout, the stress distribution would simply be the line CN. The more flexible material of the interior causes a sudden drop of stress from DE to DG at the interface of A and B. The ratio DE/DG is equal to the *modular ratio n* $(= E_A/E_B)$. The stress block method easily gives the bending moment (note that the stress diagram can be regarded as triangle *NLC minus* the triangle *NCE*).

The computation, by this method, of stresses corresponding to a given bending moment is a little more indirect. For this reason a transformed area procedure is often preferred, very similar to that used for axial deformation of similar bars.

The area of material A is transformed to an equivalent area of B. That is to say it is multiplied by n. For axial deformations we saw that the transformed area had to have the same centroid as the real area. For bending deformations it is also necessary that it should have the same second moment about the neutral axis. Where the direction of the neutral axis is obvious, as it usually is in practice, this simply means that the area must be stretched out (or contracted) in a direction parallel to the neutral axis.

Example 4.10. Calculate the bending stresses in a beam having the cross-section of Fig. 4.18 (a) if the bending moment is 500 kip-in. $E_A = 20 \times 10^3$ ksi and $E_B = 5 \times 10^3$ ksi.

Solution. Transform the section to an equivalent one of material B alone.

$$n = E_A/E_B = 4$$

The areas A are extended to 4 times their real width, giving the transformed section of Fig. 4.19 (a). This section has a moment of inertia of 2240 in⁴. A homogeneous beam of

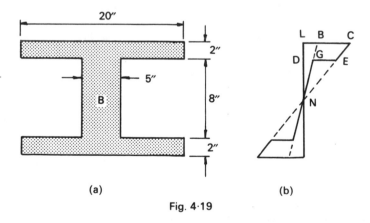

(a)

(b)

Fig. 4·19

material B with this section can now be analysed in the usual way. Stresses within the web will be correct, but stresses within the flanges are fictitious and must be multiplied by n to give the true stresses in A.

With $M = 500$ kip-in the (fictitious) extreme fibre stresses are

$$\sigma' = \frac{My}{I} = \frac{500 \times 6}{2240} = 1.34 \text{ ksi}$$

This is the ordinate LB (Fig. 4.19 (b)). The true stress is

$$\sigma = n\sigma' = 4 \times 1.34 = 5.36 \text{ ksi}$$

The stress in material B at the junction of A and B is

$$\sigma = \frac{My}{I} = \frac{500 \times 4}{2240} = 0.893 \text{ ksi}$$

This is the ordinate DC (Fig. 4.19 (b)).

It is equally valid to transform the section to an equivalent one of material A. This will simply be an I section one-quarter of the width of that in Fig. 4.19 (a). For the same bending moment, stresses will be 4 times as great. They will be true for A and must be multiplied by $\frac{1}{4}$ for B.

4.6   The Use of Step Functions

In section 4.2 the fundamental procedure was developed whereby the deformed shape of a bar subjected to bending could be calculated. Basically this consists of finding an expression for the curvature at each point along the length of the bar and then integrating this twice to obtain deflection. For linear stress distribution the curvature is given by the expression M/EI. It became evident

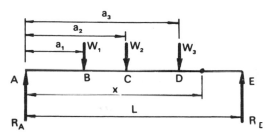

Fig. 4·20

in section 4.2 that this calculation could become very laborious even for relatively simple problems. It is therefore quite important to study methods of computation which, while based on the same principles, shorten the arithmetic work. A number of such methods have been devised of which two will be given here.

By the use of step functions it is possible to write a single expression for the M/EI function even when this is discontinuous. The introduction of extra boundary conditions at each discontinuity is then avoided.

A step function H_a is defined such that

$$H_a = 0 \quad \text{when } x < a$$

and

$$H_a = 1 \quad \text{when } x \geqslant a$$

Thus any expression which is multiplied by the function H_a takes on its normal value when x exceeds a but is suppressed when x is less than a.

For the beam of Fig. 4.20, and with origin A, the bending moment in the segment most remote from the origin (i.e., segment DE) can be written as

$$M = R_1 x - W_1[x - a_1] H_{a_1} - W_2[x - a_2] H_{a_2} - W_3[x - a_3] H_{a_3} \quad (4.37)$$

For segment DE all the step functions have unit value. For segment CD, $x < a_3$ so that H_{a_3} is zero and the last term is suppressed. The remaining expression correctly gives the value of M within CD. *The identity of the bracket terms must be preserved throughout the problem.* At no stage may the brackets be multiplied out. The term $[x - a_1]$ when integrated becomes $\frac{1}{2}[x - a_1]^2$.
To illustrate the method, example 4.5 will be re-solved.

Example 4.11. Re-solve example 4.5 (Fig. 4.21) using step functions.

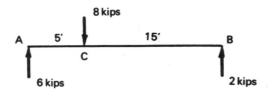

Fig. 4·21

Solution. With A as origin, the bending moment in CB can be written as

$$M = 6x - 8[x - 5] H_s$$

Then

$$EI \frac{d^2 v}{dx^2} = 6x - 8[x - 5] H_s \quad (4.38)$$

$$EI \frac{dv}{dx} = 3x^2 - \frac{8}{2}[x - 5]^2 H_s + C_1 \quad (4.39)$$

$$EIv = x^3 - \frac{8}{6}[x - 5]^3 H_s + C_1 x + C_2 \quad (4.40)$$

Note that the bracket is preserved during integration.

At A, $v = 0$ and $x = 0$. Substitution of this in equation (4.40) gives $C_2 = 0$. This is so because with $x = 0$ the term H_s has the value zero since $x < 5$.

At B, $v = 0$ and $x = 20$. Substitution of this in equation (4.40) (the term H_s is now equal to unity) gives $C_1 = -175$. The slope and deflection are therefore given by

$$EI \frac{dv}{dx} = 3x^2 - \frac{8}{2}[x - 5]^2 H_s - 175 \quad (4.41)$$

$$EIv = x^3 - \frac{8}{6}[x - 5]^3 H_s - 175x \quad (4.42)$$

These expressions will be found to agree with those obtained previously in example 4.5 (p. 60). To obtain the slope and deflection at the centre we substitute $x = 10$ and note that the bracket term is operative.

$$\frac{dv}{dx} = \frac{25}{EI} \text{ radian}$$

$$v = \frac{-917}{EI} \text{ ft}$$

The method can be used also with distributed loads. For instance, the bending moment in the beam of Fig. 4.22 can be written in the form

$$M = R_A x - \frac{w}{2}[x - a]^2 H_a$$

This applies throughout the beam.

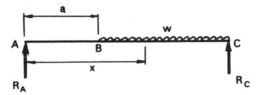

Fig. 4·22

The bending moment function can be formed by starting with the left-hand segment and adding what might be called a correction term at each discontinuity. In the beam of Fig. 4.23 (a), the bending moment in the segment AB is

$$M_{AB} = 9.5x$$

This is the equation of line 1 in Fig. 4.23 (b). When integrated twice, and without integration constants, this will give the deflection curve 1 of Fig. 4.23 (c), tangential to the x axis at A.

The bending moment in segment BC is obtained by the addition of a term

$$M_{BC} = 9.5x - \frac{1.5}{2}[x - 4]^2 H_4$$

The second term, which is zero at B and neglected to the left of B, represents the difference between lines 1 and 2 of Fig. 4.23 (b). Upon integration this second term modifies deflection curve 1 to deflection curve 2 to the right of B. The form of the expression ensures that the curves 1 and 2 are tangential at B.

In a similar way, another term modifies curve 2 to curve 3 past C.

$$M_{CD} = 9.5x - \frac{1.5}{2}[x - 4]^2 H_4 - 8[x - 9] H_9$$

This expression assumes the presence of the distributed load $w = -1.5$. Beyond D this must be eliminated by the addition of a distributed load $w = +1.5$. Hence

$$M_{DE} = 9.5x - \frac{1.5}{2}[x - 4]^2 H_4 - 8[x - 9] H_9 + \frac{1.5}{2}[x - 12]^2 H_{12}$$

The deflection curve will pass through some point E'. The integration constants will rotate (and translate when necessary) the composite curve AE' to satisfy the given support conditions.

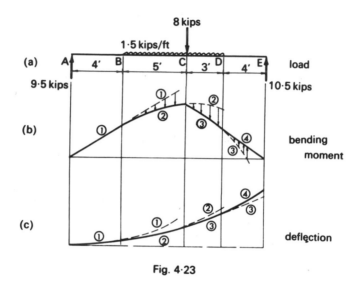

Fig. 4·23

The basis of the method can also be described algebraically. We refer to Fig. 4.20.

For segment AB,

$$EI\frac{d^2v}{dx^2} = R_A x$$

$$EI\frac{dv}{dx} = R_A\frac{x^2}{2} + c_1$$

$$EIv = R_A\frac{x^2}{6} + c_1 x + c_2$$

The condition that $v = 0$ when $x = 0$ gives $c_2 = 0$. The value of c_1 (which is the slope at A) is as yet undetermined.

For segment BC,

$$EI\frac{d^2v}{dx^2} = R_A x - W_1[x - a_1] H_{a_1}$$

$$EI\frac{dv}{dx} = R_A \frac{x^2}{2} - \frac{W_1}{2}[x - a_1]^2 H_{a_1} + d_1$$

$$EIv = R_A \frac{x^3}{6} - \frac{W_1}{6}[x - a_1]^3 H_{a_1} + d_1 x + d_2$$

The condition that at point B (where $x = a_1$) $dv/dx_{AB} = dv/dx_{BC}$ gives $c_1 = d_1$. Moreover, the condition $v_{AB} = v_{BC}$ then gives $d_2 = 0$ (i.e., $d_2 = c_2$). The effect of leaving the terms of the slope and deflection function in this particular form is that the integration constants for the various segments become identical.

Proceeding in a similar manner we find that for segment CD,

$$EIv = R_A \frac{x^3}{6} + c_1 x - \frac{W_1}{6}[x - a_1]^3 H_{a_1} - \frac{W_2}{6}[x - a_2]^3 H_{a_2}$$

and for segment DE,

$$EIv = R_A \frac{x^3}{6} + c_1 x - \frac{W_1}{6}[x - a_1]^3 H_{a_1} - \frac{W_2}{6}[x - a_2]^3 H_{a_2} - \frac{W_3}{6}[x - a_3]^3 H_{a_3}$$

When the support condition at E is substituted in this last equation, the value of c_1 is determined.

Discontinuities in the beam rigidity EI can be dealt with in a similar manner by the introduction of a new term at each discontinuity and attaching the appropriate step function to this term.

4.7 Moment-area Methods

Moment-area methods, first proposed by Otto Mohr, provide a very powerful tool for the solution of beam deflection problems. They make use of the fact that the first integral of a function can be regarded as the area under the graph of this function, while the second integral can be regarded as the moment of this area about some reference line. We are here applying these methods to the calculation of bending deformations.

The primary function here is the curvature, ρ, the first and second integrals being the slope and deflection functions. The relationships appear in their simplest form with beams of uniform section stressed within the proportional range. Then $\rho = M/EI$ and since EI is constant, the graph of ρ is the same as the graph of M to a different scale. This class of problems will be considered first.

Fig. 4.24 (a) shows a bending moment diagram for a beam of constant EI, and Fig. 4.24 (b) shows the corresponding graph of curvature $(= M/EI)$.

FIRST MOMENT-AREA THEOREM: The difference of slope between any two points E and F is equal to the area under the graph of curvature taken between these two points.

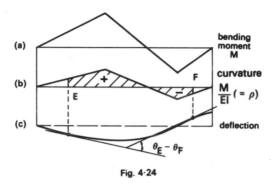

Fig. 4·24

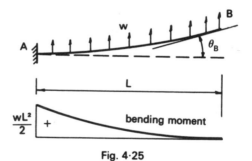

Fig. 4·25

At present we take this area to be equal to $1/EI$ times the area under the M diagram. Due account is taken of positive and negative areas. Certain problems can be solved by the use of this theorem alone. For instance in the uniform cantilever of Fig. 4.25 the slope at A is known to be zero. Consequently, the slope at the free end, B, due to the distributed load, is

$$\theta_B = \frac{1}{EI} \times \frac{1}{3} \times \frac{wL^2}{2} \times L = \frac{wL^3}{6EI}$$

(the areas and centroids of sections of parabolas are given in Appendix B).

The second theorem concerns the tangents to the beam. Let us make the tangents more realistic by fixing wire pointers to the centroidal axis to represent these tangents. The two pointers (tangents) fixed to faces C and D of an elemental slice (Fig. 4.26) are coincident before deformation, but diverge at an angle $d\theta$ afterwards, where $d\theta$ $(= M\,dx/EI)$ is the element deformation. This angle $d\theta$ is thus equal to the elemental area under the M/EI diagram which we will call dA (Fig. 4.27). The tangents at C and D will now make an intercept dY on any vertical reference line Y (Fig. 4.27 (b)) such that

$$dY = d\theta \cdot x = dA \cdot x$$

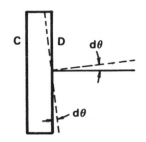

Fig. 4·26

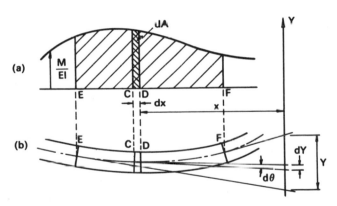

Fig. 4·27

SECOND MOMENT-AREA THEOREM: On any vertical line, the intercept Y made by the tangents at points E and F is equal to the first moment about that line of the area of the M/EI graph between the points E and F.

This follows since Y is simply $\int dY$ which in turn is $\int_{E}^{F} x\,dA$. It is most important to note that in general Y is *not* the deflection at any particular point. The use of this theorem depends largely on the careful selection of the points E and F and the position of the reference line.

The deflection of a cantilever can be readily found because the tangent at the support remains undisturbed. The loading of Fig. 4.28 (a) produces a bending moment diagram as in Fig. 4.28 (b). If we now consider the bending moment diagram between points A and C and also take our reference line through C we see that the intercept Y (Fig. 4.28 (c)) is, *in this case*, the deflection of C. From the second theorem we know that Y is equal to the moment of the M/EI diagram about C.

$$v_C = Y = \left(\frac{1}{EI}\right)\frac{(-48)6}{2}(8) = -\frac{1152}{EI}$$

Note that the discontinuous nature of the bending moment causes no difficulty.

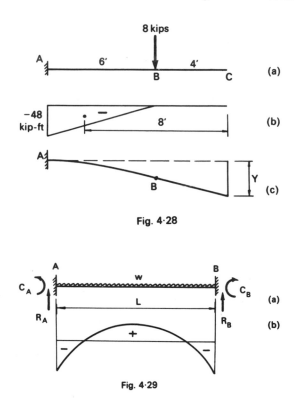

Fig. 4·28

Fig. 4·29

The direct calculation of deflections by the use of the second theorem depends on the possibility of finding by inspection a tangent whose direction is unaltered when the beam is loaded. The tangent at the centre of a symmetrically loaded and symmetrically supported beam is a case in point.

The moment-area theorems are particularly useful for beams in which one or both ends are direction-fixed. The reactions of the beam of Fig. 4.29 (a) cannot be determined by statics. They can be found by the integration method of section 4.2, but they are obtained more readily by moment-area methods.

By symmetry the reactions R_A and R_B are each equal to $wL/2$ and the support couples C_A and C_B are equal in magnitude and opposite in sign. The bending moment diagram will be as shown in Fig. 4.29 (b) with the bending moment at the ends being equal to $-C$. The only problem is to determine the value of C. Since the slopes at A and B are equal (both zero) we know from the first moment-area theorem that the area of the bending moment diagram is zero, i.e., the positive and negative areas are equal. This enables us to find C.

The most convenient procedure is to regard the loading as being the sum of the load w acting on a simply supported span and a pair of end couples C applied to a simply supported span. The first gives rise to the *free-span moment* diagram (Fig. 4.30 (a)) which can be found by statics. The second gives the

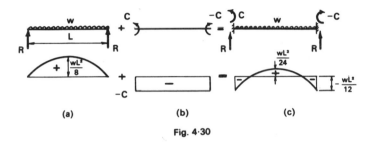

(a)   (b)   (c)

Fig. 4·30

support-moment diagram. If the sum of these areas is to be zero they must evidently be equal and of opposite sign. Thus

$$\frac{2}{3}\frac{wL^2}{8}L \quad = \quad CL$$

(area of free-span   (area of support-
moment diagram)   moment diagram)

giving

$$C = \frac{wL^2}{12}$$

The combined diagram is as in Fig. 4.30 (c).

When an unsymmetrical load acts on a fixed-ended beam both moment-area theorems are required.

Example 4.12. Find the end reactions, both forces and couples, of the beam of Fig. 4.31 (c).

Solution. There are four reactions—R_A, R_B, C_A and C_B—and only two equations of statics. Since the difference of slope between the two ends is zero, the areas of the free-span and support-moment diagrams must be equal. Also, the two end tangents are coincident. Consequently the *first moment* of the free-span and the support-moment diagrams about any point must be equal and opposite. Moments are usually taken about one end, the support-moment diagram being split into triangles for ease of calculation.

For equality of areas

$$\frac{Wab}{L}\frac{L}{2} = \frac{C_A + C_B}{2}L \qquad (4.43)$$

For equality of moments about A

$$\frac{Wab}{2}\frac{(L+a)}{3} = \frac{(C_A \times L)L}{2}\frac{L}{3} + \frac{(C_B \times L)2L}{2}\frac{2L}{3} \qquad (4.44)$$

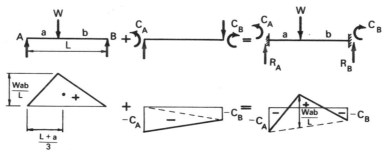

Fig. 4·31

Equations (4.43) and (4.44) give

$$C_A = \frac{Wab^2}{L^2} \quad \text{and} \quad C_B = \frac{Wa^2b}{L^2}$$

When C_A and C_B are known, the forces R_A and R_B can be found by statics.

When only one end is fixed (Fig. 4.32 (c)) we still have three reactions R_A, R_B and C_A. The end tangents of the bent beam are no longer coincident but they do make a zero intercept on the vertical line through the simple support. Hence the moment of the total bending moment diagram about the simple support is zero. This condition together with the two statical equations permits the calculation of the three reactions.

Beams in which EI varies can often be solved by these theorems. The moment-area theorems must now be applied specifically to the M/EI diagram which will be distinct in shape from the bending moment diagram (see example 4.15 on p. 85).

In Fig. 4.27 we discussed the movement of imaginary pointers attached tangentially to the beam. Sometimes it is convenient to attach the pointers at some other angle. For example, in Fig. 4.33 the loads on the bent bar $AGHB$ will clearly cause bending which will result in an increase in the distance AB. Imagine pointers $C_1 D_1$ attached to the two faces of a small element of the bar. Upon deformation of the element, these pointers will diverge by $d\theta$ ($= M\,dx/EI$). This shows clearly that the deformation will contribute

$y_1 d\theta$ to the total spreading of the supports A and B. Following the same arguments as before, we see that the total increase of AB is given by the first moment of the M/EI diagram about the line AB.

Useful as they are, the moment-area theorems are somewhat cumbersome to use for the calculation of the deflection at any point of a beam loaded in an arbitrary manner. However, this is easily accomplished if the moment-area theorems are used in conjunction with the conjugate beam concept.

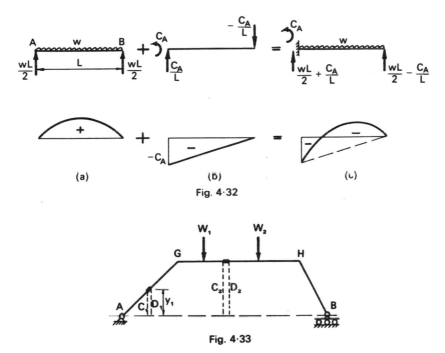

Fig. 4·32

Fig. 4·33

4.8   The Conjugate Beam

The mathematical relationships between load, shear force and bending moment (equations 4.22 and 4.23, p. 61) are the same except as regards sign as those between curvature, slope and deflection (equations 4.24 and 4.25). So when we are trying to find slopes and deflections from a known curvature function, we can imagine that we are trying to find shear forces and bending moments from a given load function. The curvature function is regarded as a distributed load of varying intensity. The beam loaded with this fictitious load is called a *conjugate beam*. The support conditions of this conjugate beam must be chosen to simulate correctly the slope and deflection boundary conditions of the real beam.

When the curvature is expressed by M/EI, then M/EI of the real beam becomes w in the conjugate beam, dv/dx of the real beam becomes $-S$ in the conjugate beam and v of the real beam becomes M in the conjugate beam. To save confusion the conjugate quantities will be denoted by w', S' and M'.

Real beam	*Conjugate beam*
ρ	w'
$\theta \ (= \int \rho \, dx)$	$-S' \ (= \int w' \, dx)$
$v \ (= \int \theta \, dx)$	$M' \ (= \int -S' \, dx)$

Let us try the method out first on the simple problem in example 4.5.

Example 4.13. Re-solve example 4.5 (Fig. 4.34 (a)) by the conjugate beam method.

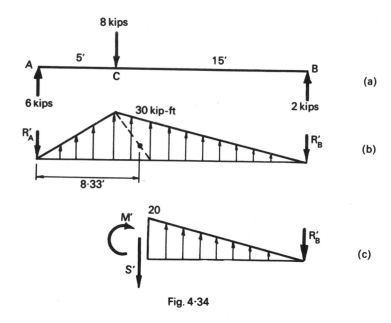

Fig. 4·34

Solution. The bending moment diagram is shown in Fig. 4.34 (b). This diagram, divided by the constant EI, becomes the "load" diagram. The bending moment is positive, so the conjugate load is also positive, i.e., upward. To make this clearer, upward arrows are drawn within the diagram. The total load is $300/EI$ and the centroid of the triangle is 8.33 ft from A. Hence the "reactions" are

$$R'_A = \frac{-300}{EI}\frac{11.67}{20} = \frac{-175}{EI} \quad \text{and} \quad R'_B = \frac{-300}{EI}\frac{8.33}{20} = \frac{-125}{EI}$$

The conjugate end shears, S'_A and S'_B, are $+175/EI$ and $-125/EI$. The real slopes at the ends, which are equal to $-S'$, are thus $\theta_A = -175/EI$ and $\theta_B = +125/EI$. We now want the conjugate shear and bending moment at the centre.

From the free-body of Fig. 4.34 (c) we see that

$$S' = -\frac{1}{EI}(125 - 100) = -\frac{25}{EI}$$

$$M' = \frac{1}{EI}\left(-125 \times 10 + 100 \times \frac{10}{3}\right) = -\frac{917}{EI}$$

Hence in the real beam

$$\frac{dv}{dx} = -S' = +\frac{25}{EI} \text{ radian}$$

$$v = +M' = -\frac{917}{EI} \text{ ft}$$

It will be seen that this solution is far simpler than that used in example 4.5.

The manner in which the conjugate beam must be supported, in order to simulate correctly the boundary conditions for the slope and deflection functions, follows from the relationships already tabulated above. If the real beam has a direction-fixed support, where $v = 0$ and $\theta = 0$, it follows that S' and M' must be zero at this point in the conjugate beam, i.e., this is a free end. The relations are reciprocal, so a free end in the real beam becomes a direction-fixed support in the conjugate beam.

At a simple end support in the real beam, $v = 0$ and θ is undefined. In the conjugate beam $M' = 0$ and S' is undefined which corresponds again to a simple support.

A beam may be supported at a point other than the end. At such an interior support, $v - 0$ and θ is undefined except that it does not change value as we pass the support. In the conjugate beam $M' = 0$ and S' is undefined but continuous. This corresponds to a hinge in the conjugate beam.

These support conditions are summarized in Fig. 4.35.

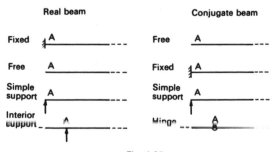

Fig. 4·35

The method will be illustrated by a fairly general problem.

Example 4.14. For the beam of constant EI supported and loaded as shown in Fig. 4.36 (a) find the slope over the support B and the slope and deflection at D.

Solution. The bending moment diagram for the real beam is first drawn (Fig. 4.36 (b)). To use this as a conjugate load diagram we draw upward arrows where the bending moment is positive and downward arrows where the bending moment is negative. If we draw the chord CE on this diagram (CE is the segment carrying the uniformly distributed load), then the mid-ordinate of the parabola measured from this chord is $wa^2/8 = 2 \times 12^2/8 = 36$ kip-ft, where a is the length of the loaded segment.

The support conditions of the conjugate beam (Fig. 4.36(c)) are determined from the table of Fig. 4.35.

Now, regarding the diagram of Fig. 4.36 (b) as the conjugate load on the beam of Fig. 4.36 (c) we can determine the conjugate reactions by statics. The load can be broken up into

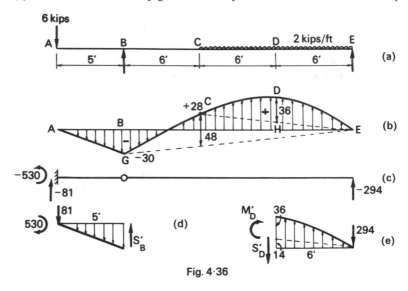

Fig. 4·36

triangles ABG (−), BGE (−), GCE (+) and the parabolic section $CDEHC$ (+). The resultants of these areas are

	Area	Centroid	
ABG	−75	−1.67′	from B
BGE	−270	+6′	from B
GCE	+432	+8′	from B
$CDEHC$	+288	+12′	from B

From these values the reactions are easily found. They are shown on Fig. 4.36 (c). The constant EI is omitted for the time being.

The real slope at B corresponds to the conjugate shear force with the sign changed. From the free-body of Fig. 4.36 (d) we see that

$$EIS'_B = +81 + \frac{30 \times 5}{2} = +156$$

The real slope at B is thus $-156/EI$ ft.

The real slope and deflection at D correspond to the conjugate shear force (with the sign changed) and bending moment at D. From Fig. 4.36 (e) we see that

$$EIS'_D - \left(\frac{14 \times 6}{2}\right) - (\tfrac{2}{3} \times 36 \times 6) + 294 = 0$$

or
$$S_D' = -108/EI$$

$$-EIM_D' + (42 \times 2) + (144 \times 2.25) - (294 \times 6) = 0$$

or
$$M_D' = 1356/EI$$

The real slope at D is $108/EI$ radian and the deflection is $1356/EI$ ft.

Example 4.15. Find the end reactions of the beam of Fig. 4.37 (a) by the conjugate beam method. (This is the same problem as that of example 4.11.)

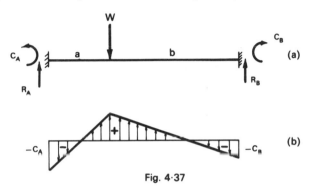

Fig. 4·37

Solution. The bending moment diagram for the real beam is drawn in general form (Fig. 4.37 (b)) although the values of the end couples C_A and C_B and hence of the bending moments at A and B, are as yet unknown.

Since each end of the real beam is fixed, each end of the conjugate beam is free. That is to say the conjugate beam is completely unsupported. This leads to the requirement that the conjugate load must be an equilibrium system. In other words, the total area of the diagram of Fig. 4.37 (b) must be zero, and its moment about any point must be zero.

These are the same requirements as we had in example 4.11. So although the point of view here is slightly different the algebra will be exactly the same as in example 4.11.

For beams of constant EI it is sufficient to work with the unfactored bending moment diagram and insert the factor $1/EI$ right at the end. If EI varies, the M/EI diagram must be specifically considered.

Example 4.16. Find the maximum deflection of the beam of Fig. 4.38 (a). For the segment AC, $EI = 12 \times 10^4$ kip-ft^2, and for CB, $EI = 3 \times 10^4$ kip-ft^2.

Solution. Fig. 4.38 (b) shown the bending moment diagram, and Fig. 4.38 (c) shows the M/EI diagram. The latter is used as the conjugate load, and the reactions are computed in the usual way.

The maximum M' occurs when $S' = 0$. This will be at a point D such that the load above DB is equal to the reaction R_B'. Thus

$$\left(20 \times 10^{-4} \frac{x}{12}\right) \frac{x}{2} = 90 \times 10^{-4}$$

$$x = 10.4 \text{ ft}$$

Then at D,

$$M' = \left(90 \times \frac{10.4}{3} - 90 \times 10.4\right)10^{-4}$$

$$= -0.0624 \text{ ft}$$

$$= -0.75 \text{ in}$$

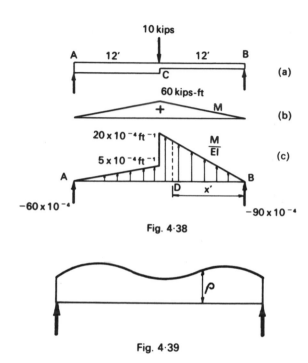

Fig. 4·38

Fig. 4·39

Moment-area and conjugate beam methods are still of some assistance in computing deformations when the bending stresses are non-linear. In such beams the curvature at a number of points must be calculated as described in section 4.4. The graph of ρ is now used as the conjugate load (Fig. 4.39). Usually it will be necessary to proceed by numerical methods. But at least the problem has been reduced to the calculation of shears and moments due to a very irregular load.

4.9 Flexibility and Stiffness

We now give some thought to the meaning of flexibility and stiffness in the case of bars deformed in bending. Discussion will be restricted here to the relation between forces (or couples) applied at the *ends* of the bar and *end* displacements. We see at once that whereas the end of a bar deformed axially can only move in the x direction, one deformed in bending can suffer displacements v and dv/dx according to the support conditions. It will thus be possible to define several stiffness or flexibility coefficients. Some typical values will be derived for straight uniform beams only.

If a single *unit* force, or *unit* couple, is applied to the end of a beam, the various end displacements are classified as flexibility coefficients. The beam of Fig. 4.40 is supported at one end only. A positive unit end force will produce the bending moments as shown. If we use moment-area methods, we see by inspection that the free end slope is $L^2/2EI$ and the end deflection is $L^3/3EI$. These expressions then are the flexibility coefficients for this type of loading.*

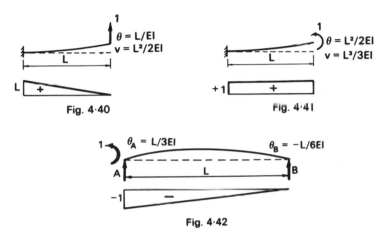

Fig. 4·40   Fig. 4·41

Fig. 4·42

The application of a unit couple (Fig. 4.41) is easier since the induced bending moment is constant. (Note that the external couple is applied in the positive sense, i.e., counterclockwise.) The area of the M/EI diagram is L/EI and its moment about the free end is $L^2/2EI$, hence these are the required displacements, or flexibility coefficients.

The beam of Fig. 4.42 is simply supported. We can apply an end couple, either at the left-hand or at the right-hand end. Applying the couple at A we obtain the bending moment diagram shown. The conjugate beam is perhaps the best tool for this problem. If the bending moment diagram is regarded as a conjugate load the total load is $L/2EI$ (downward). The end conjugate shears are

$$S'_A = -\frac{2}{3}\frac{L}{2EI} = -\frac{L}{3EI}; \quad S'_B = \frac{1}{3}\frac{L}{2EI} = \frac{L}{6EI}$$

---

* It should be noted that these are the displacements caused by the bending of the elements. Where the loading produces shear force, as in this case, a further displacement will be caused by the shearing of the elements. This is discussed in Chapter 5.

These are the values of the end rotations which are thus $\theta_A = L/3EI$ and $\theta_B = -L/3EI$.

To derive stiffness coefficients we must give the beam a *single* unit end displacement and compute the end forces induced thereby. This restricts the type of end supports we can have. The supports must be capable of inhibiting all displacements except the one being considered. We can imagine four different end displacements (Figs 4.43 (a), (b), (c) and (d)).

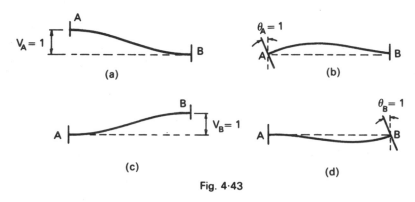

Fig. 4·43

Consider the first of these displacement modes. Our objective is to find the reactions R_A, R_B, C_A and C_B (Fig. 4.44 (a)) resulting from a unit translation of A. We could use the basic integration method of section 4.2. Starting with the fact that $w = 0$ we find that

$$S = c_1$$

$$M = -c_1 x + c_2$$

$$EI\,dv/dx = -c_1 x^2/2 + c_2 x + c_3$$

$$EIv = -c_1 x^3/6 + c_2 x^2/2 + c_3 x + c_4$$

With boundary conditions $v = 1$ and $dv/dx = 0$ at $x = 0$, and $v = 0$ and $dv/dx = 0$ at $x = L$, we can evaluate the four constants from the last two equations.

$$c_1 = -\frac{12EI}{L^3}; \quad c_2 = -\frac{6EI}{L^2}; \quad c_3 = 0; \quad c_4 = EI$$

From the first two equations we can now determine the shear force and bending moment at each end. At B the external forces will be equal to S_B and M_B. At A the external forces will be equal in magnitude to S_A and M_A but opposite in sign. We find that

$$R_A = \frac{12EI}{L^3}; \quad R_B = -\frac{12EI}{L^3}; \quad C_A = \frac{6EI}{L^2}; \quad C_B = \frac{6EI}{L^2}$$

Suppose we solve the same problem by the moment-area method. The bending moment diagram will be as shown in Fig. 4.44 (b). The difference of end slopes is zero, since tangents at A and B remain horizontal. Hence the total area of the diagram is zero, showing that $C_A = C_B$ (which could have been deduced by the symmetry of the configuration). The intercept of the end tangents on a vertical line through A is 1. So the moment of M/EI about A is 1.

$$\frac{1}{EI}\left(\frac{CL}{4} \times \frac{5L}{6} - \frac{CL}{4} \times \frac{L}{6}\right) = 1$$

$$\therefore \quad C = 6EI/L^2$$

By statics $R_A = -R_B = 2C/L = 12EI/L^3$.

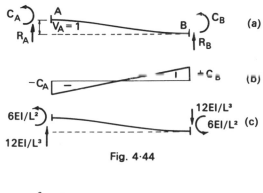

Fig. 4·44

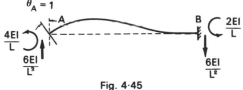

Fig. 4·45

For the deformation of Fig. 4.43 (b) the solution follows the same lines. The results are shown in Fig. 4.45. It might be noted that the value $4EI/L$, for instance, represents the constant of proportion between the rotation of A and the couple induced. If we write

$$C_A = k\theta_A$$

then

$$k = 4EI/L$$

It does not mean that the couple is expected to be $4EI/L$ when A is rotated through 1 radian, because in practical beams we should not expect the stresses to remain in the proportional range if we attempted such an enormous rotation of the end.

The results for deformations c and d of Fig. 4.43 can be deduced by inspection from the values given in Figs 4.44 and 4.45. Apart from turning the beam end for end, a few sign changes are required.

Problems

4.1   A beam has a rectangular cross-section 3 in wide and 10 in deep. The material has a linear stress–strain relationship. (*a*) What is the depth of the neutral axis below the top of the beam? If the extreme fibre stress is 8000 psi what is (*b*) the stress 1 in from the N.A. (*c*) the stress 3 in from the N.A. (*d*) the bending moment at the section?

4.2   A beam has an I section of overall dimensions 6 in wide and 15 in deep. The flanges are 1 in thick and the web is $\frac{1}{2}$ in thick. What bending moment can this section sustain if the stress is not to exceed 20,000 psi? When the maximum bending moment is applied, what is the bending stress at the junction of the web and flange?

4.3   A beam having the cross-section shown in Fig. P4.3 is subjected to a bending moment about the axis *zz*. The stress distribution is assumed to be linear. Can the extreme fibre stress be reduced by removing material from the corners as indicated?

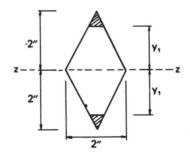

Fig. P4·3

If so, what value of y_1 gives the smallest maximum stress for a given bending moment?

4.4   Find the bending moment which can be resisted by a cast iron pipe 6 in external diameter and $4\frac{1}{2}$ in internal diameter when the greatest stress due to bending is 1500 psi.

4.5 A beam of rectangular cross-section is required to resist a bending moment of 30,000 lb-ft. The material has a linear stress–strain relation and can safely take a stress of 1800 psi. If the depth of the beam is twice the width, calculate the dimensions of the beam.

4.6 A beam has a rectangular cross-section 3 in wide and 10 in deep. The stress–strain curve is parabolic and follows the equation $\sigma = 8 \times 10^3 \epsilon \pm 8 \times 10^5 \epsilon^2$ where σ is in ksi. The curve is shown in Fig. P4.6, point F representing the failure point of the material. The curve is the same shape in tension and compression.

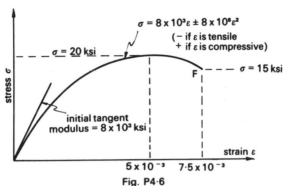

$$\sigma = 8 \times 10^3 \epsilon \pm 8 \times 10^8 \epsilon^2$$
(– if ϵ is tensile
+ if ϵ is compressive)

$\sigma = 20$ ksi

$\sigma = 15$ ksi

F

stress σ

initial tangent modulus = 8×10^3 ksi

strain ϵ

5×10^{-3} $7 \cdot 5 \times 10^{-3}$

Fig. P4·6

When the extreme fibre stress is 20 ksi (a) sketch the stress distribution on the cross-section, (b) find the magnitude and position of the resultant compressive force and the resultant tensile force, (c) find the bending moment, (d) find the extreme fibre strains, and (e) find the curvature of the beam at this section due to bending.

4.7 An I beam has a cross-section with overall dimensions 20 in × 8 in, a flange thickness of 0.8 in and a web thickness of 0.3 in. It is made of material which has the stress–strain curve given in Fig. P4.6.

When the extreme fibre stress is 20 ksi, find approximately the bending moment on the section.

4.8 A beam of rectangular section 10 in × 3 in is made of material whose stress–strain curve in compression is shown in Fig. P4.6. In tension the stress–strain curve is linear with a modulus of 8×10^7 ksi.

If the beam is subjected to a bending moment which causes a stress of 20 ksi at the compression face, find the depth of the neutral axis below the compression face. Find also the strains at the extreme fibres, and the curvature due to bending.

Would you expect the neutral axis to be at the same depth for other values of the bending moment?

4.9 A beam has a rectangular cross-section 4 in wide and 12 in deep. For a depth of 1 in at the top and bottom of the beam, the material has an elastic modulus of 30 × 10⁶ psi. For the remainder of the depth the material has an elastic modulus of 15 × 10⁶ psi. At a certain cross-section, the bending moment causes a stress of 12,000 psi at the top surface A (Fig. P4.9).

(*a*) Find the strain at level A.
(*b*) Find the strain at level B.
(*c*) Sketch the distribution of stress on the cross-section.
(*d*) Find the bending moment.

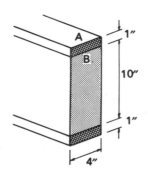

Fig. P4·9

4.10 In a beam of rectangular section of width 6 in and depth 20 in, the upper and lower halves are made of different material joined so that there is no slip at the interface. Both materials have a linear stress–strain relation and the elastic modulus of the lower is twice that of the upper material.

For a given stress σ at the top surface find the depth to the neutral axis. Hence, show that the neutral axis does not vary with the bending moment. Find the extreme fibre stresses when the bending moment is 100 kip-ft.

4.11 A beam has a rectangular section 10 in deep and 4 in wide. At a certain cross-section the distribution of longitudinal stress due to a given loading is as shown by the full line in Fig. P4.11.

(*a*) Find the bending moment at this section.

(*b*) If, for simplicity of calculation, it is assumed that the stress at every fibre is 10,000 psi as shown by the dotted line, what percentage error would this cause in the value of the bending moment?

(*c*) Find the curvature of the beam due to bending if $E = 12 \times 10^6$ psi.

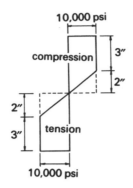

Fig. P4·11

4.12 A steel pipe is 6 in external diameter and 5 in internal diameter. Find the plastic moment of the section if the yield stress of the steel is 35,000 psi.

4.13 A steel I beam has flanges 6 in × $\frac{1}{2}$ in and a web 11 in × $\frac{1}{2}$ in, the total depth being 12 in. The yield stress of the steel is 35,000 psi. At a certain section the bending moment causes a strain equal to the yield strain at the junction of the flange and web. Calculate the value of the bending moment.
Find the curvature due to bending at this section if $E = 30 \times 10^6$ psi.

4.14 The beam of problem 4.13 is subjected to a bending moment which produces yielding to a point 4 in from the neutral axis. If the unloading curve follows the initial tangent modulus, determine the residual stress distribution when the beam is unloaded.

4.15 A beam of rectangular section, width b and depth h, is made of an elastic-plastic material of yield stress σ_y. It is stressed partially into the plastic range by bending. After removal of the loading the residual stress in the top fibre is $\sigma_y/3$ compression. Find the magnitude and sign of the applied bending moment.

4.16 Prove that, for a beam made of material with a linear stress–strain law, $M/EI = d^2y/dx^2$. State any limitations which must be observed when using this equation.

4.17 A 12 in × 15 in beam, span 20 ft, $I = 261$ in^4, is simply supported at each end and carries a uniformly distributed load of 1.5 kips per foot run. Find the deflection of the beam at the centre and the angle of slope at each end. The material is elastic and $E = 30 \times 10^3$ ksi.

4.18 A beam AB of span 12 ft is cantilevered from A. It carries a distributed load (downward) the intensity of which varies linearly from 1200 lb/ft at the fixed end to zero at the free end. Find the deflection at the free end. $I = 288$ in^4 and $E = 20 \times 10^6$ psi.

4.19 A simply supported beam AB of span 20 ft carries a distributed load (downward) the intensity of which varies linearly from zero at A to 1200 lb/ft at B. Find the maximum deflection. $I = 400$ in^4 and $E = 15 \times 10^6$ psi. (Use direct integration.)

4.20 A uniform beam of length $L + 2a$ rests on two supports L feet apart and overhangs for a distance a at each end. It is loaded with concentrated loads W at each extremity and a concentrated load $2W$ at the centre of the span. Find the deflection at each end. (*Note*: Use the conjugate beam method.)

4.21 A uniform beam of length L, simply supported at each end, carries a distributed upward load the intensity of which is given by $w = w_0 \sin \pi x/L$, where w_0 is the load intensity at the centre and x is measured from a support.
   Derive expressions for the slope and deflection of the beam.

4.22 A horizontal cantilever AB of uniform cross-section and length L is built into a wall at the end A. The beam carries a uniformly distributed load of intensity W/L along its full length and a concentrated load $5W$ at a point $L/3$ from A. Show that the deflection of the free end B is $\dfrac{241}{648} \dfrac{WL^3}{EI}$.

4.23 A simply supported beam ACB of span 20 ft carries a uniformly distributed load of 500 lb/ft run extending from A to C, a distance of 5 ft. Derive an expression for the deflection and determine the maximum deflection. $E = 30 \times 10^6$ lb/in^2 and $I = 91.4$ in^4.

4.24 A uniform beam is simply supported over a span of 20 ft. It carries loads of 5 kips and 10 kips at 6 ft and 12 ft respectively from the left-hand support. Find the deflection of the mid-point of the beam in terms of EI. State the units of EI in your answer.

4.25 A simply supported beam spanning 24 ft is loaded as shown in Fig. P4.25. Find the central deflection. $I = 100$ in^4 and $E = 30 \times 10^6$ psi.

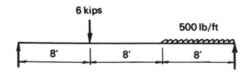

Fig. P4·25

4.26 A simply supported beam AB of length L (Fig. P4.26) is subjected to a couple C at E. Determine the deflection curve of the beam and find the slope at A, B and E and the deflection at E.

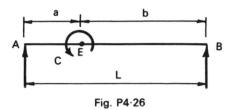

Fig. P4·26

4.27 A simply supported beam with overhang BC (Fig. P4.27) carries a uniformly distributed load w. Find an expression for the deflection curve from A to B and find the slope at B.

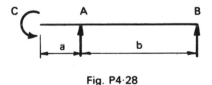

Fig. P4·27

4.28 A simply supported beam with overhang a is shown in Fig. P4.28. At the end of the overhang a couple C is applied. Determine the equation of the deflected shape of the portion AB of the beam.

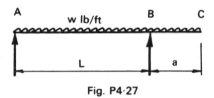

Fig. P4·28

4.29 A beam AB of constant EI spans 20 ft and is direction-fixed at each end. At 6 ft from each end the beam supports a load of 8 kips. By any method, find the end support couples on the beam and draw the bending moment and shear force diagrams.

4.30 A beam AB spans 30 ft. It is direction-fixed at A and simply supported at B. It carries a distributed load which varies from 1500 lb/ft at A to zero at B. Obtain an expression for the bending moment at a section x feet from A.

4.31   A uniform beam of span 24 ft is simply supported at A and built in at B. It is loaded as shown in Fig. P4.31. Use the step function method to find the deflection at the mid-point. $I = 240$ in^4; $E = 20 \times 10^6$ psi.

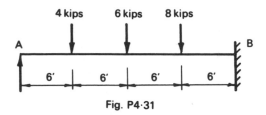

Fig. P4·31

4.32   The beam of Fig. P4.32 has constant EI. Sketch the bending moment and shear force diagrams indicating the main values on each.

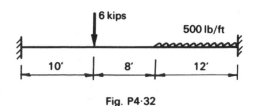

Fig. P4·32

4.33   A beam AB is 12 ft long. It is completely fixed at A. There is a support at B, but in the unloaded position there is a gap of 0.48 in between the beam and the support. For the loading shown in Fig. P4.33 find the reactions to the beam. $E = 24 \times 10^6$ psi; $I = 180$ in^4.

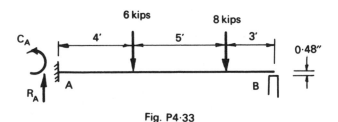

Fig. P4·33

4.34   A beam AB, spanning 20 ft, is fixed at A and supported at B on an elastic support which will deflect 0.02 feet per kip of force applied to the support. It carries a uniformly distributed load of $\frac{1}{2}$ kip/ft over the whole span. Find the force on the elastic support and draw the bending moment diagram for the beam. $I = 150$ in^4 and $E = 26 \times 10^3$ ksi.

4.35 The support couples for an elastic beam fixed at both ends and loaded by a single concentrated load are

$$C_A = +\frac{Pab^2}{L^2} \quad \text{and} \quad C_B = -\frac{Pa^2 b}{L^2}$$

where a and b are the distances of the load P from A and B respectively, and the span is L. Use this information to find the reactions and support moments for a beam fixed at A and B and loaded by a uniformly distributed load w between C and B, where C is the centre of the span.

4.36 Find the strain energy due to bending in the beam of problem 4.19.

4.37 Derive an expression for the total bending strain energy in the beam of problem 4.21.

4.38 A simply supported beam of constant EI has a span L. A couple C is applied to one end. Find the bending strain energy of the beam and use this to obtain an expression for the slope of the beam at the loaded end.

4.39 A simply supported beam spanning 20 ft carries a 12 kip load 12 ft from one end. If $E = 30 \times 10^6$ psi and $I = 150$ in^4 calculate the total strain energy stored in the beam due to bending. Hence, calculate the deflection under the load.

4.40 Solve problem 4.17 by moment-area methods.

4.41 Solve problem 4.22 by moment-area methods.

4.42 Solve problem 4.24 by moment-area methods.

4.43 A uniform beam AB is simply supported over a span of 20 ft. It carries a load of 8 kips at the centre. Use the conjugate beam method to find the slope at A and the deflection at the quarter point D. Use kip and feet units.

4.44 State the relationship between the bending moment and the variation in slope between two points in a beam. In the cantilever ABC shown in Fig. P4.44, the slope at B is zero. Determine the ratio of W_2 to W_1 given that the moment of inertia and Young's modulus are constant.

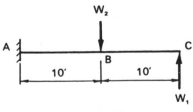

Fig. P4·44

4.45   The beam ABC of Fig. P4.45 is direction-fixed at A, simply supported at B and cantilevered to C. Use the conjugate beam method to obtain the shear force and bending moment diagrams for the member, $E = 20 \times 10^6$ psi and $I = 14,400$ in⁴.

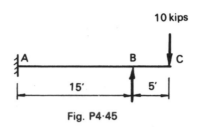

Fig. P4·45

4.46   Figure P4.46 shows a beam of constant EI spanning 36 ft. It is simply supported at A and fixed at B. Use the conjugate beam method to obtain the bending moment diagram and also to obtain the slope at A and the deflection at C.

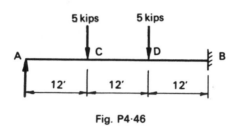

Fig. P4·46

4.47   The beam of Fig. P4.47 is direction-fixed at A and simply supported at C. The EI value of AB is three times that of the part BC. Use the conjugate beam method to find the reaction at C for the given load.

Indicate how you would locate, from your final diagrams, the position of maximum deflection.

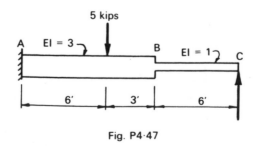

Fig. P4·47

4.48   For the beam shown in Fig. P4.48 find the difference in level of A and E. State the units of EI and the units of your answer.

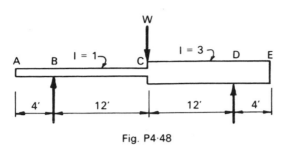

Fig. P4·48

4.49   A uniform beam AB is simply supported at A and direction-fixed at B. The length is L and the flexural rigidity EI. Find the four stiffness coefficients associated with the rotation of end A, i.e., the end couple and end force at both A and B induced by unit rotation of A.

4.50   A straight beam CD 20 ft long is cantilevered from C. The section is rectangular, the width at C being 15 in and the depth 9 in. The beam tapers linearly to D where the dimensions are one-third of those at C.

Find (a) the two flexibility coefficients for a unit force at D and (b) the two flexibility coefficients for a unit end couple at D (use inch, lb units). Take $E = 25 \times 10^6$ psi.

4.51   A straight linearly tapered beam PQ has a rectangular section and is 25 ft long. It is direction-fixed at both ends. The width is constant at 10 in and the depth varies from 20 in at P to 10 in at Q. Find the stiffness coefficients associated with vertical displacement at Q, i.e., find the couple and force induced at both P and Q when the end Q is given a unit vertical displacement without rotation. Take $E = 2.3 \times 10^6$ psi. Use lb, inch units.

4.52   A beam has a rectangular section $b \times d$. It is made of material whose Young's modulus is E up to a yield stress of σ_y (and yield strain ϵ_y). For strains in excess of ϵ_y the stress is constant at σ_y. Derive expressions for the relation between the bending moment, M, and the curvature ρ, both before and after yielding.

Plot the M-ρ curve if $b = 3$ inches, $d = 10$ inches, $E = 30 \times 10^6$ psi and $\sigma_y = 40,000$ psi.

CHAPTER 5

Shear

5.1 Stresses due to Shear

As before we consider an elemental slice of the bar lying between cross-sections
C and D. If we were to proceed as we did in Chapters 3 and 4 we should now
assume that face D moves without distortion either parallel to the y axis or
parallel to the z axis. The analysis is the same for each alternative so we shall
consider here only the movement parallel to y. The assumption that faces C
and D remain plane (Fig. 5.1) would lead immediately to a uniform distribution
of shear strain over the depth of the bar, and consequently to a uniform
distribution of shear stress. Unfortunately, such a shear stress distribution
violates the laws of equilibrium. Consequently, the basic assumption that
faces C and D undergo relative movement while remaining plane and undistorted proves to be untenable for shearing types of deformation. On the
other hand, the laws of equilibrium, which prohibit this assumption, themselves
provide a means of obtaining a partial solution of our problem.

First we must see why equilibrium is violated by the assumption of a
deformation as shown in Fig. 5.1.

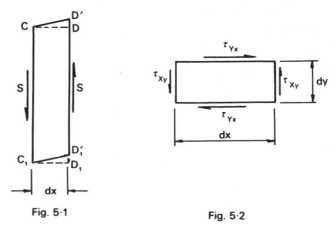

Fig. 5·1   Fig. 5·2

Fig. 5.2 shows an element of material with sides dx and dy and of length b normal to the paper. The total shear force on the top and bottom faces is $\tau_{Yx} b \, dx$ and these two forces constitute a couple of moment $(\tau_{Yx} b \, dx) dy$. Similarly, the shear forces on the vertical faces produce a couple of moment $(\tau_{Xy} b \, dy) dx$. If the element is small enough, the possible moment about the centre, O, of the normal stresses on the faces can be neglected. Rotational equilibrium then requires that the above two couples be equal and opposite.

$$(\tau_{Yx} b \, dx) \, dy = (\tau_{Xy} b \, dy) \, dx$$

or

$$\tau_{Yx} = \tau_{Xy} \tag{5.1}$$

This is known as the *Theorem of Complementary Shear Stresses.* On any two planes at right-angles the shear stresses in directions normal to the common edge of the planes are equal.

The exterior surfaces of a beam are nearly always free from shear stress. Therefore, close to the boundary of any cross-section, the component of shear stress at right-angles to the boundary must be zero (Fig. 5.3). If, on a plane normal to an external unloaded surface, there is a shear stress close to the boundary it must be parallel to the boundary, as at E in Fig. 5.3. At a protruding corner of the cross-section, such as B (Fig. 5.3), the shear stress components normal to AB and to BC must both be zero. Consequently, the shear stress must be zero at such a corner. Near any protruding corner there will be a region in which the shear stress is small. On the other hand, at a *re-entrant* corner, such as C, the shear stress tends to be high. A re-entrant corner gives rise to *stress concentration.*

These conclusions all stem from the complementary shear stress theorem, and since this is based entirely on equilibrium considerations, we can say

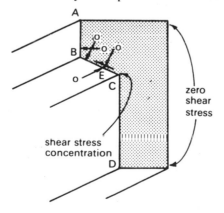

Fig. 5·3

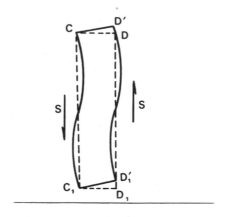

Fig. 5·4

that these are equilibrium requirements. Reverting to Fig. 5.1, we see that the uniform shear stress which results from our hypothesis would mean non-zero shear stresses at right-angles to the top and bottom boundaries of the section, which we know is impossible.

In fact, the shear strain must be zero at the top and bottom, which means that even after deformation of the elemental slice (Fig. 5.4) the corners of the element must remain right-angles. Thus the cross-sections CC and DD must become warped to an S-curve the exact shape of which will depend on the shape of the cross-section.

From these general requirements we can form some idea of the stress distribution on cross-sections of various shapes, and this will be a guide to the validity of certain assumptions. Sketches of such distributions are given in Fig. 5.5. In each example it is assumed that the plane of loading contains the vertical axis of symmetry. The sum of the *y components* of the shear stresses must be equal to the total shear force on the cross-section in the y direction.

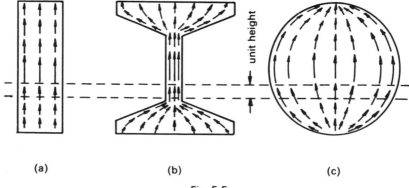

(a)   (b)   (c)

Fig. 5·5

In thin sections, such as the web portion of Fig. 5.5 (b), it is reasonable to assume that the stresses are all vertical since no point is far from a vertical boundary. In the circle and in the flange portions of the I beam, the stresses are clearly not vertical in general. In the rectangle (Fig. 5.5 (a)) the stresses are vertical at the sides and at the centreline, and it might not be unreasonable to assume that all stresses are vertical.

Across each of the shapes of Fig. 5.5 two lines have been drawn defining a band of unit height. The total vertical component of shear force resisted within such a band will be called the vertical *shear flow* at that level of the cross-section. The dimensions of shear flow are thus FL^{-1}. The shear flow will be denoted by q, and the integral of q from top to bottom of the section must equal the shear force S. At any level, the shear flow is the quantity frequently required for design purposes. It can be calculated from the bending stresses by equilibrium considerations alone.

If the actual shear *stress* at any point is required, as it sometimes is, its calculation requires an assumption about the distribution of stress across the width of the beam. As can be seen from Fig. 5.5 this might be difficult in some instances. In other cases the assumption of uniform distribution is reasonable, and then the stress can be found by dividing the shear flow by the width of the beam at the particular level.

Directly from the bending stresses we can find the shear flow on a horizontal plane of the beam, such as the shear flow q_H on the plane EG of Fig. 5.6. At any point across the width of the beam, the longitudinal shear stress on the horizontal plane EG is equal and opposite to the vertical shear stress on the

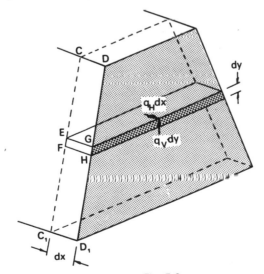

Fig. 5·6

plane *GH*. The shear flow on each of these planes is merely the summation of shear stresses across the width of the beam; hence the shear flow q_H on *EG* is also equal and opposite to the shear flow q_V on *GH*. In this way we can find q_V.

Example 5.1. A beam has a rectangular cross-section, 3 in wide and 8 in deep. At a certain cross-section the bending moment is 200,000 lb-in while at a section 1.5 in further along, the bending moment is 224,000 lb-in. Find the vertical shear flow on a cross-section between CC_1 and DD_1 at a depth of 2.8 in from the top of the beam. Assume that the bending stresses are distributed linearly.

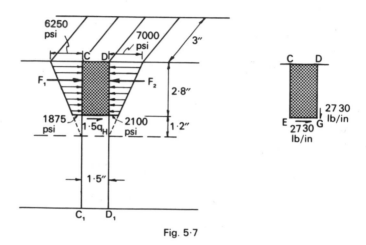

Fig. 5·7

Solution. We first calculate the bending stresses on the sections CC_1 and DD_1 (Fig. 5.7).

$$I = \frac{bd^3}{12} = \frac{3 \times 8^3}{12} = 128 \text{ in}^4$$

On section CC_1:
   At the top,

$$\sigma = \frac{My}{I} = \frac{200,000 \times 4}{128} = 6250 \text{ psi}$$

2.8 in below the top,

$$\sigma = \frac{200,000 \times 1.2}{128} = 1875 \text{ psi}$$

On section DD_1:
   At the top,

$$\sigma = \frac{224,000 \times 4}{128} = 7000 \text{ psi}$$

2.8 in below the top,

$$\sigma = \frac{224,000 \times 1.2}{128} = 2100 \text{ psi}$$

We now consider the equilibrium of the block of material whose side elevation is *CEGD* (Fig. 5.7) and which runs the full width of the beam.

The forces on the ends of this block are:

$$F_1 = \left(\frac{6250 + 1875}{2}\right) \times 3 \times 2.8 = 34,125 \text{ lb}$$

and

$$F_2 = \left(\frac{7000 + 2100}{2}\right) \times 3 \times 2.8 = 38,220 \text{ lb}$$

For horizontal *equilibrium*, the difference between these two forces is balanced by the total shear on the face EG, which is $q_H \times 1.5$

$$\therefore \quad 1.5q_H = 38,220 - 34,125 = 4095 \text{ lb}$$

and

$$q_H = 2730 \text{ lb/in}$$

The shear flow q_V on the vertical cross-section is also 2730 lb/in at a depth of 2.8 in below the top of the beam.

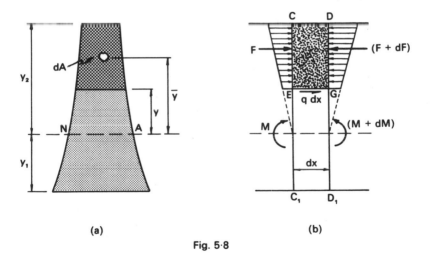

(a)   (b)

Fig. 5·8

The above calculation can be generalized to give an expression for the shear flow at any level of a cross-section of any shape. Suppose that the shear flow is required at a distance y from the neutral axis (Fig. 5.8). First consider the section CC_1 where the bending moment is M. Provided the bending stresses vary linearly, then on an element of area dA at a distance $\bar{y}$ from the neutral axis the bending stress is

$$\sigma = \frac{M\bar{y}}{I}$$

The force on the element is

$$dF = \frac{M\bar{y}\,dA}{I}$$

The total force between the top and the level y is then

$$F = \int dF = \frac{M}{I} \int_{y}^{y_2} \bar{y}\,dA$$

The integral represents the first moment of the dark area about the neutral axis. If this is denoted by Q we can write

$$F = \frac{MQ}{I} \tag{5.2}$$

This expression corresponds to the force 34,125 lb in the previous example.

In a similar way, for the section DD_1 where the bending moment is $(M + dM)$ we find that

$$(F + dF) = \frac{(M + dM)Q}{I} \tag{5.3}$$

The difference of these two forces is balanced by the shear on face EG which is $q\,dx$.

$$\therefore \quad q\,dx = (F + dF) - F$$

$$= \frac{dM\,Q}{I}$$

Hence

$$q = \frac{dM}{dx} \times \frac{Q}{I} = \frac{SQ}{I} \tag{5.4}$$

The quantity dM/dx is equal to the shear force on the cross-section, while Q/I is a function of the shape of the section and determines the distribution of shear flow from top to bottom of the beam. The same result would be obtained if we considered the equilibrium of the block C_1EGD_1 below EG. The first moment of the whole cross-section about the neutral axis is zero, therefore the Q of the light area (Fig. 5.8 (a)) is the same as the Q of the dark area. At the top and bottom of the beam, where y is equal to y_1 or y_2, the quantity Q/I becomes zero. The equation (5.4) thus gives $q = 0$ at the top and bottom which agrees with the requirements of equilibrium. This is only to be expected as equation (5.4) was derived from equilibrium considerations.

Example 5.2. Determine the distribution of shear flow for a beam of rectangular cross-section.

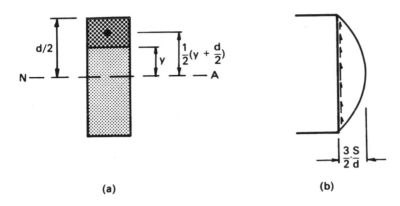

(a)   (b)

Fig. 5·9

Solution. We find the shear flow at y from the neutral axis. The area above y (Fig. 5.9 (a)) is $b(d/2 - y)$. Its centroid is $\frac{1}{2}(y + d/2)$ from the neutral axis.

$$\therefore \quad Q = b\left(\frac{d}{2} - y\right)\frac{1}{2}\left(\frac{d}{2} + y\right)$$

$$= \frac{b}{2}\left(\frac{d^2}{4} - y^2\right)$$

$$I = \frac{bd^3}{12}$$

and

$$\frac{Q}{I} = \frac{12}{bd^3}\frac{b}{2}\left(\frac{d^2}{4} - y^2\right)$$

$$= \frac{6}{d}\left(\frac{1}{4} - \left(\frac{y}{d}\right)^2\right)$$

$\therefore$ at y from N.A.

$$q = \frac{dM}{dx}\frac{6}{d}\left(\frac{1}{4} - \left(\frac{y}{d}\right)^2\right) \tag{5.5}$$

This indicates a parabolic variation of q.
When $y = \pm d/2$,

$$q = 0$$

and at the centre, $y = 0$ and

$$q = \frac{dM}{dx}\frac{6}{4d}$$

$$= \frac{2}{3}\cdot\frac{S}{d}$$

The variation is often represented graphically as in Fig. 5.9 (b). It should be noted that although the ordinates as plotted are normal to the section they represent stresses *parallel* to the section. The area of this graph must represent the total shear force, and this result satisfies this requirement.

For a rectangular beam it is reasonable to assume that the stress is uniform across the width. The maximum shear stress is then obtained by dividing the shear flow at mid-depth by b.

$$\tau_{max} = \frac{q_{max}}{b} = \frac{3\ S}{2\,bd} \tag{5.6}$$

The expression S/bd is the *average* vertical shear stress on the cross-section.

Example 5.3. A rolled-steel beam is 18 in deep. It is proposed to increase its bending strength by welding an 8 in $\times$ ¾ in plate to each flange (Fig. 5.10). Find the shear to be resisted by 1 in of the fillet weld at a location where the total shear force in the beam is 60,000 lb. The I of the unplated beam is 840 in⁴.

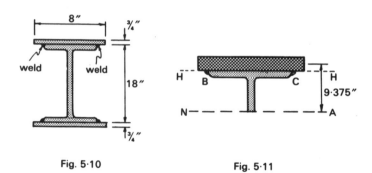

Fig. 5·10   Fig. 5·11

Solution. A length of 1 in of weld at B and C (Fig. 5.11) together have to resist the shear on the horizontal plane HH for a length of 1 in along the beam. This force is the shear flow at level HH and is given by

$$q = S \times \frac{Q}{I}$$

where Q is the first moment of the shaded area about N.A.

$$Q = (8 \times 0.75)\,9.375^2 = 56.25 \text{ in}^3$$

I of unplated beam $\qquad\qquad\qquad = 840 \text{ in}^4$

I of two plates about N.A. $= 2(8 \times 0.75)\,9.375^2 = 1054 \text{ in}^4$

$\qquad\qquad\qquad\qquad\qquad$ Total $I = 1894 \text{ in}^4$

$$\therefore \quad q = 60,000 \times \frac{56.25}{1894} = 1780 \text{ lb per in}$$

Force resisted by 1 in of one weld $= 890$ lb

In this problem, the function of the welds is clear when we realize that an increment of bending moment from one cross-section to another implies an increase in the force in the flange plate. This increase can be brought about only by the force exerted on the plate by the fillet welds.

In the previous chapters we considered an elemental slice of beam of length dx. By making an assumption in regard to the distribution of strains across this element we avoided the necessity of further subdividing it. In the present chapter, such an assumption about strains is found to be untenable, and as a result we have had to subdivide our strip further by horizontal cuts in order to achieve a solution. Even so, in order to evaluate *stresses* from the shear flow we have to make an assumption of strain distribution across the width of the beam. If such an assumption is unreasonable, as in the flanges of I beams, we must either abandon the attempt to obtain stresses or further subdivide our element laterally. The latter procedure is beyond the scope of the present book.

The present analysis thus enables us to evaluate shear stresses if we can assume that these are uniform across the width of the beam. If this assumption cannot be made, we must be content with the value of the shear flow, which is frequently sufficient for design purposes.

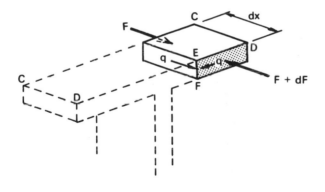

Fig. 5·12

So far we have discussed the shear flow on a *horizontal* plane of the beam. This is equal to the *vertical* shear flow in the cross-section at the corresponding level. However, the method of procedure is of more general application. We consider as usual a length of beam dx between two cross-sections CC and DD. Any block of this material can be isolated by a cutting plane EF parallel to the beam axis (Fig. 5.12). Provided the bending stresses on the ends of the block $CDEF$ are different, equilibrium of the block parallel to the beam axis will reveal a shear flow in the cutting plane in a direction parallel to the beam axis. There will be a shear flow of equal magnitude in the cross-section at right-angles to the cutting plane. The reader should check that, if the bending stress distribution is linear, the shear flow is given by equation (5.4), in which Q still represents the first moment of the shaded area about the neutral axis. The shear flow in this cutting plane is accompanied by an equal shear flow in the cross-section as shown.

5.2   Stresses in Thin-walled Sections

We have seen that close to the boundary of a cross-section the shear stress must be parallel to the boundary. It follows that if the cross-section is thin-walled, the shear flow will everywhere be predominantly parallel to the boundary of the wall, irrespective of whether this is parallel to the external loading or not. This can best be illustrated by a numerical example.

Example 5.4. A beam has the cross-section shown in Fig. 5.13. Calculate both the horizontal and vertical shear force resisted by each of the elements of the cross-section if the total shear force is 20,000 lb and acts in the plane of symmetry.

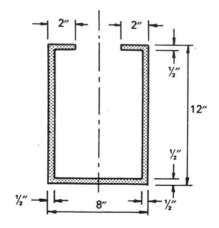

Fig. 5·13

Solution. The position of the centroid and the moment of inertia of the section about the neutral axis are first determined (see Appendix B). The centroid is 5.323 in above the base and 6.677 in below the top. The moment of inertia is 301.606 in⁴.

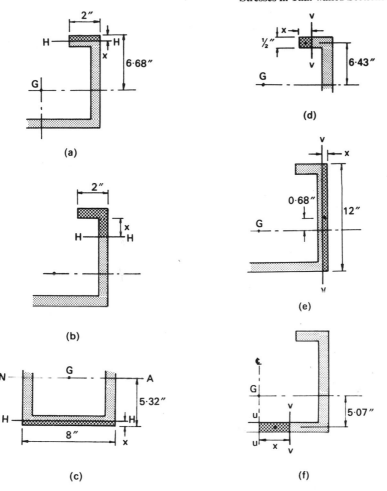

Fig. 5·14

(a) Vertical shear force in the top flange

Consider a cutting plane such as *H–H* (Fig. 5.14 (a)) at *x* inches from the top. For the area above *H–H*,

$$Q = ?r \left(6.677 - \frac{x}{2} \right)$$

$$q = \frac{S}{I} Q = \frac{20,000}{301.606} (13.352x - x^2)$$

$$V_1 = \int_0^{0.5} q \cdot dx = 66.311 \times 1.628 = 108 \text{ lb}$$

(b) Vertical shear force in the web
Consider the plane *H–H* (Fig. 5.14 (b)) to cut the web.

$$Q = (2 \times \tfrac{1}{2})6.426 + \frac{x}{2}\left(6.177 - \frac{x}{2}\right)$$

$$q = \frac{S}{I}Q = 66.311\left(6.427 + 3.088x - \frac{x^2}{4}\right)$$

$$V_2 = \int_0^{11} q \cdot dx = 9721 \text{ lb}$$

(c) Vertical shear force in bottom flange
For the lower flange it is convenient to consider the area below, rather than above, a plane such as *H–H* (Fig. 5.14 (c)).

$$Q = 8x\left(5.323 - \frac{x}{2}\right)$$

$$q = \frac{S}{I}Q = 66.311(42.584x - 4x^2)$$

$$V_3 = \int_0^{0.5} q \cdot dx = 342 \text{ lb}$$

(d) Horizontal shear force in top flange
To find the horizontal shear flow, a cutting plane such as *V–V* is considered. For the shaded area (Fig. 5.14 (d)),

$$Q = \frac{x}{2}6.427$$

$$q = \frac{S}{I}Q = 66.311(3.213x)$$

$$H_1 = \int_0^{1.5} q \cdot dx = 240 \text{ lb}$$

(The integration can only be taken as far as $x = 1.5$ otherwise the plane *V–V* will enter the vertical web.)

(e) Horizontal shear force in web
For the shaded area in Fig. 5.14 (e),

$$Q = 12x \times 0.677$$

$$q = \frac{S}{I}Q = 66.311(8.124x)$$

$$H_2 = \int_0^{0.5} q \cdot dx = 67 \text{ lb}$$

(f) Horizontal shear force in bottom flange

Cutting the bottom flange with a plane V–V (Fig. 5.14 (f)), we can consider the area to the right of this plane. However, it is more convenient to consider the area between the planes V–V and U–U. Since the horizontal shear flow on U–U (the centreline) is zero by symmetry, equation (5.4) gives the shear flow on plane V–V

$$Q = \frac{x}{2}5.073$$

$$q = \frac{S}{I}Q = 66.311(2.536x)$$

Since the stresses on the section are symmetrical, the total horizontal shear in the bottom flange is zero. The shear on each half separately will be found

$$H_3 = \int_0^{3.5} q \cdot dx = 1030 \text{ lb}$$

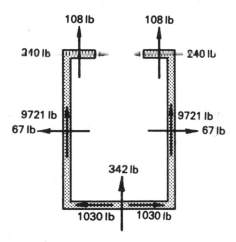

Fig. 5·15

The total shear forces carried by the various portions of the cross-section are summarized in Fig. 5.15. It should be noted that every portion sustains both a horizontal and a vertical shear component. The total of the vertical components is 20,000 lb, which is the shear force on the cross-section. The direction of these components is governed by the sign of the shear force, S. We also notice that the shear in each rectangle is predominantly in the direction of the larger dimension, irrespective of whether the rectangle lies vertically or horizontally. However, as noted above, the transverse shear component is not zero.

5.3   Shear Centre

In this chapter, we have so far considered beams whose cross-sections have a vertical axis of symmetry, and where the plane of loading contains this axis. The majority of beams, in practice, have this form. The vertical components of shear stresses on the section have a resultant equal to the shear force, S, and this resultant evidently acts along the axis of symmetry and therefore passes through the centroid. The horizontal shear stresses are symmetrical and cancel out.

In the previous chapter, it was shown that provided the plane of loading was parallel to a principal axis of the cross-section, the deflection of the beam is wholly in the direction of loading. The principal axis is not necessarily an axis of symmetry. If a beam with the cross-section of Fig. 5.16 (a) is loaded parallel to Cy it will deflect in this direction. The shear stresses will have a resultant S in the vertical direction but in general this resultant will not pass through the centroid. For a given section there is a certain point O through which the resultant shear force will always pass. This point is called the *shear centre*. If there is one axis of symmetry (as in Fig. 5.16 (a)) the shear centre will lie on this axis. If the plane of loading does not pass through the point O, the beam will twist even though there is no deflection in the direction Cz.

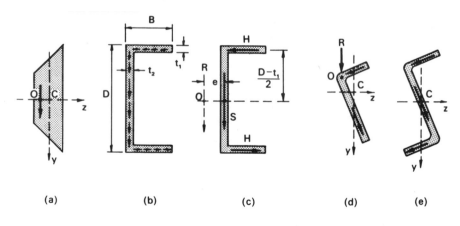

(a)   (b)   (c)   (d)   (e)

Fig. 5·16

The location of the shear centre requires the determination of the position of the resultant of the shear stresses. In general this is beyond the scope of the simple theory of bars since this theory does not yield the distribution of shear stress on the section except in special instances.

In thin-walled sections, the shear stresses transverse to the wall can be neglected, and a reasonable estimate can then be made of the location of the shear centre. The shear flow parallel to the wall of the section at any point can be found from equation (5.4), the resultant is known to be a vertical force of magnitude S, hence the position of the resultant can be found by taking moments about any convenient point.

Perhaps the most common cross-section of this type is the channel section shown in Fig. 5.16 (b). By the method used in example 5.4 we find that the total horizontal shear force in each flange is given approximately by

$$H = \frac{St_1(D - t_1)}{I} \frac{B^2}{4}$$

If the vertical shear forces in the flanges are neglected, then the total vertical shear, S, must be carried by the web. These total forces are shown in Fig. 5.16 (c). If the shear centre, O, is at a distance e from the centre of the web, the resultant of H, S, and H must pass through O. Taking moments about O, we have

$$2H\left(\frac{D - t_1}{2}\right) - Se = 0$$

$$e = \frac{H}{S}(D - t_1)$$

$$= \frac{t_1(D - t_1)^2 B^2}{4I} \tag{5.7}$$

For this section, I is approximately equal to $(D^2/4)(Dt_2/3 + 2Bt_1)$. With this value for I, equation (5.7) becomes

$$e = B\left(\frac{D - t_1}{D}\right)^2 \bigg/ \left(\frac{Dt_2}{3Bt_1} + 2\right) \tag{5.8}$$

Since this is, after all, only an approximate value for e, it may be reasonable to simplify the expression further by putting $(D - t_1) = D$. Equation (5.8) then becomes

$$e = \frac{B}{\left(\dfrac{Dt_2}{3Bt_1} + 2\right)} \tag{5.9}$$

In some sections, the location of the shear centre can be seen by inspection. For any angle section (Fig. 5.16 (d)), the lines of action of both of the internal shear forces pass through the corner of the angle, which must therefore be the shear centre. The equal Z section of Fig. 5.16 (e) has an anti-symmetry which determines that the shear centre coincides with the centroid.

5.4   Deformation due to Shear

When we considered the effects of axial force and bending moment, we started with an assumption about the deformation of an elemental slice of the beam, and from this we deduced the stress distribution. The deformation of the whole beam could easily be derived from the deformation of the individual elements.

In the present chapter we have not been able to make an assumption about the deformation of the element. Now, therefore, we must first study the deformation of the element before proceeding to consider the deformation of the beam as a whole.

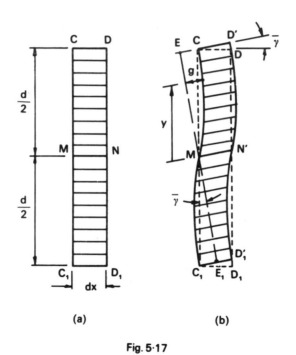

Fig. 5·17

Strictly speaking, for most shapes of cross-section we cannot, by the present elementary theory, determine the *stress* distribution, and in these cases we cannot accurately determine the strains or the beam deformation. For beams of normal proportions, the shear deformations are usually small compared with deformations due to bending, so that an approximate estimate of the former is quite sufficient.

Consider *a beam of rectangular cross-section*. Fig. 5.17 (a) shows the side elevation of an elementary slice of the beam before deformation. The shear force causes this element to deform to an S-shape which is shown much exaggerated in Fig. 5.17 (b). Vertical strains are very small and variation of vertical strains between sections C and D are negligible. Hence we may assume that layers parallel to the beam axis MN before deformation will still be parallel to the axis MN' after deformation. The deformation considered here is due to shear alone. Usually, bending deformation will also occur, as discussed in Chapter 4. This will take the form of a rotation about N' of the section DD_1 relative to CC_1.

We first compute the shape of the warped cross-section CMC_1. Line EME_1 is drawn perpendicular to the axis MN' (and perpendicular also to the top and bottom surfaces).

The shear strain at $\bar{y}$ from the neutral surface is obtained from equation (5.5)

$$\gamma = \frac{q}{Gb} = \frac{6S}{Gbd}\left\{\frac{1}{4} - \left(\frac{\bar{y}}{d}\right)^2\right\}$$

Then at y from the neutral surface the distance g (Fig. 5.17 (b)) is given by

$$g = \int_0^y \gamma \, d\bar{y}$$

$$= \frac{6S}{GA}\int_0^y \left\{\frac{1}{4} - \left(\frac{\bar{y}}{d}\right)^2\right\} d\bar{y}$$

$$= \frac{6S}{GA}\left\{\frac{y}{4} - \frac{y^3}{3d^2}\right\} \tag{5.10}$$

The greatest value of g occurs at $y = d/2$. Then

$$g_{max} = CE = \frac{Sd}{2GA} \tag{5.11}$$

The angle $\bar{\gamma}$ between the chord CMC_1 and EME_1 is

$$\bar{\gamma} = \frac{CE}{d/2} = \frac{S}{GA} \tag{5.12}$$

The angle $\bar{\gamma}$ is the change in the angle between the cross-section and the centroidal axis, and may be regarded as an overall shear strain at the given cross-section.

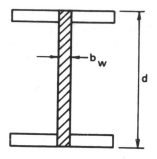

Fig. 5·18

For a beam of I section (Fig. 5.18) the corresponding quantity is given approximately by

$$\bar{\gamma} = \frac{S}{GA_w}$$

where

$$A_w = b_w d$$

and

$$b_w = \text{the web thickness}$$

In general we may say that the "shearing" deformation of a beam element is given by

$$\bar{\gamma} = \frac{S}{GA_y} \tag{5.13}$$

where A_y is called the *shear area* of the section associated with shear parallel to the y axis. A similar quantity can be defined in relation to shear parallel to the z axis. The value of these quantities is discussed further on page 123.

Along a segment of a beam where the shear force is constant, every cross-section undergoes the same degree of warping. The shear deformations induce no curvature of the beam axis (that is, no curvature additional to that induced by bending).

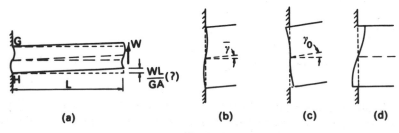

(a)   (b)   (c)   (d)

Fig. 5·19

Fig. 5.19 (a) shows a cantilever with an end force W. Any shear deflection will depend entirely on the boundary conditions assumed at the support. The statement that the beam is direction-fixed at the support may be interpreted in several ways. It may mean that the chord GH does not rotate (Fig. 5.19 (b)) in which case the beam axis rotates by the angle $\bar{\gamma}$ (equation 5.13). It may signify that the tangent to the section at mid-depth does not rotate (Fig. 5.19 (c)) in which case the axis rotates by the angle γ_0, the strain at the neutral axis. It could mean that the beam axis itself does not rotate (Fig. 5.19 (d)).

The meanings illustrated in Figs 5.19 (c) and (d) probably represent extreme values. Clearly the true condition depends on the nature of the support, i.e., of the adjacent structure. The situation is further confused by the fact that the strain distribution in the vicinity of the support cannot be computed by elementary theory, so the shape of the cross-section at the support is not well defined.

In these circumstances, the rotation of the beam axis is often taken as $\bar{\gamma}$ ($= S/GA_y$) and this leads to an end deflection of WL/GA_y (Fig. 5.19 (a)).

We consider now the consequences of a variable shear force. The warping of adjacent sections will now be different. Longitudinal strains and stresses will inevitably be introduced. The limitations of the simple theory are now apparent. Not only does the shear force induce longitudinal stresses as well as shear stresses, but our initial assumption that the longitudinal stresses vary linearly is no longer tenable. As mentioned previously these inconsistencies are negligibly small in most practical problems.

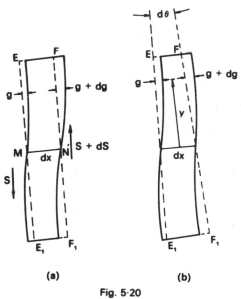

(a)   (b)

Fig. 5·20

At least we should satisfy equilibrium conditions. The longitudinal stresses induced by the shear deformation must have a zero resultant. If we assume that the beam axis remains straight, then normals to this axis at M and N' (Fig. 5.20 (a)) will be parallel. These normals, EE_1 and FF_1, should not be confused with the cross-sections. As can be seen in Fig. 5.20, the deviation of the left-hand section from EE_1 is less than the deviation of the right-hand section from FF_1. Differential warping will produce compressive strains in all fibres on one side of the neutral axis and tensile strains in all fibres on the other side. The corresponding stresses would then have a moment about the neutral axis.

We conclude therefore that the beam axis becomes curved, and the normals to the axis at M and N' are inclined at $d\theta$ to one another (Fig. 5.20 (b)). The change in length of a typical fibre is now

$$de = -g + (g + dg - y\,d\theta) = dg - y\,d\theta \qquad (5.14)$$

the strain,

$$\epsilon = \frac{de}{dx} = \frac{dg}{dx} - y\frac{d\theta}{dx}$$

the stress,

$$\sigma = E\frac{dg}{dx} - Ey\frac{d\theta}{dx}$$

If the moment of these stresses about the neutral axis is to be zero, then

$$\int_{-d/2}^{d/2} \sigma b y\,dy = 0$$

(the beam has a rectangular section). Thus

$$\int_{-d/2}^{d/2} E\frac{d\theta}{dx}by^2\,dy = \int_{-d/2}^{d/2} E\frac{dg}{dx}by\,dy \qquad (5.15)$$

Equation (5.10) gives the value of g, whence

$$\frac{dg}{dx} = \frac{dS}{dx}\cdot\frac{6}{GA}\left(\frac{y}{4} - \frac{y^3}{3d^2}\right)$$

By substituting this value into equation (5.15) and carrying out the integration, we obtain

$$\frac{d\theta}{dx} = \frac{dS}{dx}\frac{6}{5GA} = \frac{dS}{dx}\frac{1}{G(5A/6)} \qquad (5.16)$$

The quantity $d\theta/dx$ is the curvature of the beam axis. If we denote the shear deflection by v_s then

$$\frac{d\theta}{dx} = \frac{d^2 v_s}{dx^2}$$

The quantity dS/dx is the load intensity, w, at the location considered. We note that for negative w, dS/dx is positive (as shown) and $d^2 v_S/dx^2$ is positive. Thus, for a beam of any section, equation (5.16) may be written

$$\frac{d^2 v_S}{dx^2} = \frac{-w}{GA_y^*} \tag{5.17}$$

Here A_y^* is another shear area associated with shear parallel to the y axis. For a rectangle, A_y^* is equal to $\frac{5}{6}A$. The shear area A_y^* is concerned with the *curvature* of the beam.

The deformation of the beam as a whole may now be found by the procedures developed in Chapter 4. The only difference is that the curvature is now given by $(-w/GA_y^*)$ instead of (M/EI).

Example 5.5. Find the deflection due to shear at the centre of a simply supported beam carrying a uniformly distributed load of intensity w.

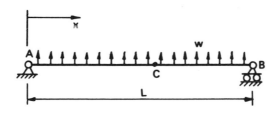

Fig. 5·21

Solution.

$$\frac{d^2 v_S}{dx^2} = \frac{-w}{GA_y^*}$$

$$GA_y^* \frac{dv_S}{dx} = -wx + c_1$$

$$GA_y^* v_S = \frac{-wx^2}{2} + c_1 x + c_2$$

At $x = 0$, $v_S = 0$

$$\therefore \quad c_2 = 0$$

At $x = L$, $v_S = 0$

$$\therefore \quad c_1 = \frac{wL}{2}$$

Hence,

$$v_S = \frac{1}{GA_y^*}\left(-\frac{wx^2}{2} + \frac{wLx}{2}\right)$$

At mid-span, $x = L/2$ and

$$v_S = \frac{wL^2}{8GA_y^*} \tag{5.18}$$

When the boundary conditions involve the beam slope we encounter the difficulty discussed previously.

Example 5.6. Find the end deflection due to shear of a cantilever carrying a uniformly distributed load (Fig. 5.22).

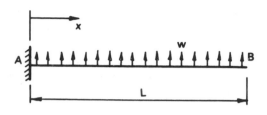

Fig. 5·22

Solution.

$$GA_y^* \frac{d^2 v_S}{dx^2} = -w$$

$$GA_y^* \frac{dv_S}{dx} = -wx + c_1$$

$$GA_y^* v_S = -\frac{wx^2}{2} + c_1 x + c_2$$

At $x = 0$, $v_S = 0$

$$\therefore \quad c_2 = 0$$

At $x = 0$ the beam is known to be direction-fixed but this is insufficient to define dv_S/dx precisely. As discussed earlier, we might assume that $dv_S/dx = \bar{\gamma} = wL/GA_y^*$. In that case

$$GA_y^* \left(\frac{wL}{GA_y}\right) = c_1$$

or

$$c_1 = \frac{wLA_y^*}{A_y}$$

We might as well take $dv_S/dx = wL/GA_y^*$ at $x = 0$ to obtain $c_1 = wL$.

For shear force in the z direction there will be another quantity A_z^*. For a rectangle this will be the same as A_y^*, namely $\frac{5}{6}A$. For thin-walled sections this is not so. In the section of Fig. 5.23, shear in the y direction is resisted very largely by the webs parallel to y, as we have seen in section 5.2. The shear area A_y^* is found to be approximately equal to the area of these walls. For such a section, A_z^* will not be the same as A_y^*.

To sum up: for shear force in the y direction there are two shear areas. The first, A_y, controls the shear deformation of beam elements, and the second,

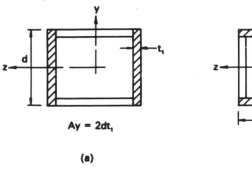

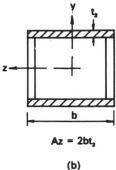

Ay = 2dt₁

(a)

Az = 2bt₂

(b)

Fig. 5·23

A_y^*, controls the curvature of the beam axis. Although these are not the same they differ by very little, and in view of the relatively small importance of shear deformation in practice, it will be sufficient to take them as equal. For simplicity, the following approximate values may be used:

(a) For thin-walled sections,
$$A_y(=A_y^*) = \text{area of walls parallel to the } y \text{ axis}$$
$$A_z(=A_z^*) = \text{area of walls parallel to the } z \text{ axis}$$

(b) For solid sections,
$$A_y = A_y^* = A_z = A_z^* = \text{area of section}$$

The following example will give some indication of the relative importance of shear and bending deflections.

Example 5.7. A beam 12 in deep is simply supported over a span of 12 ft, and carries a uniformly distributed load. Find the ratio of shear deflection to bending deflection for each of the sections shown in Fig. 5.24. Take $E = 5 \times 10^3$ ksi and $G = 2 \times 10^3$ ksi.

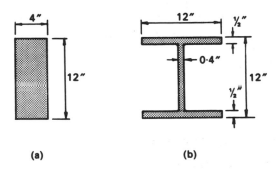

(a)   (b)

Fig. 5·24

Solution.

(a)
$$I = 576 \text{ in}^4$$

$$EI = \frac{5 \times 10^3 \times 576}{144} = 2 \times 10^4 \text{ kip-ft}^2$$

$$A_y = 48 \text{ in}^2$$

$$GA_y = 2 \times 10^3 \times 48 = 9.6 \times 10^4 \text{ kips}$$

Let v_M and v_S be the deflections due to bending and shear respectively.

$$v_M = \frac{5}{384} \frac{wL^4}{EI} = \frac{5}{384} \frac{w \times 12^4}{2 \times 10^4} = 1.35 \times 10^{-2} \text{ ft}$$

$$v_S = \frac{wL^2}{8GA_y} = \frac{w \times 12^2}{8 \times 9.6 \times 10^4} = 1.87 \times 10^{-4} \text{ ft}$$

$$\frac{v_S}{v_M} = \frac{1.87 \times 10^{-4}}{1.35 \times 10^{-2}} = 1.4 \times 10^{-2} = 1.4\%$$

(b)
$$I = 231 \text{ in}^4$$

$$EI = \frac{5 \times 10^3 \times 231}{144} = 8.02 \times 10^3 \text{ kip-ft}^2$$

$$A_y = 12 \times 0.4 \doteq 4.8 \text{ in}^2$$

$$GA_y = 2 \times 10^3 \times 4.8 = 9.6 \times 10^3 \text{ kips}$$

$$v_M = \frac{5}{384} \frac{w \times 12^4}{8.02 \times 10^3} = 3.37 \times 10^{-2} \text{ ft}$$

$$v_S = \frac{w \times 12^2}{8 \times 9.6 \times 10^3} = 18.7 \times 10^{-4} \text{ ft}$$

$$\frac{v_S}{v_M} = \frac{18.7}{3.37} \times \frac{10^{-4}}{10^{-2}} = 5.6 \times 10^{-2} = 5.6\%$$

It will be seen that the error introduced by taking A_y as 48 instead of 40 in solution (a) is of no importance.

The value of v_S/v_M is proportional to $1/L^2$, so the shear deformation becomes increasingly significant as the beam becomes shorter. It will only be of practical importance for short beams with thin webs.

Deflection due to bending and shear combined could be calculated simultaneously from the equation

$$\frac{d^2v}{dx^2} = \frac{M}{EI_z} - \frac{w}{GA_y} \tag{5.19}$$

5.5   Strain Energy due to Shear

We have seen in the previous section that, in general, an element of the bar of length dx and height dy undergoes both a change of shape and a change of length (unless the shear is constant). It is to be expected that each of these types of deformation will be accompanied by strain energy. As before, expressions will be derived for a beam of rectangular section.

Fig. 5.25 (a) shows the forces acting on such a typical element, while Fig. 5.25 (b) shows the element deformation.

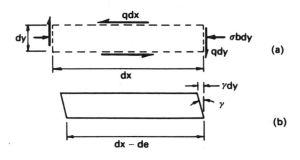

Fig. 5·25

(a) Energy due to shearing of elements

The value of the shear flow, q, in a beam of rectangular section is given by equation (5.5).

$$q = \frac{6S}{d}\left\{\frac{1}{4} - \left(\frac{y}{d}\right)^2\right\}$$

The shear strain, γ, is then

$$\gamma = \frac{q}{Gb}$$

The energy in the element $dx\,dy$ is

$$\tfrac{1}{2}(q\,dx)(\gamma\,dy) = \frac{q^2}{2Gb}dx\,dy$$

and the energy in the complete slice of the beam of length dx is

$$dU = \int_{-d/2}^{d/2} \frac{dx}{2Gb} \times \frac{36S^2}{d^2}\left\{\frac{1}{4} - \left(\frac{y}{d}\right)^2\right\}^2 dy$$

$$= \frac{6S^2\,dx}{10GA}$$

$$dU = \frac{S^2\,dx}{2GA_y} \qquad\qquad (5.20)$$

(b) Energy due to change in length of elements

The change in length of an element is given by equation (5.14)

$$de = dg - y\, d\theta = \left(\frac{dg}{dx} - y\frac{d\theta}{dx}\right) dx$$

The longitudinal force on the element is $(\sigma b\, dy)$ where σ is the bending stress My/I. The energy in the element is

$$\frac{1}{2}\left(\frac{My}{I}b\, dy\right)\left(\frac{dg}{dx} - y\frac{d\theta}{dx}\right) dx$$

and the energy in the complete slice of length dx is

$$dU = \int_{-d/2}^{d/2} \frac{1}{2}\left(\frac{My}{I}b\, dy\right)\left(\frac{dg}{dx} - y\frac{d\theta}{dx}\right) dx$$

$$= \frac{M\, dx}{2I} \int_{-d/2}^{d/2} \left(\frac{dg}{dx}by - \frac{d\theta}{dx}by^2\right) dy$$

Now this integral is equal to zero according to equation (5.15). Hence the change in length of the elements gives rise to no strain energy.

In a length dx of the beam the strain energy due to shear is thus

$$dU = \frac{S_y^2\, dx}{2GA_y}$$

and in the whole beam

$$U_S = \int_0^L \frac{S_y^2\, dx}{2GA_y} \tag{5.21}$$

This expression applies equally to beams of other section if an appropriate value for A_y is known. For the component of shear force in the z direction, a similar expression applies involving S_z and A_z.

5.6 Non-linear Stress-Strain Conditions

The shear stress distributions discussed so far in this chapter have been predicated on the assumption of a linear distribution of the bending stresses. If these are non-linear, a different distribution of shear stresses will naturally follow.

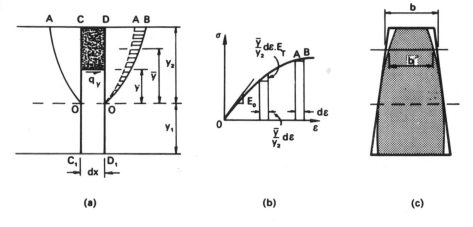

Fig. 5·26

In the present section, it will be assumed that the bending stresses have already been determined. On the cross-section CC_1 (Fig. 5.26 (a)) these stresses are shown by the graph OA, which corresponds to the portion OA of the stress–strain curve (Fig. 5.26 (b)). On the section DD_1 the stresses have increased to the values given by the curve OB. The stresses between OA and OB represent the incremental stresses and the summation of these will determine the shear flow distribution.

The strain distribution is linear; therefore, if the increase of strain from C to D on the top surface is $d\epsilon$, the strain increase at any level $\bar{y}$ is $(\bar{y}/y_2)d\epsilon$. The stress increment at level $\bar{y}$ is then $(\bar{y}/y_2)d\epsilon E_t$ where E_t is the tangent modulus at level $\bar{y}$. The differential force, dF, between sections CC_1 and DD_1 from the top down to a level y is given by

$$dF = \int_y^{y_2} \frac{\bar{y}}{y_2}(d\epsilon)\, E_t\, b\, d\bar{y}$$

The tangent modulus can be expressed in terms of some chosen value, for instance, the initial tangent modulus E_0. The shear flow at level y is then

$$q = \frac{dF}{dx} = \int_y^{y_2} \frac{\bar{y}}{y_2}\left(\frac{d\epsilon}{dx}\right) E_0 \left(\frac{E_t}{E_0}\right) b\, d\bar{y}$$

If we put

$$b\left(\frac{E_t}{E_0}\right) = b'$$

and

$$\left(\frac{d\epsilon}{dx}\right)\frac{E_0}{y_2} = C, \text{ a constant}$$

then

$$q = C \int_y^{y_2} b' \bar{y} \, d\bar{y} \qquad (5.22)$$

The quantity b' can be regarded as an effective width which defines a transformed section (Fig. 5.26 (c)) which can be used in the usual way to determine the shear flow. The integral in equation (5.22) is the first moment, Q', about the neutral axis of the portion of the transformed section above level y. Thus

$$q = CQ' \qquad (5.23)$$

The total vertical shear on the cross-section is

$$S = \int_{y_1}^{y_2} q \, dy = \int_{y_1}^{y_2} CQ' \, dy = CI'$$

where I' is the second moment of the transformed area about the neutral axis. This determines C as being S/I', so that equation (5.23) can be written

$$q = \frac{SQ'}{I'} \qquad (5.24)$$

Thus the shear flow at any level is found in exactly the same way as if the bending stresses were distributed linearly, except that the calculation is based on a transformed section. The width of this section at any level is $b(E_t/E_0)$ where E_t is the tangent modulus at that level. At the neutral axis the width remains unchanged.

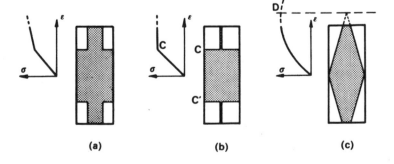

(a)   (b)   (c)

Fig. 5·27

Fig. 5.27 shows the type of transformed section which would result from some simple stress–strain curves. In each example a rectangular cross-section has been taken and the stress–strain relationship has been assumed to be the same in compression as in tension. Other assumptions present no problem. In Fig. 5.27 (a) there is a sudden reduction in slope of the stress–strain curve at the level C, and there is a corresponding sudden reduction in the width of the transformed section. A special instance of this is the elastic-plastic stress–strain curve of Fig. 5.27 (b). At level C the slope of the stress–strain curve drops to zero, and so does the effective width of the section. This indicates that for a section on which the yield stress exists down to level C (and up to C'), the whole of the shear force is resisted by the portion of the section between C and C'.

In the Fig. 5.27 (c) the stress–strain curve is parabolic with a maximum at D. Since the slope of the parabola is proportional to the distance from D, the width of the effective section decreases linearly as shown.

Example 5.8. A beam has a rectangular section 4 in wide and 10 in deep. The material has a tangent modulus of 10×10^6 psi up to a stress of 8000 psi and a tangent modulus of 6×10^6 psi thereafter. At cross-section A the bending moment is 400 kip-in and at section B the bending moment is 1100 kip-in. At each of these sections the shear force is 24 kips. Compare the distribution of shear stress and the maximum shear stress at the two sections. Assume that the shear stress is uniform across the width of the beam.

Solution. At section A, we find that the extreme fibre stress due to bending is 6000 psi, so that the bending stress distribution is linear (Fig. 5.28 (a)). The distribution of shear flow is therefore computed from the whole cross-section (Fig. 5.28 (b)).

$$I = \frac{4 \times 10^3}{12} \; 333.3 \text{ in}^4$$

At level C, 3 in below the top

$$Q_C = 4 \times 3 \times 3.5 = 42 \text{ in}^3$$

$$q_C = \frac{24{,}000 \times 42}{333.3} = 3020 \text{ lb/in}$$

and, assuming a uniform stress distribution across the width,

$$\tau_C = \frac{3020}{4} = 755 \text{ psi}$$

Similarly, at the mid-depth, D

$$\tau_D = \frac{24{,}000 \times 50}{4 \times 333.3} = 900 \text{ psi}$$

The shear stress distribution is parabolic (Fig. 5.28 (c)).

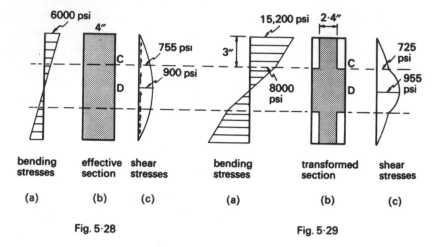

bending   effective   shear   bending   transformed   shear
stresses   section   stresses   stresses   section   stresses

(a)   (b)   (c)   (a)   (b)   (c)

Fig. 5·28   Fig. 5·29

At section B we find that the extreme fibre stress exceeds 8000 psi so that the bending stresses are not linear. By the method of section 4.4 we find that the maximum bending stress is 15,200 psi and the stress of 8000 psi occurs at C, 3 in below the top (Fig. 5.29 (a)). Above C the tangent modulus is constant and equal to 0.6 times the initial tangent modulus. The effective section is therefore as shown dotted in Fig. 5.29 (b). For this section

$$I' = \frac{2.4 \times 10^3}{12} + \frac{1.6 \times 4^3}{12} = 208.5 \text{ in}^4$$

At level C,

$$Q'_C = 2.4 \times 3 \times 3.5 = 25.2 \text{ in}^3$$

$$q_C = \frac{24,000 \times 25.2}{208.5} = 2900 \text{ lb/in}$$

This is the shear flow in the *real* beam and must therefore be divided by 4 in (not 2.4 in) to obtain the shear stress.

$$\tau_C = \frac{2900}{4} = 725 \text{ psi}$$

The distribution of stress down to C is parabolic since the width of the effective section is uniform.

Similarly, at D

$$Q'_D = 33.2 \text{ in}^3$$

and

$$\tau_D = \frac{24,000 \times 33.2}{4 \times 208.5} = 955 \text{ psi}$$

The shear stress distribution at section B is shown in Fig. 5.29 (c).

It sometimes happens that the maximum bending stress occurs slightly away from the extreme fibre, so that close to the top and bottom of the beam the tangent modulus is negative. The width of the effective section will then be taken as negative in the calculation of Q' and I'.

5.7 Flexibility and Stiffness

Usually shear deformation and bending deformation both occur as a result of a given load system. It is possible to have bending without shear deformation, namely, when the bending moment is constant (so that $S = 0$). On the other hand, we cannot have shear force without bending moment (except at an isolated section) and consequently shear deformation is always accompanied by bending deformation.

Suppose we apply a unit force to the end of a uniform cantilever AB of length L. In Chapter 4 we found that the deflection of end B due to bending of the elements is

$$v_M = \frac{L^3}{3EI}$$

The deflection due to shear may be taken as

$$v_S = \frac{L}{GA_y}$$

The total deflection at B is therefore

$$v_B = \frac{L^3}{3EI} + \frac{L}{GA_y} \tag{5.25}$$

This expression is a flexibility coefficient in which shear and bending deformations are both included. Usually the bending deformation pre-dominates and it is convenient to regard the shear term as a modification of the bending term. If we define α by

$$\alpha = \frac{6EI}{L^2} \cdot \frac{1}{GA_y}$$

then

$$\frac{L}{GA_y} = \frac{L^3 \alpha}{6EI}$$

and

$$f = \frac{L^3}{3EI} + \frac{L^3 \alpha}{6EI}$$

$$= \frac{L^3}{3EI}\left(1 + \frac{\alpha}{2}\right) \tag{5.26}$$

If shear deformation is neglected, $\alpha = 0$ and the flexibility coefficient just becomes $L^3/3EI$. If shear deformation is included, the flexibility is increased by the factor $(1 + \alpha/2)$ where α is usually, although not always, fairly small.

We now consider the end rotation caused by the unit end force. Due to bending the rotation at B is

$$\theta_M = \frac{L^2}{2EI}$$

When we consider the rotation of end B due to shear deformation, a difficulty arises. Do we take the rotation of the centroidal axis or the rotation of the end face of the beam? After shearing deformation, these do not remain perpendicular to one another, hence their rotations are different. However, the rotation of the axis is the same at each end of the cantilever, and the rotation (if any) of the cross-sections at each end are also the same. Now the beam is considered as direction-fixed at the support, and it is consistent to take the free end also as having no rotation.

The flexibility coefficient representing the rotation of end B due to unit vertical force then has the value

$$f = \frac{L^2}{2EI} \tag{5.27}$$

If a unit couple is applied to the end B of the cantilever AB the member sustains pure bending, i.e., $S = 0$ everywhere. Consequently, shear deformations do not occur. From Chapter 4 we have the values

$$v_B = \frac{L^2}{2EI} \tag{5.28}$$

$$\theta_B = \frac{L}{EI} \tag{5.29}$$

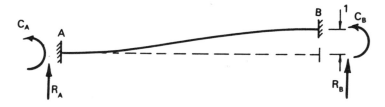

Fig. 5·30

When a unit couple is applied to one end of a simply supported beam, the end rotations are influenced to some extent by shear. The evaluation of these flexibility coefficients is left as a problem for the reader (see problem 5.17).

Stiffness coefficients are also modified when shear deformation is taken into account. Their evaluation is more indirect. Consider the beam AB of Fig. 5.30

in which B is given a unit vertical displacement, without rotation. In Fig. 5.30 all reactions are shown in the positive sense, i.e., R upward and C counter-clockwise, although clearly some will in fact be negative.

We consider the member as a cantilever fixed at A and apply at B a force R_B and a couple C_B of such magnitude that they cause unit vertical displacement of B and zero rotation. Use is made of the expressions in equations (5.26), (5.27), (5.28) and (5.29).

$$v_B = \frac{L^3}{3EI}\left(1 + \frac{\alpha}{2}\right)R_B + \frac{L^2}{2EI}C_B$$

$$\theta_B = \frac{L^2}{2EI} \cdot R_B + \frac{L}{EI}C_B$$

With $v_B = 1$ and $\theta_B = 0$, these equations can be solved to give

$$R_B = \frac{12EI}{L^3}\left(\frac{1}{1 + 2\alpha}\right) \tag{5.30}$$

$$C_B = -\frac{6EI}{L^2}\left(\frac{1}{1 + 2\alpha}\right) \tag{5.31}$$

Then from statics,

$$R_A = -\frac{12EI}{L^3}\left(\frac{1}{1 + 2\alpha}\right) \tag{5.32}$$

$$C_A = -\frac{6EI}{L^2}\left(\frac{1}{1 + 2\alpha}\right) \tag{5.33}$$

The four expressions of these last four equations are the stiffness coefficients associated with vertical displacement of end B. The shear deformation again contributes a modifying factor which tends to unity as the shear deformation becomes negligible. The stiffness coefficients for other types of displacement can be found in the same manner (see problem 5.18).

Problems

5.1 A beam having the cross-section shown in Fig. P5.1 has a shear force of 4000 lb at a certain section.

(a) Find the horizontal shear flow in the flange at the line AB.

(b) Find the vertical shear flow in the web at the line CD.

At the line AB, would you expect vertical as well as horizontal shear stresses on the cross-section?

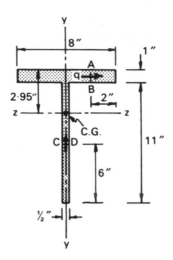

Fig. P5 1

5.2   A beam (Fig. P5.2) is built up by welding a 10 in $\times \frac{1}{2}$ in steel plate to each flange of a 12 in $\times$ 5 in R.S.J. The moment of inertia of the combined section is 598 in⁴. Each plate is attached by two welds as shown. At a section where the shear force is 20,000 lb: (*a*) Find the shear force per inch run resisted by each weld. (*b*) Find the shear flow at the mid-height of the web. (The first moment of half of the unplated R.S.J. about the neutral axis is 19.90 in³.) (*c*) Find the shear stress at this point if the stress is assumed to be uniformly distributed across the thickness of the web.

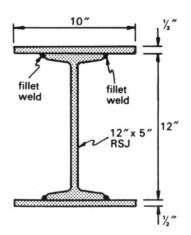

Fig. P5·2

5.3   The beam shown in Fig. P5.2, and described in problem 5.2, is built up by riveting the flange plates to the R.S.J. by $\frac{3}{4}$-in rivets. If each rivet is capable of safely transmitting 10 kips in shear, how many rivets per foot are required in each flange where the shear force in the beam is 20,000 lb?

5.4   The composite beam of Fig. P5.4 consists of a 10 in × 4$\frac{1}{2}$ in R.S.J. ($A = 7.35$ in²) with a 9 in × 3 in channel ($A = 5.14$ in²) welded to the top flange. The centroid of the channel is 0.78 in below the top of the combined section and the web thickness of the channel is 0.3 in. The moment of inertia of the composite beam about the neutral axis is 188 in⁴. Find the force per inch run resisted by each weld, at a section where $S = 18,000$ lb.

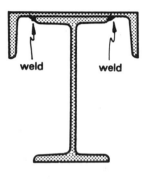

Fig. P5·4

5.5   A beam is built up by welding four 4 in × 4 in × $\frac{3}{4}$ in angles to a 24 in × $\frac{1}{2}$ in web plate as shown in Fig. P5.5. Each angle is fixed to the plate by fillet welds along its upper and lower edges, such as at A and B.

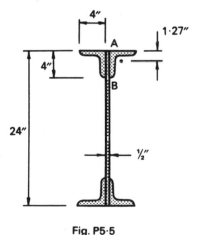

Fig. P5·5

The properties of one 4 in $\times$ 4 in $\times$ $\frac{3}{4}$ in angle are:

$$I = 7.7 \text{ in}^4 \text{ about its own axis}$$

$$A = 5.44 \text{ in}^2$$

The centroid is 1.27 in from the back of the angle. Find the shearing force on one weld per inch run at a section where the total shear force, S, is 40,000 lb. (Assume that the forces in the welds are equal.)

5.6 Four identical bars of material, each 3 in $\times$ 4 in in cross-section are placed one above the other and glued together with epoxy jointing material to form a beam of section 12 in $\times$ 4 in (Fig. P5.6). At a cross-section where the total shear force is 5000 lb, find the shear force per inch run of beam resisted by the epoxy glue at joint A. Also find the shear force per inch run resisted at joint B.

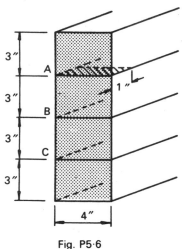

Fig. P5·6

5.7 A beam has a rectangular cross-section 3 in wide and 10 in deep. The bending stresses are distributed according to a parabolic law both in tension and compression. The stress–strain law is $\sigma = 8 \times 10^3 \, \epsilon \mp 8 \times 10^5 \, \epsilon^2$, which has a maximum value of 20 ksi. At a certain section this maximum bending stress value, 20 ksi, occurs at the extreme fibres of the beam (see problem 4.6, p. 91). At the same section the shear force is 40 kips.

(a) Find the shear flow and the shear stress 3 in below the top of the beam.

(b) Find the maximum shear stress on this cross-section. (Use the transformed area method.)

5.8 A beam has a channel section (Fig. P5.8) built up of two vertical plates 6 in $\times$ $\frac{1}{2}$ in welded to a horizontal plate 10 in $\times$ $\frac{1}{2}$ in. The beam is loaded vertically and the bending stresses are linearly distributed. At a section where the shear force is 2000 lb, find the shear flow at point A.

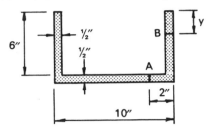

Fig. P5·8

Find the shear flow at point B, y in below the top and hence find the total shear resisted by each vertical plate, and the vertical shear resisted by the 10 in plate.

5.9   A beam of square section is placed so that the plane of bending is parallel to a diagonal. If the shear force is S and the side of the square is a, derive an expression for the distribution of shear flow.

Is it reasonable to assume that the shear stress will be uniformly distributed across the width at all sections? Give reasons.

5.10   The cross-section of a beam is a hollow square with sides vertical and loaded vertically. The square is 6 in × 6 in externally and the walls are $\frac{1}{4}$ in thick. If the bending stresses are linearly distributed, find the shear flow at a point in the top of the beam $1\frac{1}{2}$ in from the centreline. The total shear force is 3000 lb.

5.11   A composite beam has the cross-section shown in Fig. P5.11. Materials A and B are both elastic, the elastic modulus of A being twice that of B. If the joints between A and B are glued, find the shear resisted by each joint per inch run of beam if the total shear force is 4000 lb.

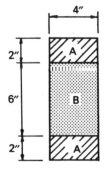

Fig. P5·11

5.12　A beam has the section shown in Fig. P5.12. Material A is homogeneous and elastic. Two 1 in × 1 in bars of material B are embedded in A and bonded thereto to prevent slip. If $E_B = 10E_A$, find the shear stress in the bonding material, assuming that the stress is distributed uniformly round the perimeter of the bar. The shear force on the cross-section is S.

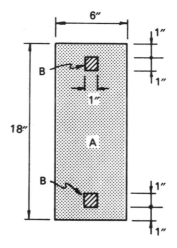

Fig. P5·12

5.13　A thin-walled beam has a cross-section in the shape of a semicircular strip. The radius to the centreline of the strip is 10 in and the strip is ¼ inch thick. Calculate approximately the position of the shear centre of this section.

5.14　A cantilever of length L has a rectangular cross-section $b \times h$. It carries a uniformly distributed load of intensity ω. Derive an expression for the ratio of the end deflection due to shear to the end deflection due to bending. At what ratio of h/L will the deflection due to shear be 10% of that due to bending? Assume $E = 2.5$ G.

5.15　A beam of I section is 12 in wide and 16 in deep overall. The flange thickness is ½ in and the web thickness is ¼ in. If the beam is simply supported and carries a uniformly distributed load, calculate the smallest span for which the shear deflection does not exceed 10% of the bending deflection. Assume that $E = 2.5$ G.

5.16　Derive an approximate expression for the strain energy due to shear in an elemental length dx of a beam which has an I section.

5.17 A uniform beam AB of length L is simply supported at each end. Derive expressions for the rotations of the end faces of the beam when a unit couple is applied at end A. Take shear deformation into account.

5.18 A uniform beam AB of length L is direction-fixed at each end. The support B is given a unit rotation. Derive expressions for the reactions induced at A and B if shear deformations are taken into account.

CHAPTER 6

Torsion

6.1 Stresses due to Torsion

We come now to the final mode of deformation of our elemental slice CD, namely, rotation about the x axis (the centroidal axis of the bar) of face D relative to face C. Will this lead to inconsistencies such as those discussed in Chapter 5, or is it a physically possible mode of deformation?

The answer is that it is quite reasonable when the bar cross-section is bounded by circles. Otherwise, difficulties are encountered—specifically, the theorem of complementary shear stresses is violated. Bars with sections of Fig. 6.1 (a) and (b) can be analysed for torsion by the simple bar theory while bars with any other section (such as Fig. 6.1 (c) or (d)) cannot. It will be easier to see where the difficulty arises if we first study the bars of circular section, to which the simple theory applies.

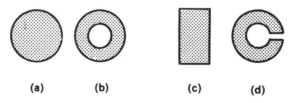

(a)   (b)   (c)   (d)

Fig. 6·1

We assume in the first instance that the stresses are within the proportional range. The elemental slice CD is isolated from the bar (Fig. 6.2 (a)). This element is shown to a larger scale in Fig. 6.2 (b). The face D is rotated about the bar axis by angle $d\phi$ relative to face C, so that point D moves to D'. This produces shear strains which are proportional to the distance from the axis. Therefore the stress also at any point is proportional to the distance from the axis.

To show this we consider the small element of material (Fig. 6.2 (b)) situated at a distance r from the axis. This element is shown to a larger scale in Fig.

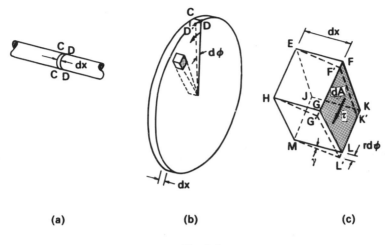

(a)   (b)   (c)

Fig. 6·2

6.2 (c). When the plane D is rotated, this element is distorted since the face $FGLK$ moves (a distance $rd\phi$) relative to the face $EHMJ$. There is a shear strain (in a plane normal to the radius) of

$$\gamma = r\frac{d\phi}{dx} \qquad (6.1)$$

This shows that the shear strain varies directly with the radius.* Furthermore, there is no strain in the radial direction.

Since

$$\tau = G\gamma$$

then

$$\tau = \left(G\frac{d\phi}{dx}\right)r \qquad (6.2)$$

which shows that the stress at any point is proportional to r (equation 6.2 should be compared with equation 4.2). For convenience we can write

$$\tau = Cr \qquad (6.3)$$

We note that the stress is normal to the radius vector at the given point.
    The total shear force on the face $FGLK$ of the element is

$$dF = \tau\,dA = Cr\,dA$$

---

* This assumes that dx is constant, i.e., that cross-sections C and D are parallel and hence that the bar is straight. If the bar is curved, dx will be smaller on the inner side of the curve and at a given radius γ is thus larger on the inside of the curve according to equation (6.1)

and the moment of this force about the bar axis is

$$dT = r\,dF = Cr^2\,dA$$

The total twisting moment is therefore

$$T = C \int r^2\,dA$$

$$= CI_p \tag{6.4}$$

where I_p is the polar moment of inertia of the section about the x axis (see Appendix B).

From equation (6.3),

$$C = \tau/r$$

$$\therefore \quad \tau = \frac{Tr}{I_p} \tag{6.5}$$

If the section is a hollow circle,

$$I_p = \frac{\pi}{2}(R_2^4 - R_1^4) = \frac{\pi}{32}(D_2^4 - D_1^4)$$

where R_1 and D_1 are the internal radius and diameter, and R_2 and D_2 are the external radius and diameter.

For a solid circle

$$I_p = \frac{\pi}{32}D^4$$

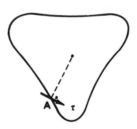

Fig. 6·3

The whole of this derivation is very similar to that used in the case of bending. In Chapter 4, the face D of the elemental slice was rotated about the z axis, and this caused direct strains and stresses proportional to the distance from the z axis. In the present instance, rotation takes place about the x axis. Shear strains and stresses are now produced which vary as the distance from the x axis.

Example 6.1. (*a*) What twisting moment (torque) could safely be transmitted by a solid shaft of 4 in diameter if the maximum shear stress is not to exceed 10,000 psi?

(*b*) If a hollow shaft, having the same sectional area, has an external diameter of 6 in, what twisting moment will it carry at the same maximum shear stress and what will be the stress at the inner surface?

Solution. (*a*) For the solid shaft

$$I_p = \frac{\pi}{32} 4^4 = 8\pi \text{ in}^4$$

The stress of 10,000 psi occurs at the outer surface where $r = 2$ in

∴ From equation (6.5)

$$10,000 = \frac{T \times 2}{8\pi}$$

and

$$T = 125,700 \text{ lb-in}$$

(*b*) If the area of the second shaft is the same as that of the first,

$$\frac{\pi}{4}(6^2 - D_1^2) = \frac{\pi}{4} \cdot 4^2$$

or

$$D_1 = \sqrt{20} \text{ in}$$

Then

$$I_p = \frac{\pi}{32}(6^4 - \sqrt{20^4}) = 28\pi \text{ in}^4$$

Now

$$\tau = 10,000 \text{ when } r = 3$$

$$10,000 = \frac{T \times 3}{28\pi}$$

and

$$T = 293,000 \text{ lb-in}$$

If the stress at the inner surface is τ_1,

$$\frac{\tau_1}{10,000} = \frac{\sqrt{5}}{3}$$

$$\tau_1 = 7450 \text{ psi}$$

Why does the theory break down if the bar has a non-circular section? Consider the section shown in Fig. 6.3. The type of deformation proposed above leads to shear stresses which at every point are normal to the radius vector. Hence at point *A* (Fig. 6.3) the shear stress is not parallel to the boundary. This violates the theorem of complementary shear stresses as was shown in Chapter 5. Evidently this theorem will be obeyed only if the boundary itself is everywhere normal to the radius.

6.2   Deformation due to Torsion

The deformation of an elemental length dx of the bar can be taken as the angular rotation, $d\phi$, of one face relative to the other. An expression for this deformation is obtained from equation (6.2)

$$d\phi = \frac{\tau}{r} \cdot \frac{dx}{G} \tag{6.6}$$

It is usually more convenient to express $d\phi$ in terms of the total torque, T. From equation (6.5)

$$\frac{\tau}{r} = \frac{T}{I_p}$$

$$\therefore \quad d\phi = \frac{T\,dx}{GI_p} \tag{6.7}$$

The total angle of twist in a bar of length L is then

$$\phi = \int_0^L \frac{T\,dx}{GI_p} \tag{6.8}$$

In many instances the bar or shaft is straight and is twisted by couples applied at the ends. If we denote the external couples by C, then the internal twisting moment, T, is equal to C at every section. In such a case equation (6.8) gives

$$\phi = \frac{CL}{GI_p} = \frac{L}{GI_p} \cdot C \tag{6.9}$$

Thus L/GI_p is the torsional flexibility coefficient of a shaft loaded in this manner. It is the angular twist for unit value of the end twisting couples.

$$f = \frac{L}{GI_p} \tag{6.10}$$

This expression is very similar to that for the axial flexibility (see Chapter 3). The torsional stiffness coefficient is given by

$$k = \frac{GI_p}{L} \tag{6.11}$$

Example 6.2. A steel rod ($G = 12 \times 10^3$ ksi) $1\frac{1}{2}$ in diameter and 50 in long fits inside a brass tube ($G = 6 \times 10^3$ ksi) of the same length. The tube has inside and outside diameters of $1\frac{1}{2}$ and 2 in respectively. The bar and the tube are fixed to rigid end plates so that they twist together. If the assembly is subjected to twisting couples of 4 kip-in at its ends, find the total angular twist and the maximum stress in each material.

Solution. The total torque is shared between the bars in proportion to their stiffnesses.
For the steel rod:

$$I_p = \frac{\pi \times 1.5^4}{32} = 0.497 \text{ in}^4$$

$$\therefore \quad k_s = \frac{12 \times 10^3 \times 0.497}{50} = 119.2 \text{ kip-in/radian}$$

For the brass tube:

$$I_p = \frac{\pi(2^4 - 1.5^4)}{32} = 1.074 \text{ in}^4$$

$$\therefore \quad k_B = \frac{6 \times 10^3 \times 1.074}{50} = 128.7 \text{ kip-in/radian}$$

Thus the torque carried by the steel is

$$T_s = 4\left(\frac{119.2}{119.2 + 128.7}\right) = 1.92 \text{ kip-in}$$

Similarly, the torque carried by the brass is

$$T_B = 4\left(\frac{128.7}{119.2 + 128.7}\right) = 2.08 \text{ kip-in}$$

The two bars may now be analysed separately.
Steel:

$$\tau_s = \frac{1.92 \times 0.75}{0.497} = 2.90 \text{ ksi}$$

The angle of twist is

$$\phi - \frac{T_s}{k_s} = \frac{1.92}{119.2} = 0.016 \text{ radian}$$

Brass:

$$\tau_B = \frac{2.08 \times 1.0}{1.074} = 1.93 \text{ ksi}$$

The angle of twist can be re-calculated as a check

$$\phi = \frac{T_B}{k_B} = \frac{2.08}{128.7} = 0.016 \text{ radian}$$

Example 6.3. A shaft tapers from 1 in diameter at end B to $\frac{3}{4}$ in diameter at end A. It is 6 in long and is made of steel ($G = 12 \times 10^6$ psi). Find (a) the maximum torque the shaft will transmit if the stress is not to exceed 6000 psi, (b) the angle of twist in the shaft at this torque and (c) the torsional stiffness of the shaft.

Solution. (a) The maximum torque is governed by the size of the shaft at the smaller end, A.
At A,

$$I_p = \frac{\pi \times (\frac{3}{4})^4}{32}$$

$$T = \frac{I_p \times \tau}{r} = \frac{\pi \times (\frac{3}{4})^4}{32} \frac{6000}{\frac{3}{8}} = 496 \text{ lb-in}$$

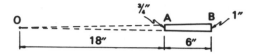

Fig. 6·4

(*b*) If *x* is measured from the origin of the taper (Fig. 6.4), then at any section,

$$d = x/24 \text{ inches}$$

$$I_p = \frac{\pi d^4}{32} = \frac{\pi x^4}{32 \times 24^4}$$

$$\phi = \int_{18}^{24} \frac{T\,dx}{GI_p} = \frac{T}{G} \frac{32 \times 24^4}{\pi} \int_{18}^{24} \frac{dx}{x^4}$$

$$= \frac{496}{12 \times 10^6} \frac{32 \times 24^4}{\pi} \frac{1}{3}\left(\frac{1}{18^3} - \frac{1}{24^3}\right)$$

$$= 4.625 \times 10^{-3} \text{ radian}$$

$$= 0.265°$$

(*c*) The stiffness is the torque per unit angular deformation.

$$k = \frac{T}{\phi}$$

$$= \frac{496}{0.265}$$

$$= 1870 \text{ lb-in/degree}$$

6.3   Strain Energy due to Torsion

In an elemental length of the bar, a twisting moment *T* produces an angular rotation *dφ* of one face relative to the other. While the stresses are within the proportional range, *dφ* is proportioned to *T* and the strain energy stored in the element is

$$dU = \tfrac{1}{2}T\,d\phi \tag{6.12}$$

This expression applies only to bars of circular section. For other bars, warping of the cross-section takes place and as a result further strain energy is stored.

Since T and $d\phi$ are related (equation 6.7), the strain energy can be expressed either in terms of T or in terms of $d\phi$. From equation (6.7)

$$d\phi = \frac{T\,dx}{GI_p}$$

$$\therefore \quad dU = \frac{T^2\,dx}{2GI_p}$$

Alternatively we can put

$$T = GI_p \frac{d\phi}{dx}$$

whence

$$dU = \tfrac{1}{2}GI_p \left(\frac{d\phi}{dx}\right) d\phi$$

$$= \tfrac{1}{2}GI_p \left(\frac{d\phi}{dx}\right)^2 dx$$

Thus for the complete bar

$$U = \int \frac{T^2\,dx}{2GI_p} \tag{6.13}$$

or

$$U = \int \frac{GI_p}{2}\left(\frac{d\phi}{dx}\right)^2 dx \tag{6.14}$$

Example 6.4. A steel bar AB, of 1 in diameter, is bent into a quadrant of a circle of radius 24 in. The bar lies in a horizontal plane. It is cantilevered from A and carries a vertical load of 40 lb at the end B. Find the strain energy stored in the bar due to torsion. Take $G = 12 \times 10^6$ psi.

Solution. It is convenient to determine the position of a cross-section by its angular distance θ from the free end.
Then at θ,

$$T = WR(1 - \cos\theta)$$

$$dx = R\,d\theta$$

$$\therefore \quad U = \int_0^{\pi/2} \frac{[WR(1 - \cos\theta)]^2\,R\,d\theta}{2GI_p}$$

$$= \frac{W^2 R^3}{2GI_p} \int_0^{\pi/2} (1 - 2\cos\theta + \cos^2\theta)\,d\theta$$

$$= \frac{W^2 R^3}{2GI_p} 0.36$$

$$I_p = \frac{\pi}{32} \text{ in}^4; \quad R = 24 \text{ in}; \quad W = 40 \text{ lb}; \quad G = 12 \times 10^6 \text{ psi}$$

$$U = \frac{40^2 \times 24^3 \times 0.36 \times 32}{2 \times 12 \times 10^6 \times \pi}$$

$$= 3.38 \text{ lb-in}$$

It might be noted that this is not the total strain energy stored in the bar. The bar is deformed also in bending and in shear, and there will be strain energy corresponding to these types of deformation.

6.4   Helical Springs

A helical spring provides a practical illustration of a bar subjected to torsion. Such springs are made from bars, normally of circular section, but sometimes of square section. The bar follows a helical path, but the ends are usually turned inwards so that the load can be applied along the axis of the helix (Fig. 6.5 (a)). It is not strictly accurate to analyse such a bar by the formulas derived earlier in this chapter, since these formulas were developed for bars which are essentially straight. However, approximate results can be obtained by this means.

The stress-resultants are the same for every point on the helix. At a typical point C they may be found in the usual way by considering the free-body AC

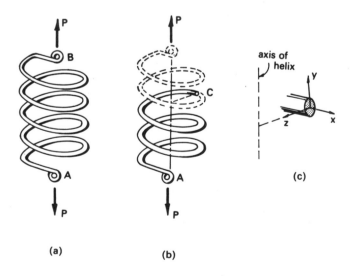

(a)   (b)

Fig. 6·5

(Fig. 6.5 (b)). The cross-section at C is shown to a larger scale in Fig. 6.5 (c). The z axis passes through the axis of the helix which will be taken as being vertical to simplify description. The x axis is tangential to the helix at C and is inclined to the horizontal by α, the helix angle. The y axis is also inclined to the vertical by α, which is usually a small angle.

The moment of the load about the x axis is $PR\cos\alpha$, where R is the radius from the axis of the helix to the centre of the bar. The moment about the y axis is $PR\sin\theta$, and about the z axis the moment is zero. Thus

$$T - PR\cos\alpha$$

$$M_y = PR\sin\alpha$$

$$M_z = 0$$

$$N = P\sin\alpha$$

$$S_y = P\cos\alpha$$

$$S_z = 0$$

In many cases it is reasonable to take α as approximately equal to zero. The spring is then said to be "close-coiled". We then have

$$T = PR; \quad S_y = P; \quad M_y = M_z = N = S_z = 0$$

The deformation due to shear is small compared with that due to torsion, so that for close-coiled springs only torsional deformation is of significance.

The strain energy in the spring can be found from equation (6.13). Suppose that R is the mean radius of the coil, while d is the diameter of the bar itself. Suppose also that the coil has n turns so that the total length of the bar is $2\pi Rn \, (= L)$.

$$T = PR \text{ (constant)}$$

$$U = \int_0^L \frac{P^2 R^2 \, dx}{2GI_p}$$

$$= \frac{P^2 R^2 L}{2GI_p}$$

$$I_p = \frac{\pi d^4}{32} \quad \text{and} \quad L = 2\pi Rn$$

$$\therefore \quad U = \frac{32P^2 R^3 n}{Gd^4} \tag{6.15}$$

This expression can be used to find the extension, e, of the spring under the axial load P.

$$U = \tfrac{1}{2}Pe = \frac{32P^2 R^3 n}{Gd^4}$$

$$\therefore \quad e = \frac{64PR^3 n}{Gd^4}$$

In regard to this type of loading the spring flexibility (extension per unit load) is therefore

$$f = \frac{64R^3 n}{Gd^4} \tag{6.16}$$

and the spring stiffness is

$$k = \frac{Gd^4}{64R^3 n} \tag{6.17}$$

The maximum stress in the material due to torsion is found from equation (6.5)

$$T = PR; \quad I_p = \frac{\pi d^4}{32}; \quad r = d/2$$

$$\therefore \quad \tau_{max} = \frac{Tr}{I_p} = \frac{16PR}{\pi d^3} \tag{6.18}$$

(In reality the maximum stress is slightly higher than this because the bar is curved.)

In addition to this there are stresses resulting from the shear force S_y. The combination of the stresses due to shear and torsion is discussed in Chapter 8. The calculation in the following example is based on torsion alone and is only approximate.

Example 6.5. A helical spring is to be made from $\frac{1}{2}$ in diameter steel bar ($G = 12 \times 10^6$ psi) to withstand a maximum load of 80 lb. The maximum shear stress is 10,000 psi and the stiffness of the spring is to be approximately 50 lb/in.

Calculate the radius of the coil and the number of turns. Also find the energy stored in the spring under full load.

Solution. The coil radius is governed by the stress

$$\tau_{max} = \frac{16PR}{\pi d^3}$$

$$\therefore \quad R = \frac{10,000 \times \pi \times (\frac{1}{2})^3}{16 \times 80} = 3.07$$

say $R = 3$ in

The stiffness depends on the number of turns

$$k = \frac{Gd^4}{64R^3 n}$$

$$\therefore \quad n = \frac{12 \times 10^6 \times (\frac{1}{2})^4}{50 \times 64 \times 3^3} = 8.7 \text{ turns}$$

The stored energy can be calculated from equation (6.15). However it is easier to calculate it from the stiffness since this is known.

$$U = \tfrac{1}{2}Pe = \tfrac{1}{2}P^2/k = \frac{80^2}{2 \times 50}$$

$$= 64 \text{ in-lb}$$

6.5   Non-linear Stress Distribution

For a bar of circular section, the type of deformation considered here leads to strains which vary linearly with the radius. In fact, equation (6.1) showed that $\gamma = r(d\phi/dx)$. We conclude that the graph of shear stress against radius will always follow the stress–strain graph of the material. Thus, while the stresses remain within the proportional range, the stress distribution can be represented graphically as in Fig. 6.6 (a).

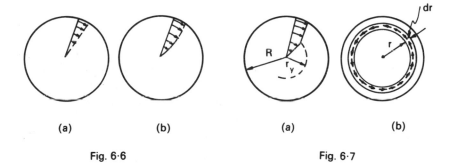

| (a) | (b) | (a) | (b) |

Fig. 6·6   Fig. 6·7

If the material has a curved stress–strain law, the stresses may be distributed as shown in Fig. 6.6 (b). We might consider first a shaft made of yielding material when the outer layers have been strained beyond the yield strain (Fig. 6.7 (a)). The yield strain occurs at radius r_y. The total torque resisted is found, as before, by summing the moments of the stresses about the axis of the shaft.

Within the circle of radius r_y we can use previous results to find the torque, T_1, resisted by this part of the shaft.

$$T_1 = \tau_y \frac{I_p}{r_y} = \frac{\pi}{2} r_y^3 \tau_y$$

Outside the radius r_y, the shear stress is constant and the force on an annular ring of radius r and width dr (Fig. 6.7 (b)) is $dF = (2\pi r\, dr)\tau_y$. The moment of this force about the axis is

$$dT = dFr = 2\pi r^2\, dr\tau_y$$

The torque T_2, resisted by the part of the shaft outside radius r_y is then

$$T_2 = \int_{r_y}^{R} \tau_y 2\pi r^2\, dr = \frac{2\pi}{3}\tau_y(R^3 - r_y^3)$$

The total torque resisted is therefore

$$T = T_1 + T_2 = \tau_y\frac{\pi}{6}(4R^3 - r_y^3) \tag{6.19}$$

The deformation of the bar can be found by the use of equation (6.1). This can be written

$$\frac{d\phi}{dx} = \frac{\gamma}{r} \tag{6.20}$$

Then if we know the strain γ at any radius r, the rate of twist can be found. In the present example, the yield strain γ_y occurs at radius r_y, hence

$$\frac{d\phi}{dx} = \frac{\gamma_y}{r_y} \tag{6.21}$$

As the twisting moment on the shaft is increased, more and more material is strained beyond yield and the radius r_y decreases. The torque approaches a limiting value which would correspond to a state of yielding over the entire cross-section. This limiting value is easily obtained by putting $r_y = 0$ in equation (6.19).

$$T_p = \tfrac{2}{3}\pi\tau_y R^3 \tag{6.22}$$

T_p is called the plastic torque. It will be noticed that this value is only $33\tfrac{1}{3}\%$ greater than the torque at which the surface of the bar first reaches the yield stress.

We consider now a bar in which the stress distribution has a more general shape. If the stresses are known, the calculation of the torque presents no problem. The torque dT contributed by an annular ring similar to that of Fig. 6.7 (b) can be expressed in terms of r, and this expression integrated to give T.

The converse problem of finding the stress distribution corresponding to a given torque is less straightforward. We know that the graph of stresses will be a portion of the given stress–strain graph—the problem is to decide how large a portion.

Example 6.6. A shaft 2 in in diameter is made of material whose stress–strain graph is given in Fig. 6.8 (a). Find the stress distribution and maximum stress in the shaft when the twisting moment is 50 kip-in. Find also the angle of twist per unit length. (The stress at the surface exceeds 24 ksi.)

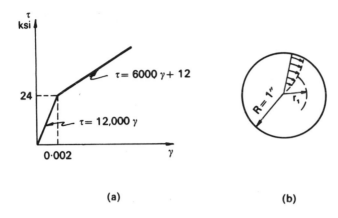

(a)   (b)

Fig. 6·8

Solution. The problem is completely solved as soon as we know the stress corresponding to some radius. We could express the total torque in terms of the maximum stress. It is easier to express it in terms of the radius at which the stress 24 ksi occurs.

Suppose that $\tau = 24$ ksi at radius r_1 (Fig. 6.8 (b)). The total torque is the sum of that part, T_1, resisted within r_1 and the part, T_2, resisted outside r_1.

For $r < r_1$

$$\tau = \frac{r}{r_1} \times 24 \text{ ksi}$$

$$T_1 = \int_0^{r_1} 2\pi r^2 \left(\frac{r}{r_1} \times 24 \right) dr - 12\pi r_1^3$$

For $r > r_1$

$$\gamma = 0.002 \times \frac{r}{r_1}$$

and

$$\tau = 6000\gamma + 12 = 12\frac{r}{r_1} + 12$$

$$T_2 = \int_{r_1}^{R} 2\pi r^2 \left(12\frac{r}{r_1} + 12 \right) dr$$

$$= \frac{6\pi R^4}{r_1} + 8\pi R^3 - 14\pi r_1^3$$

Since $R = 1$, the total torque is

$$T = 8\pi + \frac{6\pi}{r_1} - 2\pi r_1^3$$

When $T = 50$, this equation gives

$$r_1 = 0.70 \text{ in}$$

and this determines the stress distribution.
At the outer surface,

$$\tau = 12\frac{R}{r_1} + 12 = \frac{12}{0.70} + 12 = 29.1 \text{ ksi}$$

We can find the deformation since we know that the strain $\gamma = 0.002$ occurs at a radius of 0.70 in.
Then

$$\frac{d\phi}{dx} = \frac{\gamma}{r} = \frac{0.002}{0.70}$$

$$= 0.0029 \text{ radian/in}$$

6.6   Torsion in Tubes of Thin-walled Section

It has been shown that when the cross-section of the bar is not circular, the sections will not remain plane under torsion, and the foregoing analysis cannot be used. The same difficulty was encountered in Chapter 5, and a solution was obtained there by the use of equilibrium considerations. Can this method be applied to the present problem?

In general it cannot, but it can be applied to the particular problem of a hollow thin-walled tube, *provided the ends of the tube are free to warp*. It is assumed that the wall of the tube is sufficiently thin that the shear stresses are everywhere parallel to the wall (Fig. 6.9 (a)). It can then be shown that the shear flow is constant around the wall.

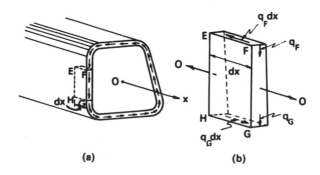

(a)   (b)

Fig. 6·9

The block of material *EFGH* (Fig. 6.9 (a)) is isolated from the tube. The length *EF* is *dx*, and no restriction is placed on the distance *FG*. Suppose that the shear flow at *F* is q_F and at *G* it is q_G. The block is shown enlarged in Fig. 6.9 (b). The shear flow on the top face, *EF*, is equal to q_F and the shear force on this face is thus $q_F dx$. Similarly, the force on the lower face *HG* is $q_G dx$. Since the ends of the tube are free to warp there are no longitudinal stresses. (If the ends are restrained against warping considerable longitudinal stresses may be present and the analysis is consequently invalid.) Equilibrium in the *x* direction then gives

$$q_F dx = q_G dx$$

or

$$q_F = q_G \tag{6.23}$$

At any point around the wall, the shear stress is equal to q divided by the wall thickness at that point. The stress will thus be greater where the wall thickness is less. At corners it cannot be assumed that the stress is uniform across the wall thickness—it will be greater on the inside.

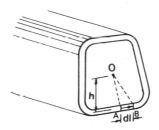

Fig. 6·10

The resultant of the shear flow constitutes a couple equal to the applied twisting moment. The value can be found by taking moments about any point. Let *A* and *B* be two points on the median line of the wall and distance *dl* apart (Fig. 6.10). The total shear force on this segment is $q\,dl$ and may be assumed to act along *AB*. The moment about any point *O* is $h(q\,dl)$ which is equal to twice the area of the triangle *OAB* multiplied by q. It follows that the total twisting moment is twice the area enclosed by the median line of the wall times the shear flow.

$$T = 2qA_0 \tag{6.24}$$

Note that A_0 is quite different from the cross-sectional area of the tube.
Thus

$$q = \frac{T}{2A_0}$$

and

$$\tau = \frac{q}{t} = \frac{T}{2A_0 t} \tag{6.25}$$

where t is the wall thickness, which is not necessarily constant.

Example 6.7. A hollow tube has the cross-section shown in Fig. 6.11. Find the maximum torque which this tube will withstand if the shear stress is not to exceed 12,000 psi.

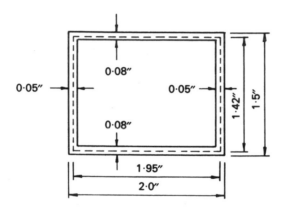

Fig. 6·11

Solution. The maximum shear stress will occur in the thinner wall. Hence the maximum permissible shear flow is

$$q = \tau t = 12{,}000 \times 0.05 = 600 \text{ lb/in}$$

The area enclosed by the median line of the wall is

$$A_0 = 1.42 \times 1.95 = 2.77 \text{ in}^2$$

Then

$$T = 2A_0 q$$
$$= 2 \times 2.77 \times 600$$
$$= 3324 \text{ lb-in}$$

Fig. 6.12 shows a small element removed from the tube wall, of dimensions dx along the tube and dl around the wall. On the face AB the shear force is $q\, dl$. The shear strain, γ, is q/Gt.

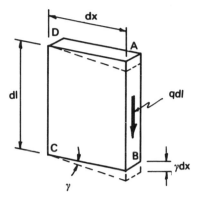

Fig. 6·12

Hence the strain energy in this element is

$$\tfrac{1}{2}(q\,dl)\,(\gamma\,dx)$$

$$= \tfrac{1}{2}(q\,dl)\left(\frac{q\,dx}{Gt}\right)$$

$$= \frac{q^2\,dx}{2G} \times \frac{dl}{t}$$

If this expression is integrated around the length of the wall, a value is obtained for the energy dU in an elemental slice of the tube of length dx.

$$dU = \frac{q^2\,dx}{2G} \int_0^l \frac{dl}{t}$$

If t is constant, then $\int_0^l dl/t$ is simply l/t where l is the length of the median line. If t is not constant, the integral is usually simple to evaluate. We denote $\int_0^l dl/t$ by Ω.
Then

$$dU = \frac{q^2\,\Omega\,dx}{2G} \tag{6.26}$$

However, from equation (6.24), $q = T/2A_0$, hence in terms of T we can write

$$dU = \frac{T^2}{4A_0^2} \frac{\Omega \, dx}{2G}$$

$$= \frac{T^2 \, dx}{2G(4A_0^2/\Omega)}$$

$$dU = \frac{T^2 \, dx}{2GJ} \tag{6.27}$$

For the whole tube

$$U = \int \frac{T^2 \, dx}{2GJ} \tag{6.28}$$

where

$$J = 4A_0^2/\Omega \tag{6.29}$$

If we compare this expression with that of equation (6.13) we see that I_p, the polar moment of inertia of the circular section, has merely been replaced by J. It is convenient to take equation (6.28) as the more general expression. The quantity J is often called the *torsion constant*. It is always a geometrical property of the cross-section and has dimensions (length)^4. In the case of circular sections J becomes equal to I_p, the polar moment of inertia, but in general its value is quite different from that of I_p. In the common case of a tube of constant wall thickness, $J = 4A_0^2 t/l$ where l is the length round the wall.

We can use the energy expression to obtain a value for the angle of twist per unit length of this type of shaft. If the angle of twist is $d\phi$ in a length dx, and the stresses are within the range of proportionality, the strain energy is

$$dU = \tfrac{1}{2}T \, d\phi$$

Then from equation (6.27)

$$\tfrac{1}{2}T \, d\phi = \frac{T^2 \, dx}{2GJ}$$

and

$$\frac{d\phi}{dx} = \frac{T}{GJ} \tag{6.30}$$

As we should expect, this is of the same form as equation (6.7) with J substituted for I_p.

Example 6.8. The tube of example 6.7 (Fig. 6.11) is 20 in long. Find the total angle of twist if the tube sustains the maximum torque of 3324 lb-in. Take $G = 4 \times 10^6$ psi.

Solution. In this problem the wall thickness is not constant, but it is constant for each individual side of the rectangle. So we can write

$$\Omega = \sum \frac{1}{t} = \frac{1.95}{0.08} + \frac{1.42}{0.05} + \frac{1.95}{0.08} + \frac{1.42}{0.05}$$

$$= 105.55$$

$$A_0 = 2.77 \text{ in}^2$$

$$J = \frac{4A_0^2}{\Omega} = \frac{4 \times 2.77^2}{105.55}$$

$$= 0.291 \text{ in}^4$$

$$\phi = \frac{TL}{GJ} = \frac{3324 \times 20}{4 \times 10^6 \times 0.291} = 0.057 \text{ radian}$$

6.7 Bars of Rectangular Section

For bars whose section is neither circular nor thin-walled, torsional stresses and deformations cannot be calculated by elementary principles. Bars of rectangular cross-section are so common in engineering that it may be desirable to give formulas without derivation for bars of this type.

The general problem of torsion was first studied by St Venant* for bars of any shape of cross-section, provided that the sections are free to warp. For any bar, the angle of twist, $d\phi$, in a small length dx, can be expressed as

$$d\phi = \frac{T\,dx}{GJ} \tag{6.31}$$

provided that the appropriate expression is used for the torsion constant J.

For a rectangular section of width b and depth d ($b < d$)

$$J = \beta b^3 d \tag{6.32}$$

The coefficient β depends on b/d and some values are given in Table 6.1.

Along any edge of the rectangle, the shear stress varies from zero at the corners to a maximum at the middle of the edge. The stress at the middle of the longer side, A, is greater than the stress at the middle of the shorter side, B. These maximum stresses can be expressed in the form

$$\tau_A = \frac{T(\lambda_1 b)}{J} \tag{6.33}$$

$$\tau_B = \frac{T(\lambda_2 b)}{J} \tag{6.34}$$

---

* Saint-Venant, Mémoire sur la Torsion des Prismes, *Mémoires des Savants Étrangers*, Vol. 14, 1856, p. 233.

TABLE 6.1

Torsion of Bars of Rectangular Section

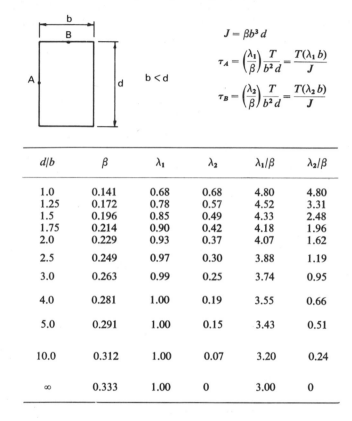

$$J = \beta b^3 d$$

$$\tau_A = \left(\frac{\lambda_1}{\beta}\right)\frac{T}{b^2 d} = \frac{T(\lambda_1 b)}{J}$$

$$\tau_B = \left(\frac{\lambda_2}{\beta}\right)\frac{T}{b^2 d} = \frac{T(\lambda_2 b)}{J}$$

d/b	β	λ_1	λ_2	λ_1/β	λ_2/β
1.0	0.141	0.68	0.68	4.80	4.80
1.25	0.172	0.78	0.57	4.52	3.31
1.5	0.196	0.85	0.49	4.33	2.48
1.75	0.214	0.90	0.42	4.18	1.96
2.0	0.229	0.93	0.37	4.07	1.62
2.5	0.249	0.97	0.30	3.88	1.19
3.0	0.263	0.99	0.25	3.74	0.95
4.0	0.281	1.00	0.19	3.55	0.66
5.0	0.291	1.00	0.15	3.43	0.51
10.0	0.312	1.00	0.07	3.20	0.24
∞	0.333	1.00	0	3.00	0

Where values of λ_1 and λ_2 are given in Table 6.1. These expressions are both in the same form as that for the maximum stress in a circular section. In the circular section we simply have r, the radius, in place of $\lambda_1 b$ or $\lambda_2 b$.

If we substitute for J from equation (6.32), the stresses at A and B can be expressed as

$$\tau_A = \left(\frac{\lambda_1}{\beta}\right)\frac{T}{b^2 d} \tag{6.35}$$

$$\tau_B = \left(\frac{\lambda_2}{\beta}\right)\frac{T}{b^2 d} \tag{6.36}$$

For convenience, the values of λ_1/β and λ_2/β are also given in Table 6.1.

Although these values are based on the condition that the sections are free to warp, the Principle of St Venant can be used since the bar is solid. This tells us that should the end of the bar be restrained against warping, the stresses and deformations will differ sensibly from the above values only within a distance b from the end of the bar. The principle should not be applied to the thin-walled tubes discussed in section 6.6.

Problems

6.1 A hollow propeller shaft is 16 in external diameter and 8 in internal diameter. Find the maximum shear stress in the material when the shaft is subjected to a pure twisting moment of 4000 kip-in. If the shear modulus of the material is 11,600 ksi, find the angle of twist, in degrees, which will occur in a length of twenty external diameters.

6.2 Find the diameter of a solid circular shaft required to withstand a torque of 80,000 lb-in if the maximum shear stress is not to exceed 8500 psi. Find the angle of twist of a 10 ft length of this shaft. Take $G = 11.6 \times 10^6$ psi. Find the torsional stiffness of this shaft.

6.3 A hollow propeller shaft has an internal diameter of 6 in. If it is required to transmit a torque of 3500 kip-in, show that the external diameter which will limit the shear stress to 10,000 psi is given by the equation

$$D^4 - 1782D = 1296$$

6.4 A hollow circular shaft 20 ft long is required to transmit a twisting moment of 270 kip-in. If the maximum shear stress must not exceed 8000 psi and the total angle of twist must not exceed 3 degrees, find the internal and external diameters. Take $G = 11.6 \times 10^6$ psi.

6.5 A brass rod of circular section is 40 in long. For half the length the diameter is $\frac{3}{4}$ in and for the other half it is $\frac{1}{2}$ in. Find the torsional stiffness and flexibility of each part separately and of the whole bar. Take $G = 6 \times 10^6$ psi.

6.6 An aluminium bar ($G = 4 \times 10^6$ psi) is 20 in long and has a diameter of $\frac{1}{2}$ in. It fits inside a brass tube ($G = 6 \times 10^6$ psi) of the same length. The outside diameter of the tube is 1 in and the wall thickness 0.1 in. The bar and the tube are connected to rigid end plates, so that they twist together.

Find the torsional stiffness and flexibility of each component and of the assembly. Use this information to find the angle of twist produced by a torque of 1000 lb-in and the maximum stress in each component due to this torque.

6.7 A horizontal cantilever AB of circular cross-section and length L is fixed at A and is subjected to a gradually applied torsional moment T at the free end B. The diameter decreases uniformly from $2d$ at A to d at the midlength. The remaining portion is cylindrical and has a diameter d.

Find the total strain energy stored in the bar due to torsion and hence determine the rotation at the free end. The shear modulus is G.

6.8 A close-coiled helical spring is required to withstand a maximum load of 890 lb and to have an extension of 13.5 in at this load. It is to be made of high-tensile steel ($G = 12 \times 10^6$ psi) with a maximum permissible shear stress of 80,000 psi.

(a) Find the dimensions of the spring if the mean diameter of the coil is 10 times the diameter of the rod from which the spring is made.

(b) Find the spring stiffness.

6.9 A close-coiled helical spring is made of steel wire 0.1 in diameter. It has 10 turns and the maximum shear stress is 22,400 psi when the load on the spring is 11 lb. If $G = 12 \times 10^6$ psi find:

(a) the mean diameter of the coil

(b) the spring stiffness

(c) the extension under the load of 11 lb

(d) the strain energy in the spring at this load.

6.10 (a) Derive an expression for the strain energy due to torsion in an open-coiled helical spring in terms of the load (P), the mean coil radius (R), the number of turns (n), the shear modulus (G), the bar diameter (d) and the helix angle (α).

(b) Derive an expression for the strain energy due to bending in terms of the same variables, given that $E = 2.5G$.

(c) Neglecting the strain energy due to axial force and shear, write an expression for the total strain energy in the open-coiled spring. Hence find the ratio of the stiffness of a close-coiled spring to the stiffness of a similar spring (i.e., all other variables have the same values) having a helix angle of $20°$.

6.11 A solid circular shaft, of diameter 4 in, is made of steel with a yield stress in shear of 18,000 psi. Find the torque carried by the shaft when the outer $\frac{1}{2}$ in of the material has yielded. What is the angle of twist per foot of shaft? Take $G = 12 \times 10^6$ psi.

6.12 A hollow circular shaft is made of material with a yield stress of 40,000 psi. The outer diameter is twice the inner diameter. Find the dimensions of the shaft if the material becomes fully plastic at a twisting moment of 587 kip-in.

6.13 A copper bar, 1 inch in diameter, fits tightly inside a steel tube of outer diameter 2 in so that no slip can take place between the two metals. What torque applied to the composite bar will cause a maximum shear stress of 6000 psi in the steel if both materials behave in a linear manner and the elastic modulus of the copper is half that of the steel? What is the maximum shear stress in the copper? What is the angle of twist per inch? Find the ratio of the strain energy in the copper to the strain energy in the steel. Take G for steel as 11.6×10^6 psi.

6.14 A hollow tube of square cross-section has an outside dimension of 4 in and a wall thickness of 0.12 in. It is subjected to a twisting moment of 20,000 lb-in. Find the shear flow and the shear stress in the material. Would you expect that when the tube is twisted the four corners of a given cross-section will remain co-planar, and if so would this plane be normal to the longitudinal axis of the tube? Give reasons.

6.15 A thin-walled tube 20 in long having the section shown in Fig. P6.15 is subjected to a torque of 1200 lb-in. Find the maximum shear stress in the material. Also find the strain energy in the bar. The ends of the bar are free to warp. $G = 12 \times 10^6$ psi.

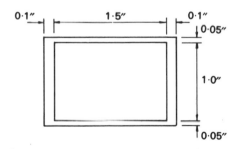

Fig. P 6·15

6.16 A hollow tube has a cross section in the shape of an equilateral triangle of side 2 in. (This dimension refers to the median line of the wall.) The tube is 6 ft long and has a wall thickness of 0.06 in. It is subjected to a twisting moment which causes a shear stress of 4000 psi in the material. Find the twisting moment and the total angle of twist in the tube. Take $G = 11.6 \times 10^6$ psi.

6.17   A circular tube has an outside diameter of 2 in and a wall thickness of 0.2 in. If the material is behaving linearly, what would be the percentage difference in the value obtained for the maximum stress if the tube were analysed

(*a*) as a thin-walled tube and

(*b*) as a hollow circular bar?

6.18   A hollow tube of square cross-section has an outside dimension of 4 in and a wall thickness of 0.05 in. What is the plastic twisting moment of this section if the material has a yield stress of 20,000 psi?

CHAPTER 7

Stresses and Strains at a Point

Chapters 3 to 6 were concerned with the stresses in bars, which were described as one-dimensional structures. The analysis was based on consideration of the behaviour of a "slice" which was infinitesimal only in the direction of the bar.

In the present chapter we shall discuss stresses and strains at a point in a body. The analysis will be based on consideration of the behaviour of an element of material of dimensions $dx \times dy \times dz$.

7.1 Stress Fields

In Chapter 1, reference was made to the fact that at a given point in a body under a given state of stress, the stresses will differ according to the direction of the plane on which they are computed. It was pointed out that in the case of a bar in simple tension, the stress at any point Q would be different on the planes AA and BB (Fig. 7.1 (a)). On the normal cross-section AA through Q, the stress appears as a normal stress (Fig. 7.1 (b)). However, if we cut the bar by the oblique plane BB, both shear stress and normal stress are present on this plane (Fig. 7.1 (c)).

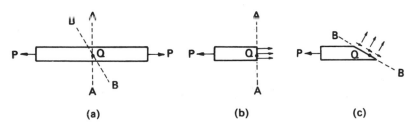

(a)   (b)   (c)

Fig. 7·1

In the bar of Fig. 7.1 we may imagine that lines of force are produced in the material by the end loads *P*. Irrespective of the cutting plane, these lines of force, for the major portion of the bar, run parallel to the length of the bar. The density of these lines of force can be taken as a measure of the stress, and we can say that there is a *stress field* parallel to the length of the bar. The presence of shear stress on a plane such as *BB* merely indicates an obliquity of the plane relative to the stress field. A plane which is chosen at right-angles to the stress field will suffer normal stress.

This is a particularly simple situation in which the stress field is one-dimensional. Even in a straight bar in axial tension, such a simple stress field will not usually apply throughout the bar. At the end of the bar, if the external force *P* is applied over a small area of the end face, the lines of force will converge to this point (Fig. 7.2 (a)).

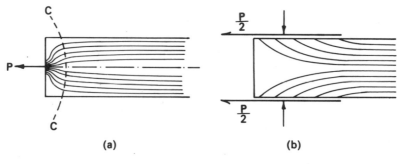

(a)   (b)

Fig. 7·2

In many practical situations a tensile force is applied to a bar through attachments which grip the external faces of the bar near its ends. In this case the lines of internal force will diverge as shown in Fig. 7.2 (b). In each case equilibrium conditions must prevail throughout the material.

Now if we think of these lines of force as strings carrying tensile force, it is clear that they will not remain in this curved position unless they are restrained by forces in the transverse direction. Where they are concave inwards they will produce compression along a line such as *CC*. The same process will occur in a direction normal to the plane of the paper. Thus a three-dimensional state of stress will exist in this region. It is convenient to imagine that in a general field of stress there are three sets of stress lines. Although these lines are curved, the lines from one set will intersect the lines of the other two sets at right-angles. Each line, or path, is called a *stress trajectory*. At a given point in the material the directions of the three stress trajectories are called the *principal axes of stress* at that point. The three principal axes at any point are mutually perpendicular, and the three sets of stress trajectories are said to be orthogonal.

Around a given point we can draw a small block $dx \times dy \times dz$. If the sides of this block are oriented so that they are normal to the principal axes of stress at that point (point A in Fig. 7.3), the faces of the block will experience normal stresses only. The faces are then called *principal planes* and the normal stresses on them are called *principal stresses*. The principal stresses will be denoted by σ_1, σ_2 and σ_3 and the principal planes will be labelled P_1, P_2 and P_3. If the block is oriented differently (point B in Fig. 7.3) its faces will be subjected to tangential stresses as well as normal stresses. The faces will then be called *oblique planes*.

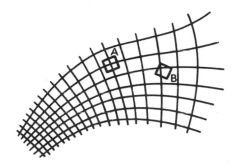

Fig. 7·3

If in any region one of the principal stresses is zero, the stress field is said to be two-dimensional in this region. At the surface of a body, the stress normal to the surface is determined by the external loading. In regions where this loading is zero we have a two-dimensional stress system, the principal stress normal to the surface being zero, and the other two principal stresses (which may or may not be zero) lying in a plane tangential to the surface.

This fact is important since, in practice, the experimental analysis of a body is usually made by the measurement of surface strains which are then used to determine surface stresses.

In a thin flat plate, subjected to loads in its own plane, it is often reasonable to assume that a condition of two-dimensional stress parallel to the surface persists throughout the thickness. Such a condition is referred to as a state of plane stress.

7.2   Oblique Stresses in Terms of Principal Stresses

At a given point in a body, the stresses on any plane are related to the principal stresses by the requirements of *equilibrium*. For simplicity this will be demonstrated with reference to a two-dimensional stress system.

In what follows, the third principal stress, σ_3, will be taken as zero, and the third stress axis will be normal to the plane of the paper as will also the z co-ordinate axis. The planes considered will be those planes parallel to z, and thus normal to the paper. As before, directions will be denoted by small letters and planes normal to these directions by the corresponding capital letter. Thus plane X is normal to the direction x.

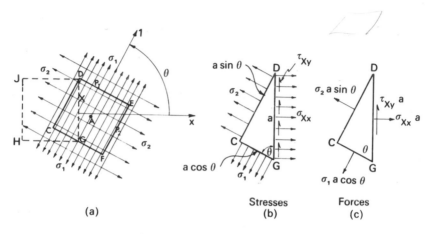

| | Stresses | Forces |
| (a) | (b) | (c) |

Fig. 7·4

Fig. 7.4 (a) shows a small element $CDEF$ at a point A and oriented similarly to the element at A in Fig. 7.3. The faces of the element are normal to the axes of stress $A1$ and $A2$ at the point A and they are, therefore, acted upon by the principal stresses σ_1 and σ_2, both of which are assumed to be tensile. Another element $DGHJ$ could be drawn with faces normal to the x and y directions. Since the elements are infinitesimally small, the two elements shown in Fig. 7.4 (a) can be considered to be at the same point in the body. Suppose that the stress axis $A1$ makes an angle θ with axis Ax. Then the face X (DG) will sustain stresses σ_{Xx} and τ_{Xy}. Consider the equilibrium of the element CDG (Fig. 7.4 (b)) which is a triangular prism with faces normal to the paper. If the area of face DG is a, then the area of CD is $a\sin\theta$ and the area of CG is $a\cos\theta$. The forces on the prism (Fig. 7.4 (c)) are found by multiplying each stress by the area of the face on which it acts. These forces must be in equilibrium.

By resolving forces in the x direction we obtain

$$\sigma_{Xx}a - \sigma_1 a\cos^2\theta - \sigma_2 a\sin^2\theta = 0$$

from which

$$\sigma_{Xx} = \sigma_1\cos^2\theta + \sigma_2\sin^2\theta \tag{7.1}$$

By resolving forces in the y direction we obtain

$$\tau_{Xy}\, a - \sigma_1\, a \cos \theta \sin \theta + \sigma_2 a \sin \theta \cos \theta = 0$$

from which

$$\tau_{Xy} = (\sigma_1 - \sigma_2)\cos \theta \sin \theta \qquad (7.2)$$

For later calculations it is more convenient to express both of these equations in terms of 2θ. Noting that $\cos^2 \theta = \frac{1}{2}(1 + \cos 2\theta)$ and $\sin^2 \theta = \frac{1}{2}(1 - \cos 2\theta)$, we can write equation (7.1) in the form

$$\sigma_{Xx} = \tfrac{1}{2}\sigma_1(1 + \cos 2\theta) + \tfrac{1}{2}\sigma_2(1 - \cos 2\theta)$$

or

$$\sigma_{Xx} = \left(\frac{\sigma_1 + \sigma_2}{2}\right) + \left(\frac{\sigma_1 - \sigma_2}{2}\right)\cos 2\theta \qquad (7.3)$$

Noting that $\cos \theta \sin \theta = \frac{1}{2}\sin 2\theta$, we can write equation (7.2) in the form

$$\tau_{Xy} = \left(\frac{\sigma_1 - \sigma_2}{2}\right)\sin 2\theta \qquad (7.4)$$

Equations (7.3) and (7.4) can be used to find the normal and tangential components of stress on any plane by an appropriate choice of the angle θ.

From equation (7.4) it can be seen that the shear stress is a maximum when $\theta = 45°$ or $135°$, i.e., on the planes which bisect the angles between the principal planes.

These two planes of maximum shear stress are called *principal shear planes*. The shear stress on these planes [equal to $(\sigma_1 - \sigma_2)/2$] is called the *principal shear stress*. For reasons which will be explained in section 7.5 this principal shear stress will be denoted by τ_3.

7.3   Principal Stresses in Terms of Oblique Stresses

More frequently in practice we know the stresses on two planes at right-angles, which may be called the X and Y planes, and require to find the principal stresses and their directions.

We consider an element $CDEF$ (Fig. 7.5 (a)) with faces normal to x and y. The stresses on this element are assumed known. Another element (at the same point) $DGHJ$ has axes $\bar{x}$ and $\bar{y}$ where $\bar{x}$ is at θ to axis x. The stress condition on face $\bar{X}$ can be obtained from the equilibrium of the wedge CGD. Fig. 7.5 (b) shows the stresses on this element and Fig. 7.5 (c) shows the forces.

By resolving these forces in the direction $\bar{x}$ we obtain

$$\sigma_{\bar{X}\bar{x}}\, a - \sigma_{Xx}\, a \cos^2 \theta - \sigma_{Yy}\, a \sin^2 \theta - \tau_{Xy}\, a \cos \theta \sin \theta - \tau_{Yx}\, a \sin \theta \cos \theta = 0$$

Noting that $\tau_{Yx} = \tau_{Xy}$, and rearranging, we obtain

$$\sigma_{\bar{X}\bar{x}} = \sigma_{Xx} \cos^2 \theta + \sigma_{Yy} \sin^2 \theta + 2\tau_{Xy} \sin \theta \cos \theta \qquad (7.5)$$

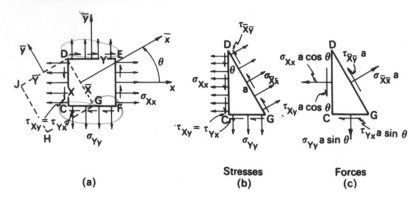

Stresses   Forces
(a)   (b)   (c)

Fig 7.5

By resolving forces in the direction $\bar{y}$ we obtain

$$\tau_{\bar{x}\bar{y}}\,a + \sigma_{Xx}\,a\cos\theta\sin\theta - \sigma_{Yy}\,a\sin\theta\cos\theta - \tau_{Xy}\,a\cos^2\theta + \tau_{Yx}\,a\sin^2\theta = 0$$

from which

$$\tau_{\bar{x}\bar{y}} = -(\sigma_{Xx} - \sigma_{Yy})\cos\theta\sin\theta + \tau_{Xy}(\cos^2\theta - \sin^2\theta) \tag{7.6}$$

As before, it is more convenient to express these equations in terms of 2θ. Equation (7.5) then becomes

$$\sigma_{\bar{X}\bar{x}} = \left(\frac{\sigma_{Xx} + \sigma_{Yy}}{2}\right) + \left(\frac{\sigma_{Xx} - \sigma_{Yy}}{2}\right)\cos 2\theta + \tau_{Xy}\sin 2\theta \tag{7.7}$$

Similarly equation (7.6) becomes

$$\tau_{\bar{x}\bar{y}} = -\left(\frac{\sigma_{Xx} - \sigma_{Yy}}{2}\right)\sin 2\theta + \tau_{Xy}\cos 2\theta \tag{7.8}$$

Plane $\bar{X}$ will be a principal plane if $\tau_{\bar{x}\bar{y}} = 0$. Equation (7.8) shows that this occurs if

$$\tan 2\theta = \frac{2\tau_{Xy}}{\sigma_{Xx} - \sigma_{Yy}} \tag{7.9}$$

For any given values of τ_{Xy}, σ_{Xx} and σ_{Yy} equation (7.9) always yields two values of 2θ which differ by $180°$, or two values of θ which differ by $90°$. Hence for any stress condition there will always be two principal planes and these are mutually perpendicular. Furthermore the normal stress is a maximum on one principal plane and a minimum on the other. If we differentiate equation (7.7) with respect to θ, we obtain

$$\frac{d}{d\theta}\sigma_{\bar{X}\bar{x}} = -(\sigma_{Xx} - \sigma_{Yy})\sin 2\theta + 2\tau_{xy}\cos 2\theta$$

$$= 2\tau_{\bar{X}\bar{y}} \tag{7.10}$$

Hence σ is a maximum or a minimum when $\tau = 0$.

The directions of the principal stresses, measured from the x direction, are given by the values of θ obtained from equation (7.9). Their magnitudes may be found by substituting these values of θ into equation (7.7). For simplicity σ_{Xx} will now be called simply σ_x.

Since

$$\tan 2\theta = \frac{2\tau_{xy}}{\sigma_x - \sigma_y}$$

then

$$\cos 2\theta = \frac{\sigma_x - \sigma_y}{\sqrt{(\sigma_x - \sigma_y)^2 + 4\tau_{Xy}^2}}$$

and

$$\sin 2\theta = \frac{2\tau_{xy}}{\sqrt{(\sigma_x - \sigma_y)^2 + 4\tau_{Xy}^2}}$$

Then from equation (7.7)

$$\left.\begin{array}{c}\sigma_1 \\ \sigma_2\end{array}\right\} = \left(\frac{\sigma_x + \sigma_y}{2}\right) \pm \sqrt{\left(\frac{\sigma_x - \sigma_y}{2}\right)^2 + \tau_{Xy}^2} \tag{7.11}$$

The positive sign gives the value of one principal stress, and the negative sign gives the value of the other.

7.4 Mohr's Circle of Stress

The relationships derived in sections 7.2 and 7.3 can conveniently be expressed by a graphical construction first proposed by Otto Mohr.

Consider first that the principal stresses are known. The stresses on the plane X are given by equations (7.3) and (7.4) or by the diagram of Fig. 7.6. The projection OB is equal to σ_x according to equation (7.3) and the projection BX is equal to τ_{xy} according to equation (7.4). We may say that the co-ordinates of the point X represent the two stress components on the plane X

The line OA is the same for all planes through the point A. The line AX is constant in length but its direction varies according to the angle between the plane X and the principal plane P_1. Consequently if the construction is performed for different positions of the plane X, a number of points will be obtained which lie on a circle with centre A. This is the Mohr stress circle, each point on which corresponds to a particular plane through the point A.

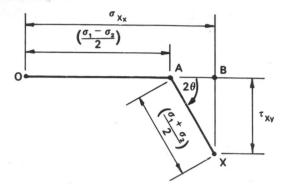

Fig. 7·6

In order that the construction can also be used to determine the principal stresses when the stresses on the X and Y planes are known, we proceed as follows. Two axes are set down. The horizontal axis represents normal stress, σ, and the vertical axis represents shear stress, τ. Along the vertical axis, τ_{Xy} is plotted positive downward, while τ_{Yx} is plotted positive upward (Fig. 7.7 (b)). We note that planes which are at an angle ϕ to one another in the actual body are represented by points which subtend an angle 2ϕ at the centre of the Mohr circle.

If we are given the stress condition on two perpendicular planes, whether they be principal planes or not, we plot the two points representing these planes. These points will be at opposite ends of a diameter of the Mohr circle,

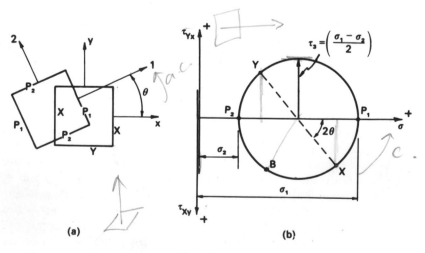

(a)   (b)

Fig. 7·7

which can thus be drawn. Points representing other planes are located with reference to the two known points.

The principal planes, having zero shear stress, will be represented by the points P_1 and P_2 on the σ axis. It is immaterial which is called P_1. The centre of the circle is at the point $\{[(\sigma_1 + \sigma_2)/2], 0\}$. The radius $[(\sigma_1 - \sigma_2)/2]$ is equal to the principal shear stress, τ_3. Since the plane X is θ clockwise from the plane P_1 in Fig. 7.7 (a), the radius AX is 2θ clockwise from radius AP_1 in Fig. 7.7 (b). If a plane is represented by some point B on the circle, we can read off the magnitude and sign of the normal stress on plane B, and the magnitude of the shear stress. We cannot specify the sign of the shear stress unless x and y directions are postulated in relation to the plane B.

(Since the angle between two planes is the same as the angle between their normals, the points on the Mohr circle could equally well be taken to represent the directions of normals, and labelled with small letters. It is important not to confuse the direction of a plane with the direction of its normal, so in what follows the points will always be taken to denote the planes.)

Example 7.1. At a certain point in a body the state of stress is two-dimensional. At this point, the principal stresses are 8 ksi tensile and 4 ksi compressive, the first acting at an angle of $+20°$ to the x axis. Find the stresses on the planes normal to x and y at the same point. What is the maximum shear stress at the point and on what plane does it occur? On what planes are the normal stresses zero?

Solution. It is most desirable to draw an element of the material (Fig. 7.8 (a)) as well as the Mohr's circle (Fig. 7.8 (b), p. 174).

One principal stress acts at $+20°$ (i.e., $20°$ anticlockwise) from x. Call this stress σ_1 and the plane on which it acts P_1 (Fig. 7.8 (a)). Plot the corresponding point P_1 at $+8$ along the normal stress axis in Fig. 7.8 (b). The other principal stress, σ_2 acts at right-angles to the first. Call the second principal plane P_2 and plot the corresponding point P_2 at -4 on the normal stress axis. With centre A midway between P_2 and P_1 draw the Mohr circle.

What point on this circle represents the X plane? Referring to Fig. 7.8 (a), we must rotate plane P_1 $20°$ clockwise to reach plane X. Consequently in Fig. 7.8 (b) we must rotate radius AP_1 in the circle, $40°$ clockwise to reach position AX. The point X of the circle corresponds to the plane X in the body. The co-ordinates of X are $(+6.60, +3.86)$—note that the shear is read $+\downarrow$ for an X plane. The first figure indicates that the normal stress on the X plane is a tensile stress of 6.60 ksi. The second figure indicates that the shear stress is $+3.86$ ksi. The plane Y is at $90°$ from plane X in the body (Fig. 7.8 (a)) so the point Y is at $180°$ from point X in the circle (Fig. 7.8 (b)). The co-ordinates of Y are $(-2.60, +3.86)$—the ordinate is $+\uparrow$ for a Y plane. Fig. 7.8 (a) shows these stresses acting on the faces of a small free-body of material isolated at the given point.

The greatest shear stress occurs on planes corresponding to the points S and S' and is equal to the radius of the circle, namely 6 ksi. These principal shear planes lie midway between the principal planes P_1 and P_2 (Fig. 7.8 (d)).

The planes of zero normal stress are represented by the points B_1 and B_2 on the vertical axis of the graph. Radius AB_1 is $70°30'$ clockwise from radius AP_2, therefore plane B_1 is $35°15'$ clockwise from plane P_2 (Fig. 7.8 (d)). Similarly B_2 is $35°15'$ anticlockwise from P_2.

The solution of this problem could have been obtained analytically by the use of equations (7.3) and (7.4).

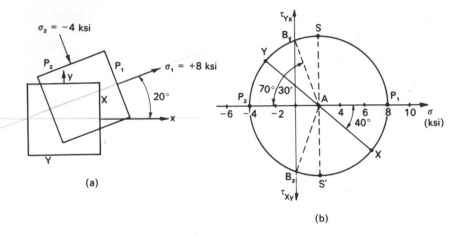

(a)

(b)

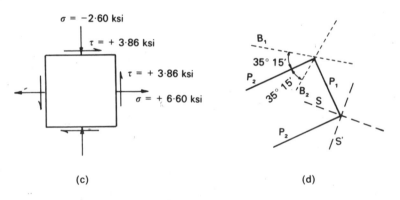

(c)

(d)

Fig. 7·8

It might be noted that while the directions of the various planes are significant, their positions within the element are not. This is because the element, unlike the drawing, really has no size. It should also be noted that while the direction of a plane such as B_1 can be *deduced* from the radius AB_1, it is important not to mistake the radius itself for the direction of the plane.

Example 7.2. At a certain point in a body, the stress on the Y plane is 5000 psi tension and +4000 psi shear. On the X plane the direct stress is 2000 psi compression. Find the magnitude and direction of the principal stresses at the point (*a*) analytically and (*b*) by means of a Mohr circle.

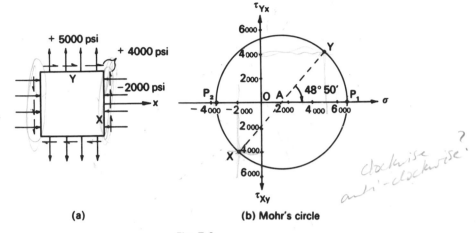

(a)

(b) Mohr's circle

Fig. 7·9

Solution. The state of stress is shown in Fig. 7.9 (a). The stress τ_{xy} is the same as τ_{yx} which is given as +4000 psi. Thus we have

$$\sigma_x = -2000 \text{ psi}$$
$$\sigma_y = +5000 \text{ psi}$$
$$\tau_{yx} = \tau_{xy} = +4000 \text{ psi}$$

(a) From equation (7.9)

$$\tan 2\theta = \frac{2(4000)}{-2000 - 5000}$$

$$= -8/7$$

hence

$$2\theta = -48° 49' \text{ or } 131° 11'$$

and

$$\theta = -24° 24' \text{ or } +65° 36'$$

From equation (7.11)

$$\sigma_1, \sigma_2 = \left(\frac{-2000 + 5000}{2}\right) \pm \sqrt{\left(\frac{-2000 - 5000}{2}\right)^2 + 4000^2}$$

$$= 1500 \pm 5315$$

hence

$$\sigma_1 = +6815 \text{ psi (tension)}$$

and

$$\sigma_2 = -3815 \text{ psi (compression)}$$

It is immaterial which is called σ_1.

(b) To construct the Mohr circle (Fig. 7.9 (b)) we first plot a point X to represent the plane X. This has abscissa -2000 and ordinate $+4000$. A point Y with co-ordinates $(+5000, +4000)$ is plotted to represent the plane Y. The circle is then drawn on XY as diameter. The points P_1 and P_2 where the circle cuts the σ axis represent the principal planes.

In this problem the points X and Y represent known planes so we take our measurements from these. In the circle, the radius AP_1 is $48°50'$ *clockwise* from radius AY. Consequently in the material, the direction of the plane P_1 is $24°25'$ clockwise from plane Y. The plane P_2 is perpendicular to plane P_1. The principal stresses σ_1 and σ_2 are normal to the planes P_1 and P_2.

In the Mohr circle, the distance OA represents the stress $[(\sigma_x + \sigma_y)/2]$ and the radius of the circle is equal to $\sqrt{[(\sigma_x - \sigma_y)/2]^2 + \tau_{xy}^2}$.

The geometry of the circle should be compared to the terms in equation (7.11).

The relationships between stresses on oblique planes and principal stresses are based on equilibrium considerations only and consequently remain true whether the body is behaving elastically or not.

7.5   Three-dimensional Stress Field

The three-dimensional stress state will not be examined in detail. However, a very brief extension of the foregoing concepts will be desirable since a pre-occupation with the two-dimensional stress state can sometimes lead to erroneous conclusions.

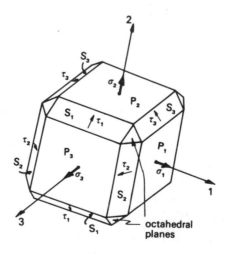

Fig. 7·10

Fig. 7.10 shows a small block of material with its main faces normal to the three principal axes of stress. These faces are then principal planes and are marked P_1, P_2 and P_3 in the figure.

The twelve edges of the block are bevelled at 45°. These bevelling planes are the principal shear planes. The four which are parallel to axis 1 are all denoted by S_1. It should be remembered that the block is infinitesimally small, and consequently the axis 1 may be said to *lie in* the planes S_1. The four planes parallel to axis 2 are labelled S_2 and so on.

Finally each of the eight corners is bevelled with a plane equally inclined to the three principal axes. These planes are called *octahedral planes*. The normal to any octahedral plane has direction cosines (with respect to axes 1, 2 and 3) which are equal in magnitude, though not necessarily equal in sign.

The stresses on the planes P_1, P_2 and P_3 are of course the normal principal stresses σ_1, σ_2 and σ_3. These planes by definition sustain no shear stress.

The principal shear planes S_1, S_2 and S_3 sustain the principal shear stresses τ_1, τ_2 and τ_3. Usually these planes also sustain normal stress. The principal shear stress τ_3 acts in a direction parallel to plane P_3. (Why is there no shear component on plane S_3 parallel to the axis 3?) The work of the preceding sections has been confined to planes which are parallel to the axis 3 (or which *contain* the axis 3), such as planes P_1, P_2 and S_3. By considering only these planes we saw (p. 169) that

$$\tau_3 = \frac{\sigma_1 - \sigma_2}{2} \qquad (7.12a)$$

We also saw that a Mohr's circle could be plotted to represent the stress conditions on all planes (at the given point) parallel to axis 3. This circle is $P_2 S_3 P_1$ in Fig. 7.11.

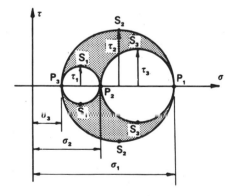

Fig. 7·11

In a similar way we can show that the principal shear stresses τ_1 and τ_2 are given by

$$\tau_1 = \frac{\sigma_2 - \sigma_3}{2} \tag{7.12b}$$

$$\tau_2 = \frac{\sigma_3 - \sigma_1}{2} \tag{7.12c}$$

Also the stress conditions on all planes parallel to axis 1 are given by the Mohr's circle $P_3 S_1 P_2$ (Fig. 7.11), while the planes parallel to axis 2 give the circle $P_1 S_2 P_3$.

It can be shown that the stress condition on any plane not containing a principal axis is represented by a point within the shaded area. The greatest shear stress at the point is therefore the largest of the three principal shear stresses.

In a two-dimensional stress state, point P_3 lies at the origin. In the event that σ_1 and σ_2 are of the same sign, the maximum shear stress at the point is then not τ_3 (the ordinate of S_3) but $\sigma_1/2$ or $\sigma_2/2$ whichever is the greater.

Finally we consider the octahedral planes. Now the shear stress on any oblique plane, whose normal has direction cosines l, m, n with the axes 1, 2, 3, is given by

$$\tau^2 = \sigma_1^2 l^2 + \sigma_2^2 m^2 + \sigma_3^2 n^2 - (\sigma_1 l^2 + \sigma_2 m^2 + \sigma_3 n^2)^2$$

For every octahedral plane, $l^2 = m^2 = n^2 = \frac{1}{3}$

$$\therefore \quad \tau_{oct}^2 = \tfrac{1}{3}(\sigma_1^2 + \sigma_2^2 + \sigma_3^2) - \tfrac{1}{9}(\sigma_1 + \sigma_2 + \sigma_3)^2$$

$$= \tfrac{1}{9}\{(\sigma_2 - \sigma_3)^2 + (\sigma_3 - \sigma_1)^2 + (\sigma_1 - \sigma_2)^2\}$$

$$= \tfrac{4}{9}(\tau_1^2 + \tau_2^2 + \tau_3^2)$$

$$\therefore \quad \tau_{oct} = \tfrac{2}{3}\sqrt{\tau_1^2 + \tau_2^2 + \tau_3^2} \tag{7.13}$$

Example 7.3. At a point in a two-dimensional stress field $\sigma_x = 13$ ksi, $\sigma_y = 7$ ksi and $\tau_{xy} = 2$ ksi. Find (a) the maximum shear stress at this point and (b) the shear stress on the octahedral planes.

Solution. We first compute the principal stresses.

$$\sigma_1, \sigma_2 = \left(\frac{\sigma_x + \sigma_y}{2}\right) \pm \sqrt{\left(\frac{\sigma_x - \sigma_y}{2}\right)^2 + \tau_{xy}^2}$$

$$= \left(\frac{13 + 7}{2}\right) \pm \sqrt{\left(\frac{13 - 7}{2}\right)^2 + 2^2}$$

$$= 10 \pm 3.606$$

Thus $\sigma_1 = 13.606$ ksi and $\sigma_2 = 6.394$ ksi. Since the stress field is two-dimensional at this point we know that $\sigma_3 = 0$.

We can now find the principal shear stresses.

$$\tau_3 = \pm \frac{\sigma_1 - \sigma_2}{2} = \pm 3.606 \text{ ksi}$$

$$\tau_1 = \pm \frac{\sigma_2 - \sigma_3}{2} = \pm \frac{6.394 - 0}{2} = \pm 3.197 \text{ ksi}$$

$$\tau_2 = \pm \frac{\sigma_3 - \sigma_1}{2} = \pm \frac{0 - 13.606}{2} = \pm 6.803 \text{ ksi}$$

(a) The maximum shear stress is 6.803 ksi

(b) $\tau_{oct} = \frac{2}{3}\sqrt{3.606^2 + 3.197^2 + 6.803^2}$

$= 5.558$ ksi

This example illustrates the point that even in a two-dimensional stress situation it is necessary to consider all three principal shear stresses in order to evaluate the maximum shear stress existing at a given point.

7.6   Some Characteristics of Stresses at a Point In a Stress Field

The foregoing consideration of the equilibrium of stresses at a point leads to certain conclusions which will now be summarized. These conclusions are probably more evident from the Mohr circle than from the analytical equations.

We refer first to a general three-dimensional field of stress:

1. At a given point in a stressed body, one of the principal stresses is the greatest normal stress acting at that point, and another principal stress is the least.

2. The three principal shear stresses are equal to the radii of the three Mohr circles at the point. The greatest of these is the maximum shear stress acting at the point.

3. When all principal stresses are of the same sign, then the normal stress on every plane at this point is of this sign also, and there is no plane on which the normal stress is zero.

4. When the principal stresses are all equal, the radius of each Mohr circle is zero and the shear stress vanishes on all planes. The normal stress on every plane is equal to the principal stresses. The stress at the point is said to be *isotropic* and the point is called an *isotropic point*.

Certain characteristics apply more particularly to a two-dimensional stress state. In the following statements, the Mohr circle for stresses σ_1 and σ_2 is referred to:

5. For all pairs of mutually perpendicular planes at the given point (all planes being normal to the plane of the stress field 1, 2) the sum of the two normal stresses is the same. That is to say

$$(\sigma_1 + \sigma_2) = (\sigma_x + \sigma_y) = (\sigma_a + \sigma_b) \text{ etc.}$$

where $(1,2)$, (x,y), (a,b) etc. represent pairs of mutually perpendicular directions.

6. In the case of a pair of mutually perpendicular normal stresses, that one which makes the smaller angle with the larger principal stress is algebraically the larger.

7. The shear stresses on any two mutually perpendicular planes are equal.

8. The normal stresses on the principal shear planes are each equal to $[(\sigma_1 + \sigma_2)/2]$.

9. Two planes which are symmetrically oriented with respect to either of the principal planes are called conjugate planes. The normal stresses on conjugate planes are equal, and the shear stresses are equal in magnitude.

10. For the given point, the abscissa of the centre of the Mohr circle, $[(\sigma_1 + \sigma_2)/2]$, and the radius of the circle, $[(\sigma_1 - \sigma_2)/2]$ are constant.

11. When σ_1 and σ_2 are numerically equal but of opposite sign, the centre of the Mohr circle is at the origin and the principal shear stress is numerically equal to σ_1 and σ_2. The stress condition at such a point is said to be one of *pure shear*.

For any point in a two-dimensional stress field, the stress condition can be specified by a scalar and a vector component.

The *scalar component* is the quantity $[(\sigma_1 + \sigma_2)/2]$. It defines the position of the centre of the Mohr circle and its effect on any plane is independent of the orientation of that plane. It gives rise to a change of volume of the element of material at the point, but to no change of shape.

The vector component is called the *deviator stress component*. It has a magnitude equal to $[(\sigma_1 - \sigma_2)/2]$ and its direction can be defined in terms of the principal stress axes at the point. It defines the radius of the Mohr circle and its effect on any plane depends on the orientation of the plane. The maximum shear stress at the point is equal to the deviator stress. This component gives rise to a change of shape of the element of material at the point, but to no change of volume.

A scalar is completely defined in terms of one quantity only, its magnitude. A vector (in a two-dimensional space) is completely defined by two quantities, its magnitude and direction (or two components). The stress at a point (in a two-dimensional field) thus requires three quantities for its complete specification and is clearly neither a scalar nor a vector. It belongs to a more extensive class called tensors, of which scalars and vectors are members of lower order.

The stress condition at a point in a three-dimensional field can be split up in a similar manner. Each principal stress can be regarded as the sum of a scalar component $\frac{1}{3}(\sigma_1 + \sigma_2 + \sigma_3)$ and a vector component which is a function of the principal shear stresses (Fig. 7.12).

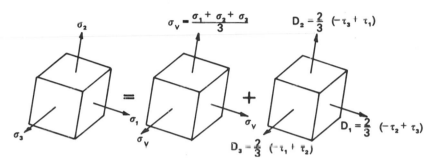

Fig. 7·12

The deviator components determine the magnitude of the three Mohr circles, while the isotropic component determines the distance of the group of circles from the origin. Moreover, the deviator components cause change of shape, while the isotropic component causes change of volume.

7.7   Equilibrium Equations of a Stress Field

Previously we considered the stress equilibrium at discrete points within the stressed body. As the position of the point is changed slightly, the values of the stresses will usually undergo variation. The stress differentials also have to obey certain relationships if equilibrium is to be maintained. To derive these relationships for a two-dimensional stress field, we consider the equilibrium of a small parallelepiped with sides dx and dy and of unit thickness in the z direction.

The system of stresses is shown in Fig. 7.13. It is assumed that all stresses are distributed uniformly along the sides on which they act. The variation of the three independent stresses over the distances dx and dy are represented by the differentials in Fig. 7.13.

The body force components per unit volume due to gravitation and the like (see p. 3) are represented by X and Y and are assumed to act at the centre of the parallelepiped.

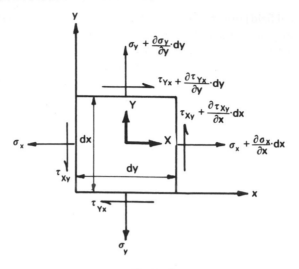

Fig. 7·13

Referring to Fig. 7.13 we may sum all the forces in the x direction to obtain the equation

$$\left(\sigma_x + \frac{\partial \sigma_x}{\partial x} dx\right) dy - \sigma_x \, dy + \left(\tau_{Yx} + \frac{\partial \tau_{Yx}}{\partial y} dy\right) dx - \tau_{Yx} dx + X \, dx \, dy = 0$$

from which

$$\frac{\partial \sigma_x}{\partial x} + \frac{\partial \tau_{Yx}}{\partial y} + X = 0 \qquad (7.14a)$$

Similarly, equilibrium in the y direction leads to the equation

$$\frac{\partial \sigma_y}{\partial y} + \frac{\partial \tau_{Xy}}{\partial x} + Y = 0 \qquad (7.14b)$$

Equations (7.14) are field equations, and are generally referred to as the differential equations of equilibrium.

In most cases the only body force is due to gravitation, and if y is vertically upward the X component is zero and the Y component is negative and represents the weight per unit volume of the material. The equations then become

$$\frac{\partial \sigma_x}{\partial x} + \frac{\partial \tau_{Yx}}{\partial y} = 0 \qquad (7.14c)$$

$$\frac{\partial \sigma_y}{\partial y} + \frac{\partial \tau_{Xy}}{\partial x} - Y = 0 \qquad (7.14d)$$

Frequently the effect of the body force on the stress values is small compared to the effect of the surface forces, and in such cases the body force can be ignored.

Equilibrium thus provides two equations containing three unknown stress components. Before a solution can be attempted a third equation is required. This arises from the fact that the material remains continuous after deformation. Neighbouring blocks of material must therefore deform in a geometrically compatible manner. This will be discussed after we have considered the geometry of strain.

7.8 Deformation of a Small Element

Just as the presence or absence of shear stresses on the faces of a small element is merely the result of the orientation of the element, so also is the presence or absence of shear strain. Thus if we imagine a small element which is rectangular before the material is strained, then after straining, the faces of this element may or may not be at right-angles, and this depends upon how we orient the element. Furthermore, the relationship between strains in various directions is entirely a matter of geometry.

Suppose that the square element $ABCD$ (Fig. 7.14) deforms to the rectangle $A'B'C'D'$. Since the corners remain right-angles, the sides of the element have suffered no shear strain. Also, by definition, the directions of the normals to the faces are the directions of *principal strain*. If, at the same point in the material, we consider the square element $EFGH$, it is clear that this element will suffer shear deformation when the material is strained.

At the given point in the material, a line PQ before distortion (Fig. 7.15) becomes $P'Q'$ after distortion. In general, $P'Q'$ is not the same length as PQ. The change of length determines the *direct strain* in the direction PQ. Also,

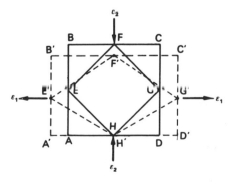

Fig. 7·14

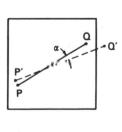

Fig. 7·15

in general, the line undergoes a rotation, α, which is called the *deviation* in the direction PQ. The deviation is taken as positive if the line rotates clockwise. For an element which was initially rectangular, the relative deviation (measured in radians) of its adjacent faces is the shear strain.

The direct strain and the deviation in any oblique direction can be expressed in terms of the principal strains. At a point A in the material, the element

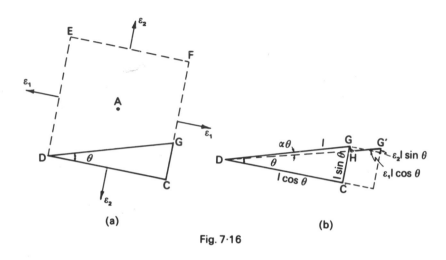

(a)   (b)

Fig. 7·16

$CDEF$ is chosen with its sides parallel to the principal axes of strain (Fig. 7.16 (a)). The line DG is drawn at an angle θ to the direction of ϵ_1. The direct strain and deviation in the direction DG are denoted by ϵ_θ and α_θ.

In the triangle CDG (Fig. 7.16 (b)), if the length of DG is l, then CD is $l\cos\theta$ and CG is $l\sin\theta$. Due to strain, G suffers displacement $\epsilon_1 l\cos\theta$ in the direction ϵ_1, and $\epsilon_2 l\sin\theta$ in the direction ϵ_2. In terms of ϵ_θ and α_θ, the displacement GG' may be expressed as $l\alpha_\theta (= GH)$ normal to DG and $l\epsilon_\theta (= HG')$ parallel to DG. It should be remembered that the angle α is extremely small.

By resolving displacements in the direction DG we obtain

$$l\epsilon_\theta = \epsilon_1 l\cos^2\theta + \epsilon_2 l\sin^2\theta$$

which gives

$$\epsilon_\theta = \epsilon_1\cos^2\theta + \epsilon_2\sin^2\theta \qquad (7.15)$$

By resolving displacements normal to DG we obtain

$$l\alpha_\theta = \epsilon_1 l\cos\theta\sin\theta - \epsilon_2 l\sin\theta\cos\theta$$

which gives

$$\alpha_\theta = (\epsilon_1 - \epsilon_2)\sin\theta\cos\theta \qquad (7.16)$$

Equations (7.15) and (7.16) are similar in form to the corresponding stress equations (p. 168). As for those equations, it is convenient to express the strain equations in terms of 2θ.

$$\epsilon_\theta = \left(\frac{\epsilon_1 + \epsilon_2}{2}\right) + \left(\frac{\epsilon_1 - \epsilon_2}{2}\right)\cos 2\theta \qquad (7.17)$$

$$\alpha_\theta = \left(\frac{\epsilon_1 - \epsilon_2}{2}\right)\sin 2\theta \qquad (7.18)$$

Because of the similarity between equations (7.17) and (7.18) and the corresponding stress equations, it is evident that the strain relations can be represented graphically by a Mohr's circle of strain. Each point on this circle (Fig. 7.17 (a)) represents a particular direction in the material and the abscissa and ordinate of the point represent respectively the direct strain and the deviation in that direction. The points D_1 and D_2 represent the principal strains ϵ_1 and ϵ_2 and the co-ordinates of any point such as Q represent the strain condition in a direction making an angle of θ with direction D_1.

Consider a rectangular element $EFGH$ (Fig. 7.17 (b)) oriented so that one of its sides, GH, makes an angle θ with D_1. Let the deviation in the direction GH be α. The side GF is at $(\theta + 90°)$ to D_1, and equation (7.18) shows that its deviation is therefore $-\alpha$. Hence the distortion of the angle FGH, which is the shear strain, is 2α. At a given point in the material the shear strain will be a maximum when the element is chosen with its sides at $45°$ to the principal axes. This is seen most easily from the Mohr circle.

The characteristics of a strain field are very similar to those of a stress field. By studying the equations and the Mohr's circle of strains, the reader should

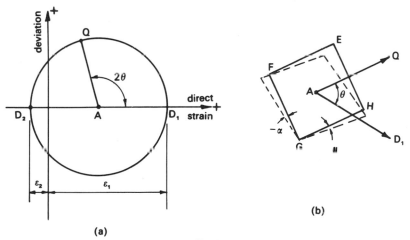

(a)

(b)

Fig. 7·17

have no difficulty in writing properties of a strain field similar to those discussed for stress in section 7.6. In fact the following transformation of symbols may be made in any of the foregoing equations.

Stress	*Strain*
σ_x	ϵ_x
σ_y	ϵ_y
σ_z	ϵ_z
τ_{xy}	$\frac{1}{2}\gamma_{xy}$
τ_{yz}	$\frac{1}{2}\gamma_{yz}$
τ_{zx}	$\frac{1}{2}\gamma_{zx}$

It should be realized that a two-dimensional stress field will usually give rise to strains in the third direction and vice versa. The preceding work of this chapter will remain true whether or not there are stresses and strains in the third direction.

7.9   Compatibility Equations of a Strain Field

We found in section 7.7 that in a stress field the stress differentials had to obey certain conditions in order to maintain equilibrium. It can be shown that strain differentials also have to obey certain relationships if compatibility is to be maintained, i.e., if the strained material is to remain continuous. To derive these relationships for a two-dimensional strain field, we consider the geometry of a small parallelepiped $ABCD$ with sides dx and dy and of unit thickness in the z direction (Fig. 7.18).

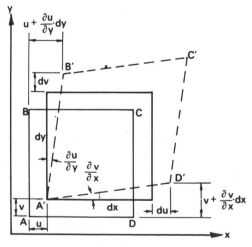

Fig. 7·18

When the material is strained the point A moves to A', the displacement components being u and v in the x and y directions respectively. If B, C and D suffer the same displacement, the block will merely experience a rigid body displacement with no straining. However, in a strain field the values of the displacements will usually undergo variation from point to point and we may suppose that B and D experience slightly greater displacements than A. *If the material in this region remains continuous*, the block $ABCD$ will become distorted. If we neglect differentials of order higher than the first (as we did when dealing with stresses), we can express the strains in terms of the corner displacements.

In the direction AD,
the direct strain,

$$\epsilon_x = \frac{\partial u}{\partial x}$$

and the deviation,

$$\alpha_x = -\frac{\partial v}{\partial x}$$

In the direction AB,
the direct strain,

$$\epsilon_y = \frac{\partial v}{\partial y}$$

and the deviation,

$$\alpha_y = \frac{\partial u}{\partial y}$$

The strains can therefore be expressed in terms of the displacement components of a point in the strain field.

$$\epsilon_x = \frac{\partial u}{\partial x} \tag{7.19a}$$

$$\epsilon_y = \frac{\partial v}{\partial y} \tag{7.19b}$$

$$\gamma_{xy} = \frac{\partial u}{\partial y} + \frac{\partial v}{\partial x} \tag{7.19c}$$

These equations depend upon the fact that the material remains continuous and they are therefore *compatibility* equations. Since the *three* strains have been expressed in terms of the *two* variables u and v, it is clear that there must be a relationship between the strains. This implied relationship is found by eliminating u and v from equations (7.19). This is achieved by differentiating

ϵ_x twice with respect to y, ϵ_y twice with respect to x, and γ_{xy} once with respect to x and once with respect to y.

$$\frac{\partial^2 \epsilon_x}{\partial y^2} = \frac{\partial^3 u}{\partial x \, \partial y^2}$$

$$\frac{\partial^2 \epsilon_y}{\partial x^2} = \frac{\partial^3 v}{\partial x^2 \, \partial y}$$

$$\frac{\partial^2 \gamma_{xy}}{\partial x \, \partial y} = \frac{\partial^3 u}{\partial x \, \partial y^2} + \frac{\partial^3 v}{\partial x^2 \, \partial y}$$

Hence

$$\frac{\partial^2 \epsilon_x}{\partial y^2} + \frac{\partial^2 \epsilon_y}{\partial x^2} = \frac{\partial^2 \gamma_{xy}}{\partial x \, \partial y} \tag{7.20}$$

This equation is called the *strain compatibility equation*, and represents the relationship which must exist between ϵ_x, ϵ_y and γ_{xy} if the material is to remain continuous at the point (x, y).

It should be noted that the strain relationships derived in sections 7.8 and 7.9 are based entirely on considerations of geometry and are applicable whether the body is behaving elastically or not.

7.10   Relationship between Stresses and Strains at a Point

In general the relationship between stress and strain in a body subjected to a two- or three-dimensional stress state can be determined only by experiment. However, when the stress and strain are proportional to one another, and when the material is isotropic, a simple relationship exists which can be expressed with a minimum of reference to experimental data.

For the uniaxial stress state, and when proportionality is obeyed, the stress and strain in the principal stress directions are related simply by the equation

$$\sigma = E \cdot \epsilon$$

as discussed in Chapter 2. Experiment is called upon only to supply the value of the elastic modulus, E.

When considering more than one direction, we have to take account of the fact that a stress in one direction causes strain not only in that direction but also in directions at right-angles to it. When a bar extends under a tensile stress, it also suffers lateral contraction. Conversely if the bar shortens under a compressive stress, it also suffers lateral expansion. Thus the direct strain is always accompanied by lateral strains of opposite sign. Furthermore, for a given material, sustaining simple uniaxial stress, the numerical ratio of these strains is a constant provided the material is within the proportional range.

The constant is called Poisson's Ratio and is denoted by μ. Numerically,

$$\text{Poisson's Ratio } (\mu) = \frac{\text{Lateral Strain}}{\text{Direct Strain}}$$

It should be noted that the same lateral strain occurs in *every direction* normal to the direct strain.

If an element of material is acted on by forces in the x, y and z directions, each of these forces produces strains in all three directions. Then the total strain in the x direction, for example, will be a combination of the strains due to all three forces. It is sufficient to find the strains due to the three forces acting separately and add them, provided that the combined effect does not stress the material beyond the range of proportionality.

Example 7.4. A cube of material of side 2 in is acted upon by a compressive force of 12,000 lb in the y direction. In the z direction it is restrained against expansion (Fig. 7.19) while in the x direction it is unconfined. If $E = 15 \times 10^6$ psi and $\mu = 0.3$ find the force exerted by the restraining walls upon the block. Also find the strain in the x direction.

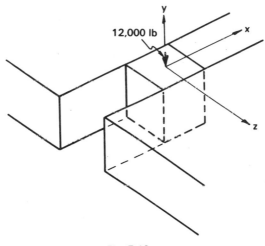

12,000 lb

Fig. 7·19

Solution. The stress in the x direction, σ_x, is zero.
    The stress in the y direction is

$$\sigma_y = -\frac{12,000}{4} = -3000 \text{ psi}$$

The stress in the z direction σ_z, is at present unknown.
    In order to compute the total strain in each direction it is convenient to tabulate the three stresses and to write opposite each the three strains to which it gives rise.

TABLE 7.1

Stress	ϵ_x	ϵ_y	ϵ_z
$\sigma_x = 0$	0	0	0
$\sigma_y = -3000$	$-0.3\left(\dfrac{-3000}{15 \times 10^6}\right)$	$\left(\dfrac{-3000}{15 \times 10^6}\right)$	$-0.3\left(\dfrac{-3000}{15 \times 10^6}\right)$
σ_z	$-0.3\left(\dfrac{\sigma_z}{15 \times 10^6}\right)$	$-0.3\left(\dfrac{\sigma_z}{15 \times 10^6}\right)$	$\left(\dfrac{\sigma_z}{15 \times 10^6}\right)$

The stress σ_x is zero and causes no strains.

The stress σ_y causes a direct strain of $-3000/(15 \times 10^6)$ in the y direction, and this is noted first under ϵ_y. The lateral strains ϵ_x and ϵ_z are then $-0.3\epsilon_y$.

The stress σ_z causes a direct strain of $\sigma_z/(15 \times 10^6)$ in the z direction and this is noted under ϵ_z. The lateral strains ϵ_x and ϵ_y are then $-0.3 \times \epsilon_z$.

By summing the values in each column we obtain the value of the total strains ϵ_x, ϵ_y and ϵ_z when all stresses are acting together. In particular, it is given that the total strain ϵ_z is zero, and from this we can calculate σ_z.

$$\text{Total } \epsilon_z = 0.3\left(\frac{3000}{15 \times 10^6}\right) + \frac{\sigma_z}{15 \times 10^6} = 0$$

$$\sigma_z = -900 \text{ psi}$$

Since the area of one face of the cube is 4 in² the force exerted by each of the restraining walls is 3600 lb compression.

The total strain in the x direction is the sum of the ϵ_x column.

$$\text{Total } \epsilon_x = -0.3\left(\frac{-3000}{15 \times 10^6}\right) - 0.3\left(\frac{-900}{15 \times 10^6}\right)$$

$$= +6.0 \times 10^{-5} + 1.8 \times 10^{-5}$$

$$= 7.8 \times 10^{-5}$$

The total strains of any small element subjected to a general three-dimensional stress system can be found in the same way.

TABLE 7.2

Stress	ϵ_x	ϵ_y	ϵ_z
x	σ_x/E	$-\mu\sigma_x/E$	$-\mu\sigma_x/E$
y	$-\mu\sigma_y/E$	σ_y/E	$-\mu\sigma_y/E$
z	$-\mu\sigma_z/E$	$-\mu\sigma_z/E$	σ_z/E

Each line of Table 7.2 lists the elastic strains due to one of the stresses σ_x, σ_y or σ_z acting alone. Provided the principle of superposition applies, these effects can be added to give

$$\epsilon_x = \frac{1}{E}[\sigma_x - \mu(\sigma_y + \sigma_z)]$$

$$\epsilon_y = \frac{1}{E}[\sigma_y - \mu(\sigma_z + \sigma_x)] \qquad (7.21)$$

$$\epsilon_z = \frac{1}{E}[\sigma_z - \mu(\sigma_x + \sigma_y)]$$

Equations (7.21) give the relations between the stresses in the x, y and z directions and the strains in these directions, in terms of the elastic constants E and μ. These equations enable us to express the change of volume of a small element in terms of the principal stresses.

In a three-dimensional field, a relationship exists between the scalar stress component and the change of volume of an element.

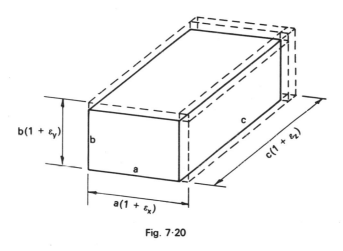

Fig. 7·20

Fig. 7.20 shows a block of material with sides a, b and c in the directions x, y and z. The volume, V, is abc. Due to strain, the sides become $a(1 + \epsilon_x)$, $b(1 + \epsilon_y)$ and $c(1 + \epsilon_z)$. The volume increases to $V + \delta V$, where

$$V + \delta V = a(1 + \epsilon_x)\, b(1 + \epsilon_y)\, c(1 + \epsilon_z)$$

$$= abc(1 + \epsilon_x + \epsilon_y + \epsilon_z)$$

if products of strains are neglected.

The change of volume is thus $V(\epsilon_x + \epsilon_y + \epsilon_z)$, and the *volumetric strain*, or *dilation*, is given by

$$\epsilon_v = \frac{dV}{V} = \epsilon_x + \epsilon_y + \epsilon_z \qquad (7.22)$$

From equations (7.21) the dilation can be expressed in terms of stresses.

$$\epsilon_v = \epsilon_x + \epsilon_y + \epsilon_z = \frac{1}{E}(\sigma_x + \sigma_y + \sigma_z)(1 - 2\mu) \tag{7.23}$$

It has been shown that in a three-dimensional stress field the scalar component is $[(\sigma_x + \sigma_y + \sigma_z)/3]$, i.e., the average of the normal stresses in any three orthogonal directions. If we call this value σ_v, equation (7.23) becomes

$$\epsilon_v = \frac{3\sigma_v}{E}(1 - 2\mu) \tag{7.24}$$

and since E and μ are constant for a given material we can write

$$\sigma_v = K\epsilon_v \tag{7.25}$$

The constant K is called the *volumetric* or *bulk* modulus of elasticity. It is the constant of proportion between the scalar stress component and the volumetric strain at a given point in a material.

A special instance of this relationship occurs when we have an isotropic stress condition. If we imagine a body to be immersed in a fluid which is then subjected to a pressure σ_p, then the stresses throughout the material are the same in all directions. At any point

$$\sigma_x = \sigma_y = \sigma_z = \sigma_1 = \sigma_2 = \sigma_3 = -\sigma_p$$

The volumetric strain is then

$$\frac{dV}{V} = \epsilon_v = \frac{-3\sigma_p}{E}(1 - 2\mu) \tag{7.26}$$

From equation (7.23) it can be seen that if a material has a Poisson's Ratio of 0.5, it can experience no volumetric strain irrespective of what stresses occur. Such a material is said to be incompressible.

7.11   Relationships between Elastic Constants

Within the range of proportionality, stresses and strains are linearly related according to Hooke's Law. In terms of E and μ, the relations between the normal stresses and strains for an isotropic material were given in equations (7.21). For the principal directions in a two-dimensional stress state these equations become

$$\epsilon_1 = \frac{1}{E}(\sigma_1 - \mu\sigma_2)$$

$$\epsilon_2 = \frac{1}{E}(\sigma_2 - \mu\sigma_1) \tag{7.27}$$

It has been shown that if an element is chosen such that its faces are not principal planes, then in general it will suffer shear stress and shear strain, each of which is a function of the orientation of the element. Because the shear stress can be expressed in terms of the principal stresses, and the shear strain can be expressed in terms of the principal strains, the shear modulus of elasticity, G, can be shown to be a function of E and μ.

We consider the case where the principal stresses are equal in magnitude but of opposite sign (Fig. 7.21 (a)). From equations 7.27 we see that the principal strains are also equal and opposite. The Mohr circles of stress and strain (Figs 7.21 (b) and 7.21 (c)) will both be circles with their centres at the origin. For such a stress state an element such as *KLMN*, drawn with its faces at 45° to the principal axes, will experience only shear stress and shear strain.

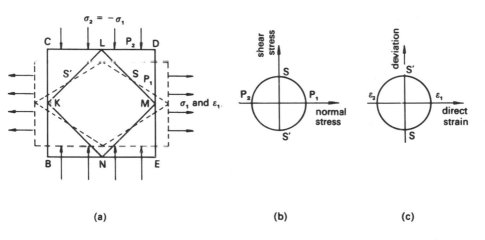

(a)   (b)   (c)

Fig. 7·21

The stresses on the faces S and S' of *KLMN* are indicated by the points S and S' in Fig. 7.21 (b), and these show that the shear stresses are equal to σ_1. Similarly the strain conditions in the direction of planes S and S' are indicated by the points S and S' in Fig. 7.21 (c), and these show that the faces suffer direct strains of zero and angular rotations equal to $-\epsilon_1$ and $+\epsilon_1$ respectively. The shear strain between the faces S and S' is thus equal to $2\epsilon_1$. For the element *KLMN* we therefore have

$$\text{shear stress, } \tau = \sigma_1 \tag{7.28}$$

and

$$\text{shear strain, } \gamma = 2\epsilon_1 \tag{7.29}$$

In terms of shear stress and strain, the relation between stress and strain may be expressed, using the shear modulus G, as

$$\tau = G\gamma \tag{7.30}$$

In terms of E and μ, the relation between normal stress and strain is given by equation (7.27). When $\sigma_2 = -\sigma_1$ this becomes

$$\epsilon_1 = \frac{1}{E}(\sigma_1 + \mu\sigma_1)$$

or

$$\sigma_1 = \frac{E}{(1+\mu)}\epsilon_1 \tag{7.31}$$

Expressing σ_1 and ϵ_1 in this equation in terms of τ and γ from equations (7.28) and (7.29) we have

$$\tau = \frac{E}{2(1+\mu)}\gamma \tag{7.32}$$

Comparing equations (7.30) and (7.32) we see that

$$G = \frac{E}{2(1+\mu)}$$

or

$$E = 2(1+\mu)G \tag{7.33}$$

In section 7.10 another elastic constant was introduced, namely the volumetric modulus K. The relationship of this constant to E and μ was made clear in equations (7.24) and (7.23). From these equations

$$K = \frac{E}{3(1-2\mu)} \tag{7.34}$$

Thus E, G, K and μ are not independent properties. When any two have been determined by experiment for a given material, the others can be calculated. Furthermore, any functions or combinations of these values will also be constant and could equally well be regarded as properties of the material. Two such combinations were proposed by Lamé, in order to simplify the equations expressing normal stresses in terms of normal strains.

Equations (7.21) express each of the strains ϵ_x, and ϵ_y and ϵ_z in terms of the normal stresses in these directions. Clearly the equation can be re-arranged to give the stresses as explicit functions of the strains. We thus obtain

$$\sigma_x = \frac{E}{1+\mu}\epsilon_x + \frac{\mu E}{(1+\mu)(1-2\mu)}(\epsilon_x + \epsilon_y + \epsilon_z)$$

$$\sigma_y = \frac{E}{1+\mu}\epsilon_y + \frac{\mu E}{(1+\mu)(1-2\mu)}(\epsilon_x + \epsilon_y + \epsilon_z) \qquad (7.35)$$

$$\sigma_z = \frac{E}{1+\mu}\epsilon_z + \frac{\mu E}{(1+\mu)(1-2\mu)}(\epsilon_x + \epsilon_y + \epsilon_z)$$

We can put $E/(1+\mu) = \Lambda_1$ and $\mu E/[(1+\mu)(1-2\mu)] = \Lambda_2$. These quantities are called Lamé's constants. The equations then appear in the form

$$\sigma_x = \Lambda_1 \epsilon_x + \Lambda_2 \epsilon_v$$

$$\sigma_y = \Lambda_1 \epsilon_y + \Lambda_2 \epsilon_v \qquad (7.36)$$

$$\sigma_z = \Lambda_1 \epsilon_z + \Lambda_2 \epsilon_v$$

It may be noted that for a material which obeys Hooke's Law, we can express the geometrical condition of strain compatibility in terms of stresses. Only the two-dimensional problem will be discussed here. In section 7.9 we found that for a material to remain continuous a relationship had to exist between the strains ϵ_x, ϵ_y and γ_{xy}, namely

$$\frac{\partial^2 \epsilon_x}{\partial y^2} + \frac{\partial^2 \epsilon_y}{\partial x^2} = \frac{\partial^2 \gamma_{xy}}{\partial x\, \partial y} \qquad (7.37)$$

Now for a two-dimensional system we have shown that

$$\epsilon_x = \frac{1}{E}(\sigma_x - \mu\sigma_y)$$

$$\epsilon_y = \frac{1}{E}(\sigma_y - \mu\sigma_x) \qquad (7.38)$$

and

$$\gamma_{xy} = \frac{1}{E}2(1+\mu)\tau_{xy}$$

Hence, we can write equation (7.37) in the form

$$\frac{\partial^2}{\partial y^2}(\sigma_x - \mu\sigma_y) + \frac{\partial^2}{\partial x^2}(\sigma_y - \mu\sigma_x) = 2(1+\mu)\frac{\partial^2 \tau_{xy}}{\partial x\, \partial y} \qquad (7.39)$$

Considering this together with the two equilibrium equations (7.14a) and (7.14b), we now have three equations all in terms of the stresses σ_x, σ_y and τ_{xy}. For certain situations it is possible to obtain analytical solutions to these equations, but a discussion of the procedures involved is beyond the scope of this book.

7.12   Elastic Strain Energy

In previous chapters expressions were derived for the strain energy stored in an element due to different kinds of deformation. In a similar way an expression can be derived for the energy stored in an element $dx \times dy \times dz$ (Fig. 7.22) in terms of the principal stresses.

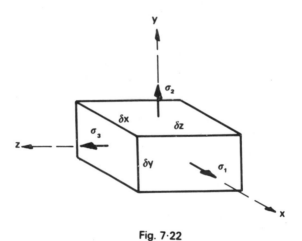

Fig. 7·22

Suppose that the principal stresses σ_1, σ_2 and σ_3 act in the x, y and z directions respectively. The forces N_1, N_2 and N_3 on the x, y and z faces are respectively $\sigma_1\,dy\,dz$, $\sigma_2\,dz\,dx$ and $\sigma_3\,dx\,dy$. Equations (7.21) give the principal strains and from these we can write the extensions in the x, y and z directions as

$$e_1 = \epsilon_1\,dx = \frac{dx}{E}\{\sigma_1 - \mu(\sigma_2 + \sigma_3)\}$$

$$e_2 = \epsilon_2\,dy = \frac{dy}{E}\{\sigma_2 - \mu(\sigma_3 + \sigma_1)\}$$

$$e_3 = \epsilon_3\,dz = \frac{dz}{E}\{\sigma_3 - \mu(\sigma_1 + \sigma_2)\}$$

The total strain energy stored in the element is $\frac{1}{2}N_1e_1 + \frac{1}{2}N_2e_2 + \frac{1}{2}N_3e_3$, i.e.,

$$dU = \frac{\sigma_1}{2E}dx\,dy\,dz\{\sigma_1 - \mu(\sigma_2 + \sigma_3)\}$$

$$+ \frac{\sigma_2}{2E}dx\,dy\,dz\{\sigma_2 - \mu(\sigma_3 + \sigma_1)\}$$

$$+ \frac{\sigma_3}{2E}dx\,dy\,dz\{\sigma_3 - \mu(\sigma_1 + \sigma_2)\}$$

$$= \frac{dx\,dy\,dz}{2E}\{\sigma_1^2 + \sigma_2^2 + \sigma_3^2 - 2\mu(\sigma_1\sigma_2 + \sigma_2\sigma_3 + \sigma_3\sigma_1)\} \qquad (7.40)$$

Since $dx\,dy\,dz$ is the volume of the element, the energy per unit volume is

$$dU = \frac{1}{2E}\{\sigma_1^2 + \sigma_2^2 + \sigma_3^2 - 2\mu(\sigma_1\sigma_2 + \sigma_2\sigma_3 + \sigma_3\sigma_1)\} \qquad (7.41)$$

It should be pointed out again that sections 7.10, 7.11 and 7.12 deal with relationships between stress and strain, and equations have only been derived for the conditions that stress and strain are proportional, i.e., for the condition that Hooke's Law applies.

Problems

In the following problems it will be assumed that the z axis is towards the viewer, so that the y axis will always appear $90°$ anticlockwise from the x axis.

7.1 A body is subjected to a two-dimensional stress state, the stresses being in the xy plane. At a given point, the principal stress σ_1 is $+1300$ psi acting at $-30°$ to the x axis (i.e., clockwise from Ox) while the principal stress σ_2 is -500 psi. Draw a small element with its faces normal to the principal stresses. Show σ_1 and σ_2 acting on the sides of the element and mark the four faces either P_1 or P_2. Determine analytically the normal stress and shear stress on the plane normal to the x axis.

7.2 For the stress condition described in problem 7.1 find by means of a Mohr circle the stresses on the planes normal to the x and y axes. Also determine the direction of the planes on which there is no normal stress. What is the magnitude of the maximum shear stress, and on what planes does this stress occur? Indicate the results by sketching an element of material.

7.3 In a two-dimensional stress system, the principal stresses are $\sigma_1 = -10,000$ psi, and $\sigma_2 = -2000$ psi. Find analytically:
(a) The stresses on an X plane at $+30°$ to plane P_1.
(b) The stresses on an X plane at $+60°$ to plane P_1.

7.4 Solve problem 7.3 by a Mohr circle.

7.5 For the stress state shown in Fig. P7.5 find analytically:
  (*a*) The direction and magnitude of the principal stresses.
  (*b*) The direction of the planes having no normal stress, and the magnitude of the shear stress on these planes.

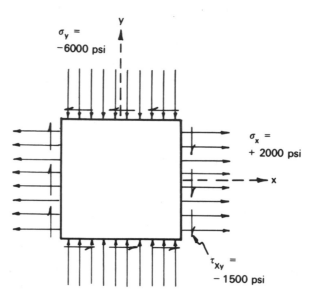

$\sigma_y = -6000$ psi

$\sigma_x = +2000$ psi

$\tau_{xy} = -1500$ psi

Fig. P7·5

7.6 Solve problem 7.5 graphically.

7.7 For the stress state shown in Fig. P7.7 (p. 199), find graphically:
  (*a*) The magnitude and direction of the principal stresses.
  (*b*) The stresses on an element whose faces are normal to $\bar{x}$ and $\bar{y}$ where axes $\bar{x}$, $\bar{y}$ are +45° from axes x, y.
  (*c*) The three principal shear stresses at the point.
  (*d*) The maximum shear stress at the point.

7.8 Solve problem 7.7 analytically.

7.9 Prove that the shear stresses on planes which make angles of $(45° - \theta)$ and $(45° + \theta)$ with the principal planes are equal.

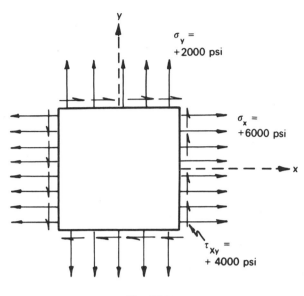

Fig. P7·7

7.10 At a point in a body the principal stresses are $\sigma_1 = +10,000$ psi and $\sigma_2 = +5000$ psi. What is the ratio of shear stress to normal stress on a plane making an angle of $30°$ with P_1? On what planes is this ratio a maximum? (Assume that the value of σ_3 lies between the values of σ_1 and σ_2.)

7.11 At a certain point $\sigma_1 = +10,000$ psi and $\sigma_2 = +6000$ psi. What are the limits between which σ_3 must lie if the ratio of shear stress to normal stress on any plane is not to exceed $\tan 30°$?

7.12 At a certain point in a beam the stresses on the plane normal to y are $\sigma_y = 0$ and $\tau_{Yx} = +400$ psi. What must be the compressive stress σ_x on the cross-section if the maximum principal tension at the point is not to exceed 200 psi? What is then the maximum compressive stress at the point?

7.13 If the largest principal stress at a point is 9500 psi and the maximum shear stress is 3000 psi, what is the smallest principal stress?

7.14 At a given point in a material, a plane normal to the y axis sustains a normal stress of $+4000$ psi and a shear stress of $+1000$ psi while a plane normal to another axis $\bar{y}$ sustains a normal stress of $+12,000$ psi and a shear stress of $+3000$ psi. What is the angle between the axes y and $\bar{y}$?

7.15   At a certain point in a two-dimensional stress field two planes X and $\bar{X}$ are at 45° to one another. Plane X has a normal stress of +12,000 psi and a shear stress of +3000 psi. Plane $\bar{X}$ has a normal stress of +2000 psi. Find the shear stress on plane $\bar{X}$ and the magnitude of the principal stresses (there are two solutions to this problem).

7.16   The stresses on a plane normal to the x axis are $\sigma = +2000$ psi and $\tau = +2000$ psi. A plane normal to $\bar{x}$, which is 60° clockwise from x, has a normal stress of −10,000 psi. Find the shear stress on plane $\bar{X}$ and the principal stresses at the point.

7.17   Three planes A, B and C are at 120° to one another. The normal stresses on the three planes are +2000 psi, +4000 psi and +8000 psi respectively. Find the magnitude of the principal stresses at the point.

7.18   At a certain point, the principal strains are $\epsilon_1 = +6 \times 10^{-4}$ at +20° to the x direction and $\epsilon_2 = -4 \times 10^{-4}$. Find, analytically, the direct strain and the deviation in the x direction and in the y direction.
   Sketch the deformation of a square element with its faces normal to x and y. What is the shear strain γ_{xy} experienced by this element?
   For the same point in the material, sketch a square element which suffers no shear strain, and show the deformation of this element.

7.19   Solve problem 7.18 by means of a Mohr circle.

7.20   In a two-dimensional stress system, two of the principal strains are $\epsilon_1 = 8 \times 10^{-4}$ and $\epsilon_2 = -2 \times 10^{-4}$.
   (*a*) What are the strain conditions in directions at +30° and +60° to the direction of ϵ_1?
   (*b*) In what direction is the direct strain zero?
   (*c*) What is the maximum deviation at this point in the material?

7.21   At a certain point, the direct strain and deviation in the x direction are $+9 \times 10^{-4}$ and $+3 \times 10^{-4}$ respectively, while in the y direction the direct strain is zero. Find analytically the magnitude and direction of the principal strains. What is the shear strain on an element with faces in the x and y directions? What is the maximum shear strain at this point in the material?

7.22   Solve problem 7.21 graphically.

7.23   A cube of material of side 3 in is subjected to stresses, normal to its faces, of $\sigma_1 = +20,000$ psi, $\sigma_2 = 5000$ psi and $\sigma_3 = -4000$ psi. If $E = 30 \times 10^6$ psi and $\mu = 0.3$, find the strains in the three directions and calculate the total change in volume of the block.

7.24 A cube of material has its faces normal to the x, y and z directions. On the x and y faces the stresses are $\sigma_x = +2000$ psi and $\sigma_y = -6000$ psi. There are no shear stresses on these faces. In the z direction the material is restrained against any expansion or contraction. If $E = 10 \times 10^6$ psi and $\mu = 0.25$ find the stress in the z direction and the total strain in the x direction.

7.25 A bar of material ($E = 15 \times 10^6$ psi and $\mu = 0.3$) is subjected to a longitudinal stress of 8000 psi. Find the percentage change in volume of the bar.

7.26 Derive an expression for the percentage change in volume of a small element in terms of the three principal stresses at the point. Assume that the material behaves elastically. Simplify the expression for an element situated at a point where the stress is isotropic.

7.27 For a two-dimensional stress state (i.e., $\sigma_3 = 0$) show that an element subjected to a state of pure shear suffers no change in volume.

7.28 For a two-dimensional state of stress, show that at a given point the sum of the normal stresses on any two planes at right-angles is a constant.

7.29 At a certain point in a material $\sigma_x = +1200$ psi, $\sigma_y = +800$ psi and $\sigma_z = 0$. Find the strain in the z direction if $E = 5 \times 10^6$ psi and $\mu = 0.25$.

7.30 At a certain point $\sigma_3 = 0$ and $\epsilon_3 = +5 \times 10^{-4}$. If the other principal stresses differ by 2000 psi find the values of these stresses. $E = 5 \times 10^5$ psi and $\mu = 0.2$.

7.31 A sheet of material is subjected to a two-dimensional stress system in the xy plane. At a given point, an element with its sides parallel to x and y is found to have a shear strain of 1×10^{-4}, while an element at 45° to the first has a shear strain of 4×10^{-4}.

Find the maximum shear strain at this point. If the shear modulus for the material is 10^7 psi find the difference between the two principal stresses.

7.32 For a certain material $E = 15 \times 10^6$ psi and $\mu = 0.25$. What is the value of the shear modulus, G?

7.33 At a certain point in a two-dimensional stress system, the shear strain between the x and y directions is 2×10^{-4}. The direct strain in a direction midway between x and y is the same as that in the x direction. Find the maximum shear strain in the xy plane at this point in the material.

7.34 For a given point in a body, show that if the normal stresses on two planes are equal, the shear stresses on these planes must be equal in magnitude.

7.35   At a point on the surface of a body, the strains are measured in three directions at 120° to one another. They are found to be $+2 \times 10^{-4}$, $+4 \times 10^{-4}$ and $+8 \times 10^{-4}$ respectively. Find the magnitude of the principal strains at the point.

If $E = 10 \times 10^6$ psi and $\mu = 0.25$ find the magnitude of the principal stresses.

7.36   At a certain point the principal stresses are $\sigma_1 = 8000$ psi, $\sigma_2 = 20,000$ psi and $\sigma_3 = 12,000$ psi. Find the maximum shear stress at this point and the shear stress on the octahedral planes.

7.37   In a simple tension test, the test specimen has a cross-section of 0.8 in². When the axial force is 27,600 lb find the values of (a) the three principal stresses, (b) the three principal shear stresses and (c) the octahedral shear stress.

CHAPTER 8

Combined Stresses

8.1 Re-combining the Stress-resultants

In previous chapters we have discussed the stresses and deformations which are caused separately by axial force, bending moment, shear force and torsion. In reality these stress resultants do not occur in isolation. The mutual interaction between the two parts of a beam separated by a given cross-section consists of stresses which vary in magnitude and direction from point to point across the section. The idea of summing the stresses to give a resultant force (or couple) and then of decomposing this force into components, is one which is adopted largely for computational purposes. It is not even a meaningful process except in the case of a beam or bar whose cross-sectional dimensions are small compared with its length.

Having adopted this procedure, we now consider the fact that the stresses so calculated at any one point really occur simultaneously. It is frequently necessary to consider the real state of stress of which these partial stresses are the components.

We note that axial force and bending moment each produce stresses normal to the cross-section. It is fairly easy to consider the combined effect of axial force, N, and bending moment, M. In fact, if the principle of superposition is valid, the stresses can simply be added algebraically. This will be studied in section 8.2.

Shear force, S, and torsion, T, on the other hand, each produce shear stresses on the cross-section. The combination of these two effects will be considered in section 8.3.

Finally, we must consider the simultaneous presence of normal stresses (due to N and M) and shear stresses (due to S and T). It may be necessary to find the principal stresses by the methods discussed in Chapter 7.

8.2 Axial Force and Bending Moment

Axial force and bending moment both give rise to stresses which are normal to the cross-section. The stresses may be added algebraically provided that the combined stress is within the proportional range at every point on the section.

Due to axial force,

$$\sigma = N/A \text{ at every point}$$

Due to bending moment M_z,

$$\sigma = -M_z y/I_z \text{ at } y \text{ from the } z \text{ axis}$$

Due to bending moment M_y,

$$\sigma = M_y z/I_y \text{ at } z \text{ from the } y \text{ axis}$$

It will be easier to study the effect of superposition if bending about the z axis only is considered. Then

$$\sigma = \frac{N}{A} - \frac{My}{I} \tag{8.1}$$

The stresses caused by N and M acting together are indicated in Fig. 8.1.

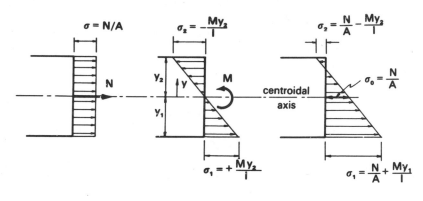

Stresses due to axial force
(a)

Stresses due to bending moment
(b)

Combined stresses
(c)

Fig. 8·1

Example 8.1. A rectangular beam is 3 in wide and 8 in deep. At a certain section $N = +12,000$ lb and $M = -10,000$ lb-ft. Find the longitudinal stress at a point 2 in above the centre of the section. What is the location of the layer of zero stress?

Solution.

$$I = 128 \text{ in}^4; \quad A = 24 \text{ in}^2; \quad M = -10,000 \text{ lb-ft} = -120,000 \text{ lb-in}$$

(a) At $y = +2$,

$$\text{stress due to } M = \sigma_M = \frac{-My}{I} = \frac{(+120{,}000) \times 2}{128} = 1875 \text{ psi}$$

$$\text{stress due to } N = \sigma_N = \frac{N}{A} = \frac{12{,}000}{24} = 500 \text{ psi}$$

$$\text{Total stress } \sigma = \sigma_M + \sigma_N = 2375 \text{ psi}$$

(b) At the layer of zero stress, we have

$$0 = \frac{12{,}000}{24} - \frac{(-120{,}000)\, y}{128}$$

Hence

$$y = -0.53 \text{ in}$$

Zero stress occurs 0.53 in below the centre of the beam.

Although this location should be called the neutral axis, it is usual to reserve this term for the position at which the stress due to bending alone is zero.

An axial force and two bending moments can be regarded as an axial force acting at a point other than the centroid. Conversely, an axial force acting through any point of the section can be replaced by an axial force at the centroid together with bending moments about the y and z axes.

Example 8.2. A beam has the cross-section shown in Fig. 8.2. The bending moment about the z axis, M_z, is 300,000 lb-in (causing compression along DC) and M_y is 240,000 lb-in (causing compression along BC).
(a) What axial force will cause the combined stress at B to be zero?
(b) If M_y, M_z and N are combined into a single equivalent axial force, where will this act?

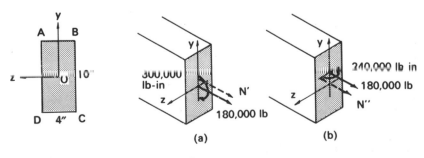

Fig. 8·2   Fig. 8·3

Solution.

$$I_y = \frac{10 \times 4^3}{12} = 53.3 \text{ in}^4; \quad I_z = \frac{4 \times 10^3}{12} = 333.3 \text{ in}^4; \quad A = 4 \times 10 = 40 \text{ in}^2$$

(a) At point B, $y = 5$ and $z = -2$. Hence

$$\text{due to } M_y, \ \sigma = -\frac{240,000 \times 2}{53.3} = -9000 \text{ psi}$$

$$\text{due to } M_z, \ \sigma = +\frac{300,000 \times 5}{333.3} = +4500 \text{ psi}$$

$$\text{due to } N, \ \sigma = \frac{N}{40} \text{ psi}$$

$$\therefore \quad \frac{N}{40} + 4500 - 9000 = 0$$

and

$$N = +180,000 \text{ lb}$$

(b) The force of 180,000 lb and the couple 300,000 lb-in (Fig. 8.3 (a)) can be replaced by a force of 180,000 at 300,000/180,000 (= 1.67 in) above the z axis. Similarly, this force N' and the couple of 240,000 lb-in (Fig. 8.3 (b)) can be replaced by a force of 180,000 lb at a distance of 240,000/180,000 (= 1.33 in) from the y axis. The two bending moments and the axial force are thus equal to an axial force of +180,000 lb at the point $y = 1.33$ in, $z = 1.67$ in.

Example 8.3. An I beam has the section shown in Fig. 8.4. It is subjected to an axial compression of 50 kips acting at a point P whose co-ordinates are $y = 5$ in, $z = -2$ in. Find the longitudinal stress at each corner of the section.

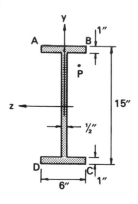

Fig. 8·4

Solution.

$$I_y = 36.135 \text{ in}^4; \quad I_z = 680.5 \text{ in}^4; \quad A = 18.5 \text{ in}^2$$

The compressive force of 50 kips is replaced by an axial force of -50 kips at the centre together with two bending moments

$$M_y = 50 \times 2 = 100 \text{ kip-in (causing compression at } B \text{ and } C)$$

and
$$M_z = 50 \times 5 = 250 \text{ kip-in (causing compression at } A \text{ and } B)$$

Then,

the corner stress due to $N = \dfrac{-50,000}{18.5} = -2700$ psi $(-$ at $A, B, C, D)$

the corner stress due to $M_y = \dfrac{100,000 \times 3}{36.135} = 8330$ psi $\left(\begin{matrix} + \text{ at } A, D \\ - \text{ at } B, C \end{matrix}\right)$

the corner stress due to $M_z = \dfrac{250,000 \times 7.5}{680.5} = 2755$ psi $\left(\begin{matrix} + \text{ at } C, D \\ - \text{ at } A, B \end{matrix}\right)$

The resultant corner stresses are

$$\sigma_A = -2700 + 8330 - 2755 = +2875 \text{ psi}$$
$$\sigma_B = -2700 - 8330 - 2755 = -13,785 \text{ psi}$$
$$\sigma_C = -2700 - 8330 + 2755 = -8275 \text{ psi}$$
$$\sigma_D = -2700 + 8330 + 2755 = +8385 \text{ psi}$$

An axial force N (assumed tensile) at a distance e from the z axis is equivalent to N at the centroid together with a bending moment Ne (which is positive if e is below the neutral axis). The extreme fibre stresses can then be expressed in terms of N and $e\ (= M/N)$, as in Fig. 8.5.

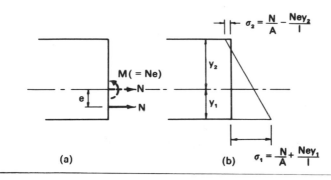

(a) (b)

Fig. 8·5

Fig. 8.6 shows how the stress distribution varies as e is changed. The stress at the centroid is always N/A, but the difference between top and bottom stresses increases as e increases. There will be some value of e for which the stress at the top of the beam is zero (Fig. 8.6 (c)). From Fig. 8.5 (b) this critical eccentricity is given by

$$\sigma_2 = \frac{N}{A} - \frac{Ne_2 y_2}{I} = 0$$

whence

$$e_2 = \frac{I}{Ay_2} \text{ below the centroid.}$$

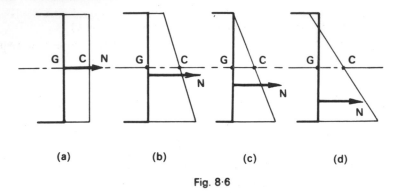

(a)   (b)   (c)   (d)

Fig. 8·6

Similarly, the stress at the bottom fibre will be zero if $e_1 = I/Ay_1$ above the centroid. If N acts on the vertical centreline between e_1 and e_2, the stress has the same sign at every point on the section. When eccentricity in the z direction is also considered, it is possible to determine a region of the cross-section within which N must lie if the stress is to be the same sign over the whole section. This region is called the *kern* of the section.

For a rectangle, the kern is a .diamond (Fig. 8.7). The limits are easily obtained by stating the condition that the stress at each corner in turn shall be zero. Different shapes of cross-sections will have different kerns.

The kern is important in the case of materials which can resist stress of one sign only. Concrete, for example, can resist compression, but not tension, with any reliability.

When the resultant stress at any point falls outside the proportional range, the method of superposition used in the previous examples is not valid. It is then necessary to consider the distribution of resultant stress directly, and to

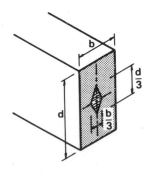

Fig. 8·7

find a distribution such that the resultant of the stresses is equal to N and the resultant of the moment of the stresses is equal to M.

This presents no difficulty if the material is elastic-plastic, but can involve lengthy algebra for some stress–strain relations.

Example 8.4. A rectangular beam, 3 in wide and 8 in deep, is made of a material which is linearly elastic up to a yield stress of 30 ksi. Find the stress distribution at a section where the axial force is −345 kips and the bending moment is +875 kip-ft.

Solution. A trial calculation on the basis of a linear stress variation shows that the top fibre is strained beyond yield while the bottom fibre is not. Clearly the top of the beam will have yielded, and it will be assumed that the bottom has not.

The stress distribution can be described in terms of two parameters. A good choice of variables makes a considerable difference in the complexity of the calculation.

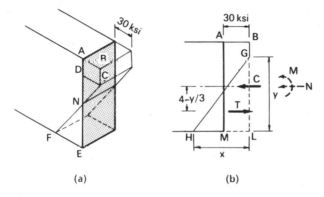

(a)   (b)

Fig. 8·8

For instance, in Fig. 8.8 (a), the distances EN and ED could be taken as the variables. In terms of these the stress blocks $ABCD$ and CDN (compression) and NEF (tension) can be expressed. However, if the stresses are considered as rectangle $ABLM$ (Fig. 8.8 (b)) (compression) and triangle GLH (tension), the stresses are then described in terms of two blocks only.

Denoting the stress ordinate HL by x and the height LG by y, we have

$$C = 30 \times 8 \times 3 = 720 \text{ kips acting at the centroid}$$

$$T = x \times \frac{y}{2} \times 3 = \frac{3}{2}xy \text{ kips acting at } \left(4 - \frac{y}{3}\right) \text{ below the centroid.}$$

For equilibrium of forces

$$720 - \frac{3}{2}xy = 345$$

from which

$$\frac{3}{2}xy = 375 \text{ kips} \tag{8.2}$$

For equilibrium of moments about the centroid

$$T\left(4 - \frac{y}{3}\right) = 875$$

i.e.,

$$\frac{3}{2}xy\left(4 - \frac{y}{3}\right) = 875$$

From equation (8.2)

$$375\left(4 - \frac{y}{3}\right) = 875$$

whence

$$y = 5 \text{ in}$$

then

$$x = 50 \text{ ksi}$$

Thus the top 3 in of the beam has yielded in compression and the stress at the bottom fibre is 20 ksi tension.

When the stress distribution is non-linear, the centroid of the section has no particular significance. The location of the axial force N is arbitrary and the value assigned to the bending moment will depend on the position specified for N. It may be easier to consider the two combined into a single eccentric force N.

Example 8.5. A beam of T cross-section has a flange 10 in × 1.5 in and a stem 16 in × 2 in. The material yields at a stress of 3000 psi both in tension and compression. Where must a compressive axial force of 75 kips act so that the whole section has yielded? (The top fibre is in compression.)

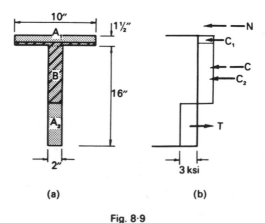

(a) (b)

Fig. 8·9

Solution. The area of the section is 47 in^2.

With the whole section in compression at σ_y the total force would be $3000 \times 47 = 141,000$ lb. Clearly the lower part of the beam must yield in tension. The section may be divided as in Fig. 8.9 (a), where regions A_1 and B have yielded in compression and region A_2 has yielded in tension. If areas A_1 and A_2 are equal, the forces on them form a couple, and the net force on the section is C_2, the force on B.

Hence

$$C_2 = N = 75 \text{ kips}$$

$$\therefore \quad \text{Area } B = \frac{75}{3} = 25 \text{ in}^2$$

Then

$$A_1 = A_2 = \frac{47 - 25}{2} = 11 \text{ in}^2$$

$$C_1 = T = 11 \times 3 = 33 \text{ kips}$$

Area A_2 extends up 11/2 in from the bottom hence T is 2.75 in above the bottom of the beam.

Forces C_1 and C_2 are equivalent to a single force C of 108 kips at the centroid of A_1 and B, i.e., 4.25 in below the top of the beam.

The resultant of forces C and T is a force of 75 kips 0.37 in above the top of the beam.

A compressive force of 75 kips 0.37 in above the top of the beam will produce a yield condition over the whole cross-section.

8.3   Shear Force and Torsion

Shear force and torsion both produce shearing stresses in the plane of the cross-section. Provided the stress is within the proportional range at all points of the section, the shear stress due to shear force and that due to torsion can be computed separately and then combined. In general, at a given point on the cross-section the two shear stresses will not act in the same direction and addition will be vectorial. In practice, stresses are mainly required at points on the boundary, and here the two stresses are in the same direction, although not necessarily in the same sense

For instance, in a bar of circular section which sustains both shear and torsion (Fig. 8.10 (a)), the shear force gives rise to a distribution of stress as

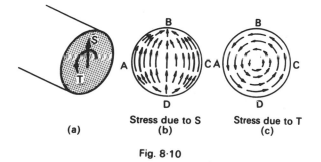

(a)

Stress due to S
(b)

Stress due to T
(c)

Fig. 8·10

in Fig. 8.10 (b), while the torsion produces stresses as in Fig. 8.10 (c). Critical points of the section are usually A, B, C and D. At A and C the stresses are added algebraically (subtracted at A, added at C). At B and D the stresses due to shear force are zero.

Example 8.6. A circular shaft is 2 in diameter. At a certain section, there is a vertical shear force of 4000 lb and a torque of 9000 lb-in. Find the resultant shear stress at the ends of the horizontal and vertical diameters.

Solution.

$$I = \frac{\pi d^4}{64} = 0.785 \text{ in}^4; \quad J = \frac{\pi d^4}{32} = 1.57 \text{ in}^4$$

Let τ_S be the stress due to shear force and τ_T be the stress due to torsion. In order to find τ_S at A and C we require the first moment of the semicircle about the diameter. This is given by

$$Q = \frac{2r^3}{3} = \frac{2 \times 1^3}{3} = 0.667$$

It will be assumed that the stress is distributed uniformly across the diameter. Then, referring to Fig. 8.10, we have

At A and C

$$\tau_S = \frac{SQ}{Id} = \frac{4000 \times 0.667}{0.785 \times 2} = 1696 \text{ psi}$$

At B and D

$$\tau_S = 0$$

At A, B, C and D

$$\tau_T = \frac{Tr}{J} = \frac{9000 \times 1}{1.57} = 6420 \text{ psi}$$

Hence

the resultant stress at $A = 6424 - 1696 = 4724 \text{ psi}$

the resultant stress at $C = 6424 + 1696 = 8116 \text{ psi}$

the resultant stress at B and $D = 6420 \text{ psi}$

Example 8.7. A beam of homogeneous material has a rectangular section 12 in × 20 in with the longer side vertical. At a section where the twisting moment is 60,000 lb-ft and the shear force (vertical) is 80,000 lb, find the shear stress at the mid-point of each face.

Solution. Consider first the stresses arising due to the shear force. At the top and bottom faces the stress is zero.
For the whole section,

$$I = \frac{bd^3}{12} = 8000 \text{ in}^4$$

At the mid-height,

$$Q = \left(\frac{bd}{2}\right)\frac{d}{4} = 600 \text{ in}^3$$

so that

$$\tau_S = \frac{SQ}{Ib} = \frac{80,000 \times 600}{8000 \times 12} = 500 \text{ psi}$$

Now consider the torsion stresses. With $d/b = 1.67$, we find from Table 6.1 (p. 160) that $\beta = 0.208$, $\lambda_1 = 0.88$ and $\lambda_2 = 0.44$.

∴ The torsion constant, $J = 20 \times 12^3 \times 0.208 = 7260 \text{ in}^4$.

At the mid-points of the vertical faces

$$\tau_T = \frac{T(\lambda_1 b)}{J} = \frac{(60,000 \times 12)(0.88 \times 12)}{7260} = 1055 \text{ psi}$$

and at the mid-points of the horizontal faces

$$\tau_T = \frac{T(\lambda_2 b)}{J} = \frac{(60,000 \times 12)(0.44 \times 12)}{7260} = 520 \text{ psi}$$

Hence the resultant shear stresses are as follows:
At the mid-points of the horizontal faces

$$\tau = \tau_T = 520 \text{ psi}$$

At the mid-points of the vertical faces,

$$\tau = \tau_T \pm \tau_S = 1055 \pm 500 = 1555 \text{ psi or } 555 \text{ psi}$$

The calculation of shear stresses in thin-walled sections has been dealt with in Chapters 5 and 6. The stresses due to shear forces and torsion both act parallel to the boundary and may be considered uniform across the thickness of the material. Hence, for these sections the combined shear force can be obtained at all points by algebraic addition (always provided the final stress is within the proportional range).

Example 8.8. A beam of hollow box section is 1.2 in deep and 2.4 in wide overall. The wall thickness is 0.04 in. If the shear force is 500 lb and the torsional moment is 2000 lb-in at a certain cross-section, find the shear stress at the mid-points A and B of the vertical portions and also at C which is 0.8 in from the centre of the top face.

Solution. The stresses due to shear force and to torsion are distributed as indicated in Figs 8.11 (a) and 8.11 (b) respectively.

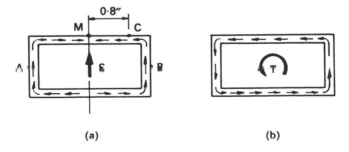

(a)   (b)

Fig. 8·11

(a) *Shear force*

For the whole section

$$I = \frac{0.08 \times 1.2^3}{12} + 2(2.32 \times 0.04 \times 0.58^2)$$

$$= 0.0740 \text{ in}^4$$

The first moment of half the section about the line AB is

$$Q = (0.6 \times 0.08 \times 0.3) + (2.32 \times 0.04 \times 0.58)$$

$$= 0.0683 \text{ in}^3$$

At A and B, $\tau_S = \dfrac{SQ}{Ib} = \dfrac{500 \times 0.0683}{0.0740 \times 0.08} = 5770 \text{ psi}$

(Since Q was calculated for the complete half beam from A round to B, the shear flow SQ/I is resisted by the section at A and B, hence b is equal to twice the wall thickness.)

To find the stress at C, isolate the portion MC.

For this portion, Q (about AB) $= 0.8 \times 0.04 \times 0.58 = 0.0186 \text{ in}^3$.

The shear flow at M is zero by symmetry, hence the shear flow SQ/I is resisted entirely at C (hence $b = 0.04$).

At C, $\tau_S = \dfrac{SQ}{Ib} = \dfrac{500 \times 0.0186}{0.0740 \times 0.04} = 1570 \text{ psi}$

(b) *Torsion*

$$A_0 = 2.36 \times 1.16 = 2.74 \text{ in}^2$$

$\therefore$ At every section

$$\tau_T = \frac{T}{2A_0 t} = \frac{2000}{2 \times 2.74 \times 0.04} = 9130 \text{ psi}$$

(c) *Combined shear stress* (Refer to Figs 8.11 (a) and (b).)

At A, $\tau = \tau_T - \tau_S = 9130 - 5770 = 3360 \text{ psi}$

At B, $\tau = \tau_T + \tau_S = 9130 + 5770 = 14{,}900 \text{ psi}$

At C, $\tau = \tau_T + \tau_S = 9130 + 1570 + 10{,}700 \text{ psi}$

8.4   Normal Stresses and Shear Stresses

At a given point on the cross-section of a beam there is usually a normal stress component (due to axial force or bending moment or both) and a shear stress component (due to shear force or torsion or both). When the total normal component has been computed (see section 8.2), and the total shear component has been computed (see section 8.3), the complete stress state at the point can be investigated by the methods of Chapter 7.

In the simple theory of bars it is assumed that the normal stress perpendicular to the beam axis is everywhere zero. This is very close to the truth except at points in the vicinity of externally applied loads of high intensity.

Example 8.9. A round bar 2 in diameter and 6 in long is cantilevered from one end (Fig. 8.12). Attached to the free end is a bracket which supports a load of 1500 lb at 4 in from the axis of the bar.

Find the principal stresses and the principal shear stresses at the points A, B, C and D of the section close to the support.

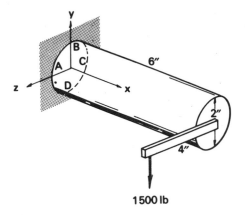

Fig. 8·12

Solution.

$$D = 2 \text{ in}; \quad I = 0.785 \text{ in}^4; \quad J = 1.57 \text{ in}^4$$

(a) *Normal stress* σ_x

The axial force is zero, and σ_x arises due to bending alone.

At A and C, $\sigma_x = 0$

At B and D, $\sigma_x = \dfrac{My}{I} = \dfrac{(1500 \times 6) \times 1}{0.785} = 11{,}450 \text{ psi}$

$(+ \text{ at } B, - \text{ at } D)$

(b) *Shear stress*

At A, B, C and D, $\tau_T = \dfrac{Tr}{J} = \dfrac{(1500 \times 4) \times 1}{1.57} = 3820 \text{ psi}$

At B and D, $\tau_S = 0$

At A and C, $\tau_S = \dfrac{SQ}{ID} = \dfrac{1500}{0.785} \dfrac{0.67}{2} = 636 \text{ psi}$

(for a semicircle, Q about the diameter $= 2r^3/3$).
   The resultant shear stresses on the cross-section are:

At A, $\tau = \tau_T + \tau_S = 3820 + 636 = 4456 \text{ psi}$

At B, $\tau = \tau_T$ $\qquad\qquad\qquad = 3820 \text{ psi}$

At C, $\tau = \tau_T - \tau_S = 3820 - 636 = 3184 \text{ psi}$

At D, $\tau = \tau_T$ $\qquad\qquad\qquad = 3820 \text{ psi}$

(c) *Principal stresses*

   At each of the points A, B, C and D we consider the equilibrium of a small element of material with its faces normal to the x, y and z directions. The elements at A and B are indicated in Fig. 8.13. The elements are shown as free-bodies in Fig. 8.14 (a), (b), (c) and (d).

For each element, the face which lies in the surface of the bar is free of shear. Hence it is a principal plane. This plane also happens to be free of normal stress. The other two principal stresses lie in planes perpendicular to the bar surface. Since the transverse normal stress is zero in each case, the principal stress will be given by

$$\left.\begin{array}{c}\sigma_1\\\sigma_2\end{array}\right\} = \frac{\sigma_x}{2} \pm \sqrt{\left(\frac{\sigma_x}{2}\right)^2 + \tau^2}$$

and the principal shear stress $\tau_3 = \dfrac{\sigma_1 - \sigma_2}{2}$

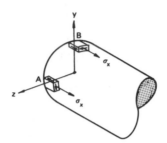

Fig. 8·13

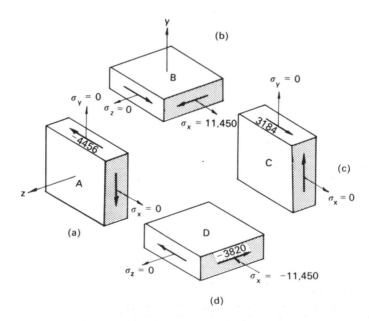

Fig. 8·14

Hence, using these formulas and the values of σ_x and τ as shown in Fig. 8.14 we have

At A, $\sigma_1 = +4456$; $\sigma_2 = -4456$; $\tau_3 = 4456$

At B, $\sigma_1 = +12{,}600$; $\sigma_2 = -1160$; $\tau_3 = 6882$

At C, $\sigma_1 = +3184$; $\sigma_2 = -3184$; $\tau_3 = 3184$

At D, $\sigma_1 = -12{,}600$; $\sigma_2 = +1160$; $\tau_3 = 6882$

It can be noted that the maximum tensile and compressive stresses in this bar are each 12,600 psi. The maximum shear stress is 6882 psi.

The calculation of principal stresses and principal shear stresses can often be important. From the bending moment and axial force we can calculate the longitudinal stress σ_x. However, it is quite possible that the maximum tensile or compressive stress does not occur in the x direction; this maximum can be a principal stress at a region where high σ_x and high τ occur at the same place.

In a similar way we can, from the shear force and torsion, compute directly the shear stresses on cross-sections of the bar. However, the maximum shear stress in the material is determined by the maximum difference between principal stresses. This can easily occur at a region where the shear stress on the *cross-section* is quite small.

Example 8.10. The beam of Fig. 8.15 has a cross-section 12 in × 6 in.
(*a*) Find the principal stresses at the mid-points of the vertical sides at the quarter point (section *D*).
(*b*) Also find the maximum shear stress in the beam.

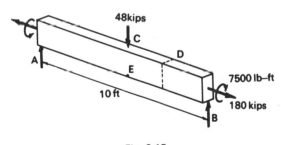

Fig. 8·15

Solution. (*a*) Stresses at Section *D*

(i) *Normal stress σ_x*

At the mid-height the bending stress is zero. Thus σ_x is due to axial force alone.

$$\sigma_x = \frac{N}{A} = \frac{180{,}000}{72} = 2500 \text{ psi}$$

(ii) *Shear stress*

At D, $S = 24,000$ lb and $T = 7500$ lb-ft

For a rectangular section, the shear stress at mid-height is $\dfrac{3}{2}\dfrac{S}{bd}$ (see p. 108). Hence

$$\tau_S = \frac{3}{2} \times \frac{24,000}{72} = 500 \text{ psi}$$

At the same points, due to torsion (see p. 160),

$$\tau_T = \frac{T(\lambda_1 b)}{\beta\, db^3} = \frac{\lambda_1}{\beta}\frac{T}{db^2} = 4.07\left(\frac{7500 \times 12}{12 \times 6^2}\right) = 845 \text{ psi}$$

On one side

$$\tau = \tau_T - \tau_S = 345 \text{ psi}$$

On the other side

$$\tau = \tau_T + \tau_S = 1345 \text{ psi}$$

(iii) *Principal stresses*

It is assumed that $\sigma_y = \sigma_z = 0$

On one side,

$$\left.\begin{matrix}\sigma_1\\ \sigma_2\end{matrix}\right\} = \frac{2500}{2} \pm \sqrt{\left(\frac{2500}{2}\right)^2 + 345^2} = \left.\begin{matrix}+2547\\ -\ \ 47\end{matrix}\right\}$$

On the other side,

$$\left.\begin{matrix}\sigma_1\\ \sigma_2\end{matrix}\right\} = \frac{2500}{2} \pm \sqrt{\left(\frac{2500}{2}\right)^2 + 1345^2} = \left.\begin{matrix}+3086\\ -\ 586\end{matrix}\right\}$$

(b) In this beam high values of σ_x do not coincide with high values of τ on the cross-section. The maximum principal shear stress is then likely to be governed by the maximum value of σ_x which in this problem is at the bottom fibre of the central section (E in Fig. 8.15).

$$\text{At } E, \ M = \frac{48,000 \times 120}{4} = 1,440,000 \text{ lb-in}; \quad I = \frac{6 \times 12^3}{12} = 864 \text{ in}^4$$

$$\sigma_x = \frac{My}{I} + \frac{N}{A} = \frac{1,440,000 \times 6}{864} + \frac{180,000}{72}$$

$$= 10,000 + 2,500 = 12,500 \text{ psi}$$

At this point,

$$\tau_S = 0 \text{ (bottom of section)}.$$

At the mid-point of the lower face,

$$\tau_S = \frac{T(\lambda_2 b)}{\beta\, db^3} = 1.62\left(\frac{7500 \times 12}{12 \times 6^2}\right) = 337 \text{ psi}$$

$$\left.\begin{matrix}\sigma_1\\ \sigma_2\end{matrix}\right\} = \frac{12,500}{2} \pm \sqrt{\left(\frac{12,500}{2}\right)^2 + 337^2} = \left.\begin{matrix}+12,510\\ -\ \ \ \ 10\end{matrix}\right\}$$

$$\sigma_3 = 0,$$

being the stress normal to the lower face of the beam (since this is an external surface it is a principal plane).

Hence the maximum shear stress at this point is

$$\tau_3 = \frac{\sigma_1 - \sigma_2}{2} = 6260 \text{ psi}$$

The shear force and torsion are the same at every section of the beam. Hence the maximum shear stress on a cross-section is 1345 psi (calculated in part (a)). So we see that the maximum shear stress in the beam is much greater than any shear stress which occurs on a normal cross-section. This is a matter which must be carefully considered in the case of members made of a material which is liable to shear failure.

8.5   Yield Criteria

The simultaneous presence of all the stress-resultants may, if the values are large enough, cause some parts of the material to be strained beyond the proportional range. Before any analysis of such a state can be carried out it is necessary to know what particular combination of stresses initiates non-proportional behaviour. For the present purposes, the commencement of such behaviour is usually termed *yielding*, irrespective of the nature of the subsequent stress–strain relation.

Material properties are usually obtained from simple uniaxial tests. From such a test, the external load at which yielding occurs is known. The test does not indicate, however, which particular characteristic of the internal stress state precipitated the onset of yielding. Perhaps the longitudinal stress, σ_x, in the test specimen reached a critical value; or perhaps it was the shear stress on an inclined plane which reached a critical value; possibly the strain energy stored in the test bar was the criterion of yielding.

In a beam in pure bending the material at any point is in a state of stress approximating to that in the uniaxial test and little difficulty is experienced in applying the test data. Suppose that in a simple tension test the stress on a normal cross-section is 50 ksi at yielding. Then, in a beam of the same material, we can reasonably expect yielding to occur when the longitudinal bending stress reaches 50 ksi. However, to call 50 ksi the yield stress is quite misleading. With a round bar of the same material stressed in torsion, the yield load is not obvious. If we suppose that yielding occurs when the principal tension reaches 50 ksi we obtain one value for the torsional load, whereas if we suppose that yielding occurs when the principal shear stress reaches 25 ksi (a condition which was also present in the tension test) then we shall predict yield at a much lower value of torsion.

It follows that in order to make use of uniaxial test data in situations where the stress state is more complex, it is necessary to know which characteristic of the stress field governs yielding, and the numerical value of that characteristic at which the phenomenon will occur. This is known as the *yield criterion*.

One obvious approach to this problem is to carry out tests under a variety of different stress states—uniaxial, biaxial and triaxial—and to see what property has the same critical value for all cases. We do not obtain the same answer for all materials. It is convenient once more to divide the stress state into its isotropic component (the average of the principal stresses) and its deviator components (which are functions of the difference between principal stresses) (see p. 181). For many materials it is found that quite large changes in the isotropic component have no effect on yielding. For these materials we may conclude that it is not the magnitude of the principal stresses which causes yield, but rather the shear stresses or some function thereof.

For such materials, two yield criteria have been suggested: (*a*) Tresca has proposed that yield occurs when the maximum principal shear stress reaches a critical value, which might be called λ_1. According to this hypothesis, yield will occur in any stress situation as soon as either τ_1 or τ_2 or τ_3 reaches the value λ_1. The value of λ_1 can be obtained from the simple tension test. (*b*) Von Mises has suggested that yield occurs when the octahedral shear stress (p. 178) reaches a critical value, say, λ_2. According to this hypothesis yield will occur as soon as $\frac{2}{3}\sqrt{\tau_1^2 + \tau_2^2 + \tau_3^2}$ reaches the value λ_2. Again, the value of λ_2 can be obtained by reference to the simple tension test.

The Von Mises criterion can be stated in terms of the energy of distortion, since the energy of distortion is a maximum when the octahedral shear stress is a maximum.

These two hypotheses lead to slightly different predictions regarding yielding for most stress conditions. Therefore, one might expect to decide between them by tests other than simple uniaxial tests. However, the difference between the predictions is fairly small and the test results generally lie between the two.

No other proposed criteria are tenable in the case of materials for which yield is found to be independent of the isotropic stress component $(\sigma_1 + \sigma_2 + \sigma_3)/3$.

On the other hand, not all materials exhibit this last characteristic. For instance, concrete, when tested in tension, is not indifferent to variations in this isotropic stress component. For this material we might conclude that yield depends on the value of the maximum principal tension.

The operation of the Tresca and the Von Mises criteria, and the difference they give in predicted values, may be illustrated by an example.

Example 8.11. A circular shaft, 4 in diameter, is tested in combined bending and torsion. At a certain cross-section the bending moment is twice the twisting moment. Determine the value of *M* and *T* at which yielding will first occur (*a*) according to Tresca's criterion and (*b*) according to the Von Mises criterion. In a simple tension test yielding occurs when the longitudinal tension is 50 ksi.

Solution. The critical point on the section is that which has the greatest bending stress, since this also has the greatest torsion stress.

For the circle, $J = 2I$
Due to torsion,

$$\tau = \frac{Tr}{J} = \frac{Tr}{2I}$$

At the same point, due to bending,

$$\sigma_x = \frac{Mr}{I} = \frac{2Tr}{I} \quad (\text{since } M = 2T)$$

For simplicity let $A = Tr/2I$
Then $\tau = A$ and $\sigma = 4A$
The principal stresses are,

$$\sigma_1 = \frac{\sigma_x}{2} + \sqrt{\left(\frac{\sigma_x}{2}\right)^2 + \tau^2} = 2A + \sqrt{4A^2 + A^2} = A(2 + \sqrt{5}) = 4.236A$$

$$\sigma_2 = \frac{\sigma_x}{2} - \sqrt{\left(\frac{\sigma_x}{2}\right)^2 + \tau^2} = A(2 - \sqrt{5}) = -0.236A$$

$$\sigma_3 = 0$$

The principal shear stresses are

$$\tau_1 = \frac{\sigma_2 - \sigma_3}{2} = -0.118A$$

$$\tau_2 = \frac{\sigma_3 - \sigma_1}{2} = -2.118A$$

$$\tau_3 = \frac{\sigma_1 - \sigma_2}{2} = 2.236A$$

(a) According to Tresca

For a given value of A, τ_{max} is $2.236A$
In the tension test, $\sigma_1 = 50$, $\sigma_2 = \sigma_3 = 0$

$$\therefore \quad \tau_{max} = \frac{50}{2} = 25 \text{ ksi}$$

Thus the critical value of $\tau_{max} = 25$
Hence yield occurs when $2.236A = 25$

$$A = 11.18$$

Since $I = 12.56$ in^4 and $r = 2$ in

$$\frac{T \times 2}{2 \times 12.56} = A = 11.18$$

Hence $T = 140.3$ kip-in and $M = 280.6$ kip-in

(b) According to Von Mises

The octahedral shear stress is

$$\tau_{oct} = \tfrac{2}{3}\sqrt{\tau_1^2 + \tau_2^2 + \tau_3^2} = \tfrac{2}{3}A\sqrt{0.118^2 + 2.118^2 + 2.236^2} = 2.056A$$

In the tension test, $\sigma_1 = 50$, $\sigma_2 = \sigma_3 = 0$

$$\tau_1 = 0, \quad \tau_2 = -25, \quad \tau_3 = 25$$

$$\therefore \quad \tau_{oct} = \tfrac{2}{3}\sqrt{25^2 + 25^2 + 0} = \sqrt{\tfrac{2}{3}}\,50 = 23.6 \text{ ksi}$$

Thus the critical value of $\tau_{oct} = 23.6$
Hence yield occurs when $2.056A = 23.6$

$$A = 11.47$$

$$\frac{T \times 2}{2 \times 12.56} = A = 11.47$$

Hence $T = 144$ kip-in and $M = 288$ kip-in

In this problem, the Von Mises criterion predicts yield loads about 2.6% higher than the Tresca criterion.

Problems

8.1   An I beam is 12 in deep and has properties $I = 500$ in^4 and $A = 14$ in^2. At a certain section the bending moment is $+25,000$ lb-in. What longitudinal force must be applied at a point 1 in above the bottom of the beam in order that the stress in the bottom fibre is just zero?
What is then the resultant stress in the top fibre?

8.2   A beam with the cross-section shown in Fig. P8.2 is subjected to an eccentrically applied axial load producing the distribution of direct stress as shown. Determine the magnitude and point of application of this load.

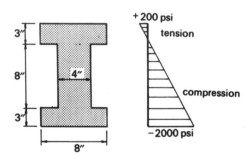

Fig. P8·2

8.3   A short column has a circular cross-section of diameter D. It is subjected to a compressive force P acting parallel to the axis. At what distance from the centre of the column must P act in order that the stress at one edge of the circle shall be zero.

8.4   A straight bar AB, 8 ft long, is cantilevered from A at an angle of $30°$ to the horizontal (Fig. P8.4). At B it carries a vertical load of 8000 lb.

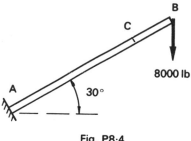

Fig. P8·4

Write down the axial force and bending moment at a section C, 2 ft from B. If the bar is 4 in deep and 1 in wide, find the total longitudinal direct stress at the upper and lower edges at C.

8.5 A beam has a rectangular cross-section $ABCD$ (Fig. P8.5). At a certain section the total normal stress at A is -7000 psi, at B it is zero, and at C it is $+16,667$ psi. Find the bending moment about each axis and the axial force. In the case of each bending moment indicate at which edge it causes compression.

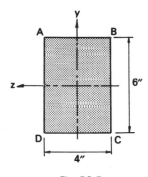

Fig. P8·5

8.6 A pulley weighing 600 lb is mounted on a shaft $2\frac{1}{2}$ in diameter midway between bearings which are 24 in apart. At every cross-section of the shaft, the torque is 5000 lb-in. Calculate the principal stresses at the extremities of (a) a vertical diameter, (b) a horizontal diameter, of a cross-section of the shaft midway between the pulley and a bearing. Also find the maximum shear stress at the same points.

8.7 The outside and inside diameters of a hollow shaft are 18 in and 16 in. At a certain section the twisting moment is 200 kip-ft, the bending moment is

50 kip-ft and the axial force is 35 kips compression. Find the maximum compressive stress in the shaft.

8.8   An I beam of overall depth 20 in, has flanges $7\frac{1}{2}$ in × 1 in and a web 18 in × 0.6 in. At a certain section the shear force is 40 kips and the bending moment is 800 kip-in.

Calculate the principal stresses (a) at the top of the beam, (b) at the top of the web, assuming that the shear stress is uniform across the thickness of the web, and (c) at the neutral axis.

8.9   A horizontal circular conical shaft is fixed at A and free at B. It tapers uniformly from 6 in diameter at A to 2 in diameter at B and the length is 8 ft. A static load of 200 lb hangs vertically on the circumference at the extremity of the horizontal diameter of the cross-section at the free end. Find the total strain energy due to torsion and bending in the rod and the vertical movement of the point of application of the load. $E = 30 \times 10^6$ psi; $G = 11.6 \times 10^6$ psi.

8.10   A cantilever has a cross-section in the form of a hollow square (Fig. P8.10) with its diagonal vertical. The properties of the section are $I = 7.25$ in⁴ and $A = 3.04$ in².

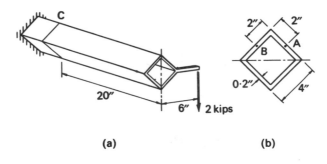

(a)   (b)

Fig. P8·10

At its free end the cantilever supports a vertical load of 2 kips applied 6 in from the axis of the beam. For a cross-section C, 20 in from the free end, find the total normal stress and shear stress on the cross-section at the mid-points A and B of the upper sides.

8.11   A beam ABCD, freely supported at A and C, is loaded as shown in Fig. P8.11. The cross-section is as shown. For the section midway between B and C find the bending stress and the shear stress at the point P on the cross-section. Assume that the shear stress is uniform across the width of the web. Find the principal stresses at P.

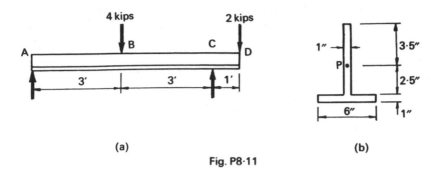

(a)   (b)

Fig. P8·11

8.12 The bent cantilever *ABCD* (Fig. P8.12) lies in the horizontal plane with right-angled bends at *B* and *C*. This cantilever, which consists of a tube 4 in external diameter and $\frac{1}{4}$ in thick, is fixed at *A* and carries a vertical load of 1000 lb at *D*.

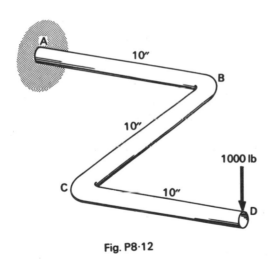

Fig. P8·12

Calculate the strain energy due to torsion and bending, and hence determine the vertical deflection at *D*. Assume that all cross-sections are free to warp. $E = 30 \times 10^6$ psi; $G = 12 \times 10^6$ psi.

8.13 A round steel rod *ABC*, 1 in diameter, is cantilevered from *A* and has a right-angle bend at *B*. *AB* is 16 in, *BC* is 6 in and both *AB* and *BC* lie in the horizontal plane. A vertical load of 100 lb is suspended from the end *C*.

Calculate the strain energy stored in the rod (*a*) due to bending and (*b*) due to torsion. Hence, find the deflection of *C*. Take $E = 30 \times 10^6$ psi and $G = 12 \times 10^6$ psi.

8.14   A cantilever is 24 in long and has a hollow square cross-section (Fig. P8.14). It is loaded at the free end by loads which act at the end of a bracket attached to the main member as shown. Point *A* lies 0.5 in above the bottom of the beam on the cross-section 12 in from the free end. Find the principal stresses at the point *A*.

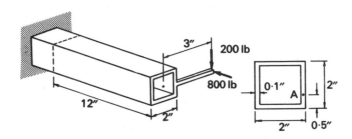

Fig. P8·14

8.15   *PQ* is a hollow tube cantilevered from *P* (Fig. P8.15). At *Q* the 1500 lb force is axial. The 75 lb vertical force is carried on a horizontal bracket perpendicular to the cantilever. At the cross-section 4 in away from *P*, find the principal stresses at the points *A*, *B*, *C* and *D* which are on the external surface of the tube.

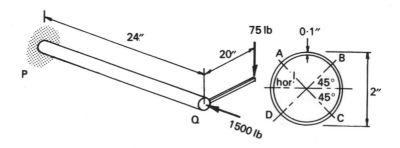

Fig. P8·15

8.16 The cross-section of a beam is a solid square of side 4 in. The side of the square is vertical. At a certain cross-section, the bending moment is 50,000 lb-in, the twisting moment is 40,000 lb-in, and the vertical shear force is 20,000 lb.

Find the three principal stresses (*a*) at the mid-point of each side and (*b*) at each corner of the square.

8.17 Find the total strain energy in the bar of problem 8.15. $E = 30 \times 10^6$ psi; $G = 12 \times 10^6$ psi.

8.18 A cantilever supports a vertical load of 2 kips at its end and also a compressive load of 15 kips. The cross-section of the member is a hollow square of side 5 in, and thickness $\frac{1}{4}$ in, with its side vertical. In the cross-section *BB*, 6 ft from the free end, find the total normal stress and the shear stress at the point *A* which is 1 in above the centre of the section. Assume that the shear stress is distributed uniformly across the thickness of the wall.

8.19 A cantilever *ABC* in the form of an L is made from a solid bar with a circular cross-section 2 in in diameter. *ABC* lies in the horizontal plane. It is supported at *A* and right-angled at *B*. *AB* = *BC* = 10 in. Find the maximum vertical load which can be applied at *C* if the principal tensile stress is not to exceed 10,000 psi at any point.

8.20 A beam which spans 25 ft has a rectangular cross-section 5 in wide and 16 in deep. It supports a uniformly distributed load of 1 kip per foot. Find the maximum shear stress in the beam.

8.21 The beam of Fig. P8.21 sustains axial force and bending moment which together cause a plastic stress distribution as shown. Find the axial force and bending moment at this cross-section. The axial force acts through the centroid of the cross-section.

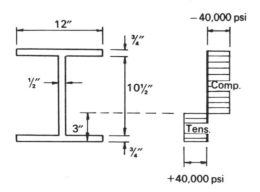

Fig. P8·21

8.22   A beam has a rectangular section 6 in wide and 12 in deep.

(a) Find the full plastic limit moment M_P which the beam will carry, if the yield stress is 36,000 psi.

(b) The beam is subjected to the limit moment M_P and the bending moment is then reduced until the stress at the extreme fibres is zero. What moment does the beam then carry? Sketch the stress distribution.

(c) In addition to this moment an axial compressive force P is now applied to the member at such a location that it produces uniform compressive strain over the whole section. What value of P will produce a compressive stress of 18,000 psi at the upper and lower edges?

8.23   A homogeneous beam has a T cross-section in which the flange and web are both 6 in × 1 in. For the material, stress and strain are proportional up to a yield stress σ_y and for higher strains the stress remains constant.

The beam is subjected to a compressive force P parallel to the centroidal axis, and acting at a point $1\frac{1}{2}$ in below the top of the beam and on the centreline of the section. The force P is gradually increased from zero.

(a) Sketch the stress distribution when the top fibre stress just reaches the value σ_y. Indicate the value of the bottom fibre stress, and find the value P_1 at which these stresses occur.

(b) Sketch the stress distribution for a value of P slightly higher than P_1. Find the value P_2 at which the bottom fibre stress becomes zero.

(c) Sketch the stress distribution for a value of P slightly larger than P_2. Find the maximum value of P which the section will withstand, assuming that the stress–strain curve is the same in tension and compression.

8.24   A beam of T-section has a flange 6 in wide by 1 in deep, and a web 1 in wide by 8 in deep. The overall depth of the beam is 9 in. At a certain section the longitudinal stress varies as shown in Fig. P8.24.

(a) Find the magnitude and position of the resultant of these stresses.

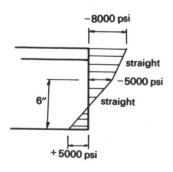

Fig. P8·24

(*b*) Express this stress-resultant as an axial force acting at the mid-depth of the section together with a bending moment.

8.25 Fig. P8.25 shows the stress distribution on the section of a rectangular beam 8 in wide and 20 in deep.

(*a*) Find the magnitude and position of the resultant of these stresses.

(*b*) Find the axial force and bending moment at this section of the beam if the axial force is supposed to act at the centre of the section.

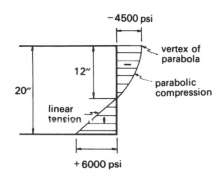

Fig. P8·25

8.26 In a tension test on a certain steel, the material is observed to yield when the longitudinal stress is 60 ksi. If a 3 in diameter shaft of the same steel is tested in torsion, at what torque would you expect yield to occur (*a*) according to the maximum shear stress criterion (Tresca) and (*b*) according to the octahedral shear stress criterion (Von Mises)?

8.27 In a tension test on a certain steel, the material is observed to yield when the longitudinal stress is 60 ksi. A bar of the same steel, 2 in × 1 in in cross-section, is subjected to an axial tension of 20 ksi and simultaneously twisted. At what torque would you expect yield to occur according to (*a*) the maximum shear stress criterion (*b*) the octahedral shear stress criterion?

8.28 A concrete specimen, tested in tension, fails at a stress of 500 psi. A 6 in diameter cylinder of the same concrete is subjected to an axial compressive stress of 2000 psi and, while sustaining this stress, is also subjected to torsion. If it is assumed that failure is governed by the maximum tensile stress, at what torque will the cylinder fail?

In view of the data, is it possible that failure of this material is due to maximum shear stress? Give reasons for your answer.

CHAPTER 9

Bar Assemblies

9.1   Statical Determinacy and Indeterminacy

The earlier chapters enable us to calculate the deformation of a simple bar under a variety of loads. The methods are approximate but, provided the length of the bar is large compared with its transverse dimensions, the errors are negligible, and these results are used in the great majority of practical structures.

Once the relation between force and deformation is known for a simple bar, it is possible to proceed to the analysis of any structure which is made up from such bars. Examples of the analysis of very simple bar assemblies will be considered in this chapter. In order to focus attention on the principles the examples will be restricted to plane pin-jointed frames in which the bars are subjected only to axial forces. Bar assemblies are called frames. When the bars are pin-jointed the frame is sometimes called a truss.

It is assumed that the student has studied the analysis of certain types of trusses previously in a course of statics. He will be familiar with the problem of finding bar forces by the method of equilibrium of joints or by the drawing of force diagrams. However, we now propose to study also the alteration of the shape of the truss. Obviously, if the various bars sustain forces they will undergo extension or contraction and this in turn will cause small but definite displacement of all the joints.

When we consider bar deformations we notice that trusses can be classified into three types.

1. Trusses which can deform as a whole without the deformation of any individual member. Fig. 9.1 (a) shows one such truss, and Fig. 9.1 (b) shows another. Since the truss deformation induces no stresses it will not resist external loads, and it is called an unstable structure or a mechanism.

2. Trusses in which each member is free to deform independently of the others, but which cannot deform as a whole unless at least one member undergoes deformation. Fig. 9.1 (c) shows such a truss, and indicates the general deformation which would result from an extension of bar *BD*, all other bars remaining

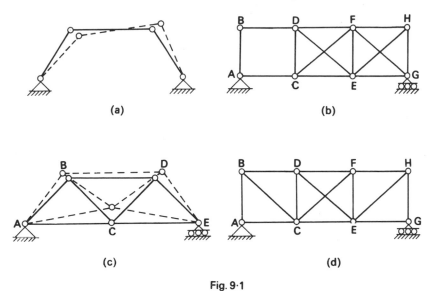

Fig. 9·1

unchanged in length. Such a structure is said to be statically determinate, because it is possible to determine all the bar forces by the application of the principles of statics alone.

3. Trusses in which there can be found some bars which cannot deform unless there is an accompanying deformation in other bars. In Fig. 9.1 (d) deformation of any bar within panel $CDEF$ must involve deformation of all other bars within this panel if the bars are to remain attached to the pins C, D, E and F. Bars outside this panel are free to deform independently of their neighbours. Such a truss is said to be statically indeterminate. Bar forces cannot be determined by statics alone but account must be taken of compatibility of bar deformations within the region where these are interdependent.

It is not possible to determine, merely by counting the bars and joints, whether or not a structure is statically indeterminate. The truss of Fig. 9.1 (b) contains more bars than necessary for stability, but these are so arranged that the structure is redundant (statically indeterminate) in panels DF and FH, while being unstable in panel BD. In elementary structures it is usually easy to see to which category the frame belongs.

9.2   Statically Determinate Trusses

In statically determinate trusses it is possible to determine the bar forces from statics alone. From the force-deformation properties (Chapter 3) the change of length of each bar can then be computed. The consequent alteration in the configuration of the structure then follows from considerations of geometry.

A simple relationship exists between the deformation of a given bar and the displacement of the joints at each end. The joint displacements are resolved in the direction of the bar, and the difference between these components of end displacement is equal to the bar extension. In the present work it is assumed that the joint displacements are sufficiently small that the change in orientation of the bar can be ignored.

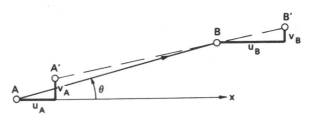

Fig. 9·2

In Fig. 9.2 the joints A and B move to A' and B'. If we arbitrarily assign a sense to the bar, from A to B for instance, we can then say that the bar direction makes an angle θ to the x direction. The x displacement ($\overset{+}{\rightarrow}$) and y displacement ($\uparrow$+) of joint A are denoted by u_A and v_A respectively, while the displacements of B are u_B and v_B. The component of the displacement AA' in the bar direction is then $(u_A \cos\theta + v_A \sin\theta)$, and the corresponding component of BB' is $(u_B \cos\theta + v_B \sin\theta)$. The extension of the bar is then

$$e = (u_B \cos\theta + v_B \sin\theta) - (u_A \cos\theta + v_A \sin\theta) \qquad (9.1)$$

When the bar lies parallel to the x or y axes the relationship can be written readily without the use of equation (9.1).

Example 9.1. In the pin-jointed truss of Fig. 9.3 calculate the bar forces and the joint displacements.

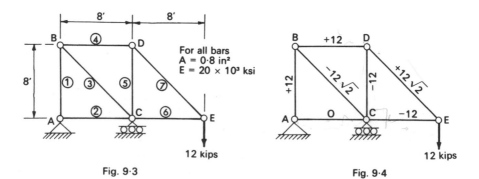

Fig. 9·3   Fig. 9·4

Solution. For convenience the bars are numbered and the joints are lettered (Fig. 9.3).

(a) By the method of joint equilibrium (or any other statical method) the bar forces are determined. These are shown in Fig. 9.4.

(b) Since the bars are uniform and the material is linearly elastic, the extension of any bar is given by $e = NL/EA$. The values are computed in Table 9.1.

TABLE 9.1

Bar	L (in)	N (kips)	$e = \dfrac{NL}{EA}$
1	96	12	0.072
2	96	0	0.0
3	$96\sqrt{2}$	$-12\sqrt{2}$	-0.144
4	96	12	0.072
5	96	-12	-0.072
6	96	-12	-0.072
7	$96\sqrt{2}$	$12\sqrt{2}$	0.144

(c) The joint displacements are now found.

$$\text{Joint } A \qquad u_A = 0 \text{ (given)}$$

$$v_A = 0 \text{ (given)}$$

$$\text{Joint } C \qquad u_C = u_A + e_2 = 0$$

$$v_C = 0 \text{ (given)}$$

Knowing the locations of A and C we can find the position of B.

$$\text{Joint } B \qquad v_B = e_1 - +0.072 \text{ in}$$

$$e_3 = \frac{-u_B}{\sqrt{2}} + \frac{v_B}{\sqrt{2}} + \frac{u_C}{\sqrt{2}}$$

$$\therefore \quad u_B = -\sqrt{2}\,e_3 + v_B + u_C = 0.204 + 0.072 + 0 = 0.276 \text{ in}$$

From B and C, D can be found.

$$\text{Joint } D \qquad e_4 = u_D - u_B$$

$$\therefore \quad u_D = e_4 + u_B = 0.072 + 0.276 = 0.348 \text{ in}$$

$$e_5 = v_D - v_C$$

whence

$$v_D = -0.072 \text{ in}$$

$$\text{Joint } E \qquad e_6 = u_E - u_C$$

whence

$$u_E = -0.072 \text{ in}$$

Writing equation (9.1) for bar DE, we have

$$e_7 = \left(\frac{u_E}{\sqrt{2}} - \frac{v_E}{\sqrt{2}}\right) - \left(\frac{u_D}{\sqrt{2}} - \frac{v_D}{\sqrt{2}}\right)$$

whence

$$v_E = -\sqrt{2}\,e_7 + u_E - u_D + v_D = -0.696 \text{ in}$$

The above procedure can be used equally well when the bars are made of a material having a non-linear stress–strain law.

Example 9.2. A truss has the same dimensions and loading as that of Fig. 9.3 but is made of material for which the stress–strain equation is

$$\epsilon = 6 \times 10^{-5}(\pm 0.04\sigma^2 + \sigma) \qquad (9.2)$$

the + and − referring to tension and compression respectively, and σ being measured in ksi.

Solution. (a) The bar forces are the same as before (Fig. 9.4) since they depend only upon equilibrium.

(b) The bar extensions are determined from equation (9.2).

TABLE 9.2

Bar	N	σ (ksi)	ϵ	L (in)	$e(= \epsilon L)$
1	12	15	144×10^{-5}	96	0.138
2	0	0.0	0	96	0.0
3	$-12\sqrt{2}$	$-15\sqrt{2}$	-235.3×10^{-5}	$96\sqrt{2}$	-0.320
4	12	15	144×10^{-5}	96	0.138
5	-12	-15	-144×10^{-5}	96	-0.138
6	-12	-15	-144×10^{-5}	96	-0.138
7	$12\sqrt{2}$	$15\sqrt{2}$	235.3×10^{-5}	$96\sqrt{2}$	0.320

(c) The joint displacements are obtained from the bar extensions by the principles of geometry. The expressions obtained in example 9.1 are therefore still valid.

Joint A $\begin{cases} u_A = 0 \text{ (given)} \\ v_A = 0 \text{ (given)} \end{cases}$

Joint C $\begin{cases} u_C = u_A + e_2 = 0 \\ v_C = 0 \text{ (given)} \end{cases}$

Joint B $\begin{cases} v_B = e_1 = 0.138 \text{ in} \\ u_B = -\sqrt{2}\,e_3 + v_B + u_C = 0.590 \text{ in} \end{cases}$

Joint D $\begin{cases} u_D = e_4 + u_B = 0.728 \text{ in} \\ v_D = e_5 + v_C = -0.138 \text{ in} \end{cases}$

Joint E $\begin{cases} u_E = e_6 + u_C = -0.138 \text{ in} \\ v_E = -\sqrt{2}\,e_7 + u_E - u_D + v_D = -1.456 \text{ in} \end{cases}$

It will be noted that the only step in this solution which is complicated by the non-linear stress–strain law is the calculation of bar extensions from bar forces.

In a statically determinate truss each member is free to deform independently of the others. Consequently, if the temperature of one member is changed, that member suffers change in length without inducing any alteration in the length of the other members. The change in truss geometry resulting from thermal changes can therefore be found in exactly the same way as described above.

Example 9.3. The truss of Fig. 9.3 is made of material which has a coefficient of linear thermal expansion of 12×10^{-6} per °F. Find the joint displacements caused by a temperature rise of 50°F.

Solution. For all members the thermal strain is

$$\epsilon = \alpha T = 6 \times 10^{-4}$$

The bar deformations are therefore

$$e_1 = e_2 = e_4 = e_5 = e_6 = 96\epsilon = +0.0576 \text{ in}$$

$$e_3 = e_7 = 96\sqrt{2}\,\epsilon \qquad = +0.0815 \text{ in}$$

From these deformations the joint displacements are found by the same expressions as before.

Joint A $u_A = v_A = 0$

Joint C $\begin{cases} u_C = u_A + e_2 = 0.0576 \text{ in} \\ v_C - 0 \end{cases}$

Joint B $\begin{cases} v_B = e_1 = 0.0576 \text{ in} \\ u_B = -\sqrt{2}\,e_3 + v_B + u_C = -0.1152 + 0.0576 + 0.0576 = 0 \end{cases}$

Joint D $\begin{cases} u_D = e_4 + u_B \doteq 0.0576 \text{ in} \\ v_D = e_5 + v_C = 0.0576 \text{ in} \end{cases}$

Joint E $\begin{cases} u_E = e_6 + u_C = 0.1152 \text{ in} \\ v_E = -\sqrt{2}\,e_7 + u_E - u_D + v_D = 0 \end{cases}$

9.3   The Work Equation

In the above examples it will be observed that the computation of joint displacements from bar deformations, although very elementary, is rather tedious. In other problems the calculation of bar forces by statics may involve long calculations. It is sometimes possible to effect considerable saving of time by making use of either the Principle of Virtual Forces or the Principle of Virtual Displacements. Before discussing these extremely useful theorems it will be advisable to consider certain relevant fundamental theorems.

First of all, we consider any rigid body in equilibrium under the action of any system of forces. The body undergoes a rigid body movement, during which equilibrium is not disturbed. It is a well-known theorem of mechanics that, during such a motion, the total work done by the various forces is equal to the work done by their resultant. Since in this case the resultant is zero, it follows that the total work done by the forces will be zero.

In the present study we are interested only in systems of concurrent forces, but the theorem is valid generally.

We now consider a pin-jointed frame. This frame will be subjected successively to two quite independent operations.

In stage 1 the frame is loaded. As a result, every pin joint and every member is in equilibrium under the action of certain forces. Displacements will have occurred, and work will have been done, but this is not important. We are interested only in the fact that force systems *in equilibrium* exist as a result of stage 1.

In stage 2 something happens as a result of which the joints of the truss are displaced. If the truss as a whole undergoes a rigid body displacement, the members will not suffer deformation. On the other hand, the members may be deformed—some other load system may be applied to the structure, a temperature change may occur, chemical change may cause expansion or contraction, and so on. During this second stage *compatibility is preserved*, i.e., bars do not come adrift from the joints or the supports. We are interested only in the bar deformations which occur during stage 2. We will stipulate, however, that during these displacements the magnitude and direction of every stage 1 force remain unchanged.

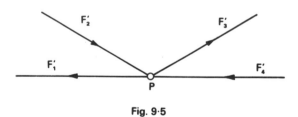

Fig. 9·5

Consider now the work done by the forces which exist at the end of stage 1 during the displacements which occur in stage 2. Any pin such as P (Fig. 9.5) is in equilibrium under a system of (stage 1) forces. Thus, during any displacement of P which may occur in stage 2 these forces do no work. One could equally well apply the same argument to any small element within the length of a bar, or any point within the structure. Hence the total work done by the stage 1 forces during the stage 2 displacements is zero. Note that this conclusion rests entirely upon the fact that the stage 1 forces were in equilibrium at every point throughout the structure.

If we denote the work done by the external forces W_1, W_2 (which will include support reactions) by U_e, and denote by U_i' the work done by the internal forces F_1', F_2', F_3' (Fig. 9.6), then we can write

$$U_e + U_i' = 0 \qquad (9.3)$$

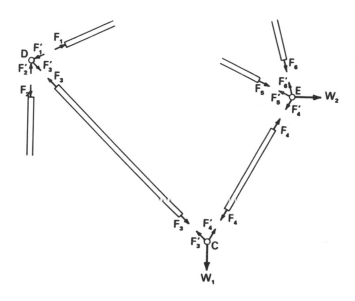

Fig. 9·6

As regards the internal forces it may often be more convenient to consider the forces F which act upon the members, rather than the forces F' which act upon the pin-joints (see Fig. 9.6). The change is merely one of sign, for if we observe the forces F_3 and F_3' at joint C (Fig. 9.6) it is obvious that when C moves, the work done by F_3 will be equal and opposite to the work done by F_3' and so on for all the internal forces. If we now think of the internal work as that done by the member forces F, and denote this work by U_i we can write

$$U_e = U_i \qquad (9.4)$$

We shall use this form of the work equation and consider forces acting on the members. It is noted that the work done by F_3 and F_3' at C are equal and opposite provided that the end of bar DC remains attached to pin C. Hence, the importance of the proviso that compatibility be preserved during the stage 2 operation.

Several points are of interest:

1. U_i is not the total internal energy of the system. It takes no account of work done during the stage 1 loading or of the work done by stage 2 forces (if any) during stage 2 displacements. It refers only to the work done during the stage 2 displacements by the internal forces existing at the end of stage 1.

2. The above arguments could be extended, without difficulty, to structures other than pin-jointed frames. In a general continuum, for example, the concept of a "member" would be replaced by the concept of "a particle of material". The "pin-joint" would become the interface between particles. The forces F would be the forces acting on the surfaces of a particle. However, such extensions will not be discussed further.

3. Throughout the discussion, no mention has been made of the material properties. The results are, therefore, valid for elastic or inelastic, for linear or non-linear material behaviour.

9.4 The Principle of Virtual Forces

We now have to put equation (9.4) to practical use. In the first place, suppose that it is required to investigate the displacement at a given point in a frame caused by various known member deformations—a purely geometric problem.

In this case our stage 1 forces are a fictitious set, called *virtual forces*. They must be in equilibrium throughout the frame, and there must be an external force at the point (and in the direction) where the desired displacement is to be calculated. Since any set will do it is often convenient to *apply a unit force* where we wish to find the displacement. The use of such fictitious forces to calculate displacements is called the Method of *Virtual Forces*. The work done by these forces is called *virtual work*.

Stage 2 is the real event which gives rise to member deformations and, consequently, to joint displacements. The relation between these deformations and displacements is easily obtained from the work equation. It is best explained by examples.

Example 9.4. In the frame of example 9.1 find the vertical and horizontal displacements of joint E by the method of virtual forces.

Solution. To find v_E we apply a unit force in the positive v_E direction (Fig. 9.7 (a)) and compute the corresponding member forces. Fig. 9.7 (a) indicates what we have called the stage 1 forces. In this problem they are virtual forces. The bar forces can be denoted by $n_1 \ldots n_7$.

Suppose that during stage 2 (the real stage in this case) bar 3 undergoes an extension of e_3. The consequent displacement v_E will be $\sqrt{2}e_3$ since n_3 is $\sqrt{2}$.

In the present problem, stage 2 consists of loading the frame with the 12 kip load at E.

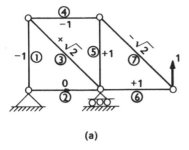

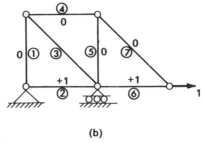

(a)   (b)

Fig. 9·7

The bar forces and deformations have been computed in Table 9.1. The internal work done by the virtual n forces during the real deformations e is

$$U_i = -1(e_1) + 0(e_2) + \sqrt{2}(e_3) - 1(e_4) + 1(e_5) + 1(e_6) - \sqrt{2}(e_7) \qquad (9.5)$$

$$= -1(0.072) + 0 + \sqrt{2}(-0.144) - 1(0.072) + 1(-0.072) + 1(-0.072) - \sqrt{2}(0.144)$$

$$= -0.696$$

The units of U_i are inches and whatever force units we imagine the virtual forces to be measured in.

Since the supports do not move, the total external work done is

$$U_e = 1 \times v_e$$

Hence from the work equation (9.4)

$$1 \cdot v_e = -0.696 \qquad (9.6)$$

$$\therefore \quad v_E = -0.696 \text{ in}$$

A non-unit virtual force δW at E would simply have caused both sides of equation (9.6) to be multiplied by δW. Such a procedure is sometimes convenient to simplify the arithmetic in computing the internal virtual force system.

To find the horizontal displacement of E, a unit force is applied in the positive u_E direction, and the corresponding internal forces calculated (Fig. 9.7 (b)).

$$U_e = U_i$$

$$1 \times u_E = 1(e_2) + 1(e_5) \qquad (9.7)$$

$$= 0 - 0.072$$

$$u_E = -0.072 \text{ in}$$

Equations (9.5) and (9.7) are valid irrespective of the cause of the member deformations.

Example 9.5. Find u_E and v_E for the frame of example 9.2 in which the material has a non-linear stress–strain law.

Solution. The virtual force systems are the same as those in the previous example. We use the appropriate e values from Table 9.2 in equation (9.5)

$$U_i = -1(0.138) + 0 + \sqrt{2}(-0.320) - 1(0.138) + 1(-0.138) + 1(0.138) - \sqrt{2}(0.320)$$

$$= -1.456$$

$$U_e = 1 \times v_E$$

$$\therefore \quad 1 \times v_E = -1.456$$

$$v_E = -1.456 \text{ in}$$

Similarly from equation (9.7)

$$1 \times u_E = 1(e_2) + 1(e_5)$$

$$= 0 - 0.138$$

$$u_E = -0.138 \text{ in}$$

Example 9.6. In the frame of example 9.1 (Fig. 9.3) find v_E and u_E caused by a temperature rise of 50°F (see example 9.3)

Solution. The virtual forces are still the same as in example 9.4. The bar deformations caused by the temperature change were computed in example 9.3. Using these values we have

$$1 \times v_E = -1(0.0576) + 0 + \sqrt{2}(0.0815) - 1(0.0576) + 1(0.0576) + 1(0.0576) - \sqrt{2}(0.0815)$$

$$= 0$$

Similarly

$$1 \times u_E = 1(e_2) + 1(e_5)$$

$$= 0.0576 + 0.0576$$

$$= 0.1152$$

$$u_E = 0.1152 \text{ in}$$

It is useful to bear in mind that in reality the problem being solved here is one of geometry. The virtual forces constitute a device for obtaining the geometrical relationships by the laws of statics. This device is possible because a relationship exists between the laws of geometry and the laws of forces.

9.5   The Principle of Virtual Displacements

The relationship just mentioned also makes it possible to solve statics problems by the laws of geometry. Although this is not likely to be an advantageous change in the case of very simple structures, it can be a powerful tool in more complex problems. For this reason it is discussed here.

Suppose that we wish to find the force in a particular member, or a reaction, caused by a given external load system.

In this case the application of the known *real* forces constitutes stage 1 of our 2-stage operation. Stage 2 consists of subjecting the member whose force we require to a fictitious deformation, called a *virtual deformation*, and computing by geometry the resulting displacements (*virtual displacements*) of any point at which a real force is acting. The work done during these fictitious displacements is again *virtual work*. The use of such displacements to solve statics problems is called the Method of *Virtual Displacements*. Displacement must be very small so that the geometry is essentially unchanged.

Example 9.7. In the truss of example 9.1 (Fig. 9.8) find the force in bar 4 due to the 12 kip load at *E*.

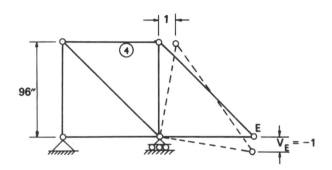

Fig. 9·8

Solution. The 12 kip load is first applied and this causes real forces in the various members.

Since we wish to find the force in bar 4 we give this bar a unit virtual extension. By geometry (Fig. 9.8) we see that the consequent vertical displacement of *E* is $v_E = -1$. During this operation the internal work done is

$$U_i = N_4 \times 1$$

while the external work done is

$$U_e = 12(-v_E) = 12 \times 1$$

Since

$$U_i = U_e$$

$$N_4 \times 1 = 12$$

The units of this equation are kips and whatever units we imagine the virtual displacements to be measured in. We then have

$$N_4 = 12 \text{ kips}$$

Example 9.8. In the frame of Fig. 9.9 (a) find the force in bar *DF* and the reaction at *G* due to the external loads. Use the method of virtual displacements.

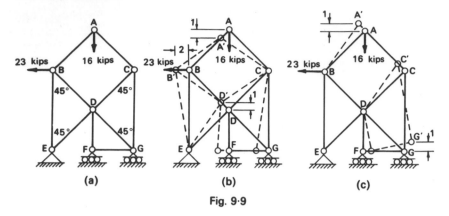

Fig. 9·9

Solution. Apply the real force system.

To find N_{DF} give member DF a unit extension (1 mm for instance), and calculate by geometry u_B and v_A (Fig. 9.9 (b)). Now D moves vertically upwards by 1 mm and since it must move normally to ED, then D must also move horizontally by 1 mm. Triangle BED rotates about E, and B therefore moves by 2 mm. Since neither B nor C moves vertically, the upward displacement of D must be accompanied by a corresponding downward displacement of A.

Thus

$$U_i = N_{DF} \times 1 \qquad U_e = 23 \times 2 + 16 \times 1 = 62$$

$$\therefore \quad N_{DF} = +62 \text{ kips}$$

To find the reaction at G we displace G vertically by unit distance (Fig. 9.9 (c)). To preserve compatibility this must cause a rigid body rotation of section $CDFG$ about D, so that CA moves in the direction of its length. Since GC and CA do not vary in length, joints G, C and A all move vertically by the same amount.

Hence

$$U_e = R_G \times 1 - 16 \times 1 + 23 \times 0 \quad \text{and} \quad U_i = 0$$

$$\therefore \quad R_G \times 1 - 16 \times 1 = 0$$

and

$$R_G = 16 \text{ kips}$$

Further examples of the use of these principles are given later in this chapter.

9.6 Statically Indeterminate Trusses—Force Method

We consider next the principle of analysis of bar forces in statically indeterminate frames, that is to say, in frames in which the bars are not free to deform independently of one another.

In the analysis of statically determinate trusses (section 9.2) it was possible (*a*) to find the bar forces from the equilibrium alone, then (*b*) to find bar deformations from the force–deformation relation of each bar and (*c*) to find the frame displacements from these bar deformations by the laws of geometry.

In the analysis of a statically indeterminate frame we cannot proceed in such a simple way because the bar forces cannot be found from the equations of equilibrium alone. We must use all three sets of equations (equilibrium, force–deformation, and geometry) before any of the variables can be evaluated explicitly.

It would be possible merely to write all the equations down and then to solve them simultaneously. But in a practical frame this would involve the solution of an unnecessarily large number of simultaneous equations. This number can be greatly reduced by using certain orderly procedures, whereby many variables are eliminated as we go along.

The two most common procedures are known as the *force method* and the *displacement method* respectively. In the former the bar forces are determined first, while in the latter method the joint displacements are determined first. In this book only very small problems will be examined in order to describe the principles of solution.

Consider first the *force method*. By the nature of the structure there are more bar forces than there are equilibrium equations. Suppose the excess is *n*. We choose *n* convenient bar forces (called redundant forces) as unknowns and, using the equilibrium equations, express all forces in terms of these *n* unknowns. Next we find the bar deformations. These are not independent and they must (in this type of structure) obey certain geometric relationships if the deformed bars are to fit together. These are called *compatibility relationships*. There will be just enough of these relationships to enable us to evaluate the *n* unknown forces. Once these are known, all bar forces can be found, and finally the joint displacements can be calculated.

Example 9.9. Find the bar forces in the structure of Fig. 9.10 (a) and the vertical displacement of joint *A*. The cross-sectional area of each bar is 0.5 in² and the material is linearly elastic with $E = 20 \times 10^3$ ksi.

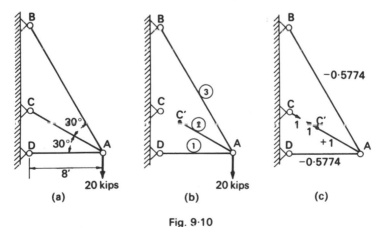

Fig. 9·10

Solution. (*a*) *Bar forces*

There are two equilibrium equations (for joint *A*) and three bar forces. Hence we must choose one unknown. Take the force in *AC* as the unknown, and call it *R* (for redundant force).

A convenient device is to imagine the bar *AC* to be detached from joint *C* while the 20 kip load is applied (Fig. 9.10 (b)). Numbering bars *AD, AC, AB* as 1, 2 and 3 respectively, we see that due to the 20 kip load

$$N_1 = -11.55 \text{ kips}, \quad N_2 = 0, \quad \text{and} \quad N_3 = +23.10 \text{ kips}$$

A unit value of the redundant force *R* is now applied in the form of a pair of forces between *C* and the detached member end *C'* (Fig. 9.10 (c)). This will cause bar forces

$$n_1 = -0.5774; \quad n_2 = +1.0; \quad n_3 = -0.5774$$

With the 20 kip load and the unknown force *R*, the total bar forces are then

$$\left.\begin{aligned}
N_1 &= -11.55 - 0.5774R \\
N_2 &= 0 + R \\
N_3 &= 23.10 \ \ - 0.5774R
\end{aligned}\right\} \tag{9.8}$$

All we have done up to now is to use the two equations of equilibrium to express the three bar forces in terms of a single unknown which we have called *R*. This force will be determined from the fact that in the real structure the gap *CC'* must be zero. Hence we must now find the value of this gap and equate it to zero.

(*b*) *Bar deformations*

With a Hookean material it is convenient to evaluate first the bar flexibility coefficients, the extension caused by unit load.

$$f_1 = \frac{L_1}{E_1 A_1} = \frac{96}{20 \times 10^3 \times 0.5} = 0.0096 \text{ in/kip}$$

$$f_2 = \frac{L_2}{E_2 A_2} = \frac{96/\cos 30°}{20 \times 10^3 \times 0.5} = 0.0111 \text{ in/kip}$$

$$f_3 = \frac{L_3}{E_3 A_3} = \frac{96/\cos 60°}{20 \times 10^3 \times 0.5} = 0.0192 \text{ in/kip}$$

Then

$$\left.\begin{aligned}
e_1 &= f_1 N_1 = 0.0096(-11.55 - 0.5774R) = -0.111 - 0.0055R \\
e_2 &= f_2 N_2 = 0.0111(R) &&= \qquad\quad 0.0111R \\
e_3 &= f_3 N_3 = 0.0192(23.10 - 0.5774R) &&= \ \ 0.444 - 0.0111R
\end{aligned}\right\} \tag{9.9}$$

(*c*) *Compatibility*

To find the displacement of *C'* relative to *C* we may use the force system of Fig. 9.10 (c) as a convenient *virtual force* system. This shows that

$$CC' = -0.5774e_1 + 1.0e_2 - 0.5774e_3$$

Hence the bar deformations will be compatible provided

$$-0.5774(-0.111 - 0.0055R) + 1.0(0.0111R) - 0.5774(0.444 - 0.0111R) = 0$$

$$-0.1929 + 0.0207R = 0 \tag{9.10}$$

$$\therefore \quad R = +9.30 \text{ kips} \tag{9.11}$$

(*d*) *Bar forces*

From equations (9.8) we now have the bar forces

$$N_1 = -16.91 \text{ kips}; \quad N_2 = +9.30 \text{ kips}; \quad N_3 = +17.74 \text{ kips}$$

(e) Bar deformations

From equations (9.9) the bar deformations are

$$e_1 = -0.163 \text{ in}; \quad e_2 = +0.103 \text{ in}; \quad e_3 = +0.341 \text{ in}$$

(f) Joint displacements

The two displacements of joint A can now be found by geometry. Obviously

$$u_A = e_1 = -0.163 \text{ in}$$

To find v_A a virtual force system is useful. We apply a unit vertical force at A, and *any set* of bar forces which make equilibrium with the vertical force. The forces shown in Fig. 9.11 will serve. Then

$$v_A = +\sqrt{3}\,e_1 - 2e_2 + 0e_3$$

$$= -0.488 \text{ in}$$

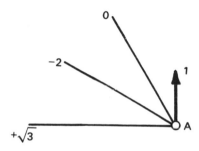

Fig. 9·11

A different choice of virtual forces would produce the same result because the e values are compatible. The only requirement of the virtual force system is that it shall be in equilibrium.

In the foregoing problem, where the principle of superposition is valid, we could have calculated the displacement of C' (relative to C) due to the 20 kip load (giving −0.192) and then, separately, the displacement of C' due to the force R (giving $0.0207R$). When the sum of these displacements is equated to zero, equation (9.10) is obtained as before. The separate calculation of the displacement due to the different load systems may give an appearance of greater simplicity, but it is not recommended since it cannot be applied to all problems. When the material has a non-linear characteristic it is essential to compute deformations with all forces acting simultaneously.

This is illustrated in the next example.

Example 9.10. Solve the frame of example 9.9 (Fig. 9.10 (a)). In the present problem the stress–strain relation is

$$\epsilon = 6 \times 10^{-5}(\pm 0.04\sigma^2 + \sigma)$$

where σ is in ksi and the + sign is to be used when σ is tensile and vice versa.

Solution. (a) *Bar forces*

We choose the same redundant force as before and hence obtain, as in example 9.8, the bar forces as

$$\left.\begin{array}{l} N_1 = -11.55 - 0.5774R \\ N_2 = 0 + R \\ N_3 = 23.10 \ - 0.5774R \end{array}\right\} \tag{9.12}$$

(b) *Bar deformations*

This time we cannot find flexibility coefficients. For bar 1 we find σ_1 from N_1

$$\sigma_1 = \frac{N_1}{0.5} = 2N_1$$

Then from the stress–strain function, and assuming σ_1 to be compressive,

$$\epsilon_1 = 6 \times 10^{-5}(-0.04\sigma_1^2 + \sigma_1)$$
$$= 6 \times 10^{-5}(-0.16N_1^2 + 2N_1)$$
$$e_1 = 96\epsilon_1$$

and using the value of N_1 from equations (9.12) we obtain

$$e_1 = 10^{-3}(-0.307R^2 - 18.95R - 256) \tag{9.13a}$$

In the same laborious way (and assuming that σ_2 and σ_3 are tensile) we find that

$$e_2 = 10^{-3}(1.066R^2 + 13.33R) \tag{9.13b}$$

and

$$e_3 = 10^{-3}(0.614R^2 - 62.5R + 1515) \tag{9.13c}$$

(c) *Compatibility*

Again using the force system of Fig. 9.10 (c) as a system of virtual forces, we see that the displacement of C' relative to C is

$$CC' = -0.5774e_1 + 1.0e_2 - 0.5774e_3$$

Hence for compatibility we require that

$$10^{-3}(-0.307R^2 - 18.95R - 256) + 10^{-3}(1.066R^2 + 13.33R) + 10^{-3}(0.614R^2 - 62.5R - 1515) = 0$$

or

$$0.889R^2 + 60.33R - 727 = 0$$

whence

$$R = +10.4 \text{ kips}$$

(d) *Bar forces*

Then from equations (9.12)

$$N_1 = -17.60 \text{ kips}; \quad N_2 = +10.48 \text{ kips}; \quad N_3 = +17.05 \text{ kips}$$

We note that our assumptions are correct concerning the signs of these forces.

(e) *Bar deformations*

From equations (9.13)

$$e_1 = -0.488 \text{ in}; \quad e_2 = +0.256 \text{ in}; \quad e_3 = +0.927 \text{ in}$$

(f) *Joint displacements*

The displacements of A are found as in the previous example.

$$u_A = e_1 = -0.49 \text{ in}$$

$$v_A = \sqrt{3}\,e_1 - 2e_2$$

$$= -1.36 \text{ in}$$

A particularly simple class of statically indeterminate bar systems are those in which the bars are parallel. The principle of solution is the same as before, but the simplification stems from the fact that the geometrical relations between bar deformations and joint displacement are obvious.

As a first example of this class of structures we take the case of two parallel bars of linearly elastic material.

Example 9.11. In the structure of Fig. 9.12 the properties of the two bars are:

$$L_1 = 60 \text{ in} \qquad L_2 = 72 \text{ in}$$

$$A_1 = 1.0 \text{ in}^2 \qquad A_2 = 1.2 \text{ in}^2$$

$$E_1 = 30 \times 10^3 \text{ ksi} \quad E_2 = 15 \times 10^3 \text{ ksi}$$

Find the location of the 50 kip load so that the beam GH remains horizontal, and find the force in each bar.

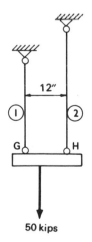

50 kips

Fig. 9·12

Solution. Regard the force in bar 1 as the unknown force R. Then

$$N_1 = R \quad \text{and} \quad N_2 = 50 - R$$

The flexibility coefficients are

$$f_1 = \frac{L_1}{E_1 A_1} = 2 \times 10^{-3} \text{ in/kip} \quad \text{and} \quad f_2 = \frac{L_2}{E_2 A_2} = 4 \times 10^{-3} \text{ in/kip}$$

The extensions of the bars must be equal
i.e.,

$$f_1 N_1 = f_2 N_2$$

$$2 \times 10^{-3} R = 4 \times 10^{-3}(50 - R)$$

hence

$$R = 33.33 \text{ kips}$$

Thus

$$N_1 = 33.33 \text{ kips} \quad \text{and} \quad N_2 = 16.67 \text{ kips}$$

It would have been sufficient to notice that bar 2 is twice as flexible as bar 1 and hence carries only half as much load.

From the rotational equilibrium of bar GH we see that the 50 kip load must be 4 in from G.

Instead of being separated, as in Fig. 9.12, the bars sometimes lie one inside another. Such an arrangement is often called a *compound bar*. Fig. 9.13 (a) shows a brass rod inside a steel tube, *the ends being maintained level*. The forces in the two materials are determined in exactly the same way as in example 9.10.

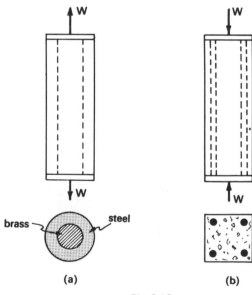

(a)   (b)

Fig. 9·13

Fig. 9.13 (b) shows a similar arrangement with steel rods embedded in concrete. This construction is called reinforced concrete. Again the forces are determined as before. Frequently it is assumed that both the steel and the concrete behave elastically. Such a solution will be approximate because the stress–strain characteristic of the concrete varies with time, and, as a consequence, the distribution of load between the two materials also varies with time. In any such compound bar, the stiffer bar will sustain the greater load. As time goes on the stiffness of the steel remains about the same, while that of the concrete decreases. Hence, in time, more and more of the load will be taken by the steel.

In all compound bar problems it is assumed that the external force is so located that uniform axial strain is produced over the whole cross-section.

Example, 9.12. A bar of aluminium, 2 in² in cross-section, and a bar of copper, 1.5 in² in cross-section, are rigidly connected at their ends (Fig. 9.14 (a)). The compound bar is subjected to a compressive force of 30,000 lb applied in such a way that the bars remain straight. The elastic moduli for aluminium and copper are $F_{al} = 10 \times 10^6$ psi and $E_c = 16 \times 10^4$ psi.

Find the force in each bar.

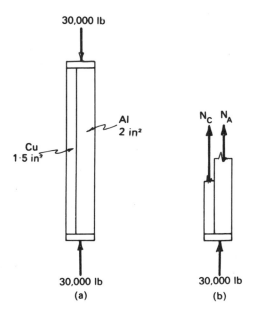

30,000 lb

Al
2 in²

Cu
1·5 in³

N_C N_A

30,000 lb   30,000 lb
(a)   (b)

Fig. 9.14

Solution. Although the bars are in contact, the problem is essentially the same as that of example 9.10.

Expressing the bar forces in terms of the force in the aluminium ($= R$) we have

$$N_A = R$$

The bar forces are assumed to be tensile, and equilibrium (Fig. 9.14 (b)) then gives $30,000 + N_A + N_C = 0$. Hence

$$N_C = -30,000 - R$$

The bar flexibilities are

$$f_A = \frac{L}{E_A A_A} = \frac{L}{20 \times 10^6} \text{ in/lb}; \quad f_C = \frac{L}{E_C A_C} = \frac{L}{24 \times 10^6} \text{ in/lb}$$

Since

$$e_A = e_C \text{ (for compatibility)}$$

then

$$f_A N_A = f_C N_C$$

$$\frac{L}{20 \times 10^6} (R) = \frac{L}{24 \times 10^6} (-30,000 - R)$$

$$R = -13,667 \text{ lb}$$

Thus

$$N_A = -13,667 \text{ lb} \quad \text{and} \quad N_C = -16,333 \text{ lb}$$

Example 9.13. Fig. 9.15 shows a rigid beam GJH supported by three bars, whose cross sections and spacing are as indicated. A load of 150 kips is attached to the mid-point of the beam.

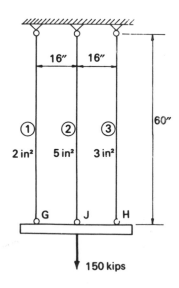

Fig. 9·15

For the stress range of this problem, bars 1 and 2 are Hookean with $E_1 = 20 \times 10^3$ ksi and $E_2 = 12 \times 10^3$ ksi. For bar 3, the stress–strain law is a straight line with a slope of 10×10^3 ksi up to a stress of 7 ksi and above this stress it is another straight line with half this slope. The equation of the second line is therefore $\sigma = 5 \times 10^3 \epsilon + 3.5$ provided that σ is in ksi.

Find the force in each bar.

Solution. The rigidity of beam GJH implies a relationship among the bar deformations. Hence the assembly is statically indeterminate. Since there are three bar forces and two equilibrium equations for beam GJH ($\sum Y = 0$ and $\sum M = 0$) we must choose one force as an unknown. Any bar force will do

(a) *Bar forces*

Let
$$N_2 = R$$

Then $\sum Y = 0$ gives
$$N_1 + R + N_3 = 150$$

$\sum M = 0$ gives
$$N_1 = N_3$$

$$\therefore \quad N_1 = N_3 = \frac{150 - R}{2}$$

(b) *Bar deformations*

$$e_1 = \left(\frac{150 - R}{2}\right) \frac{60}{20 \times 10^3 \times 2} = 7.5 \times 10^{-4}(150 - R)$$

$$e_2 = (R)\frac{60}{12 \times 10^3 \times 5} = 10 \times 10^{-4}(R)$$

For bar 3, assume that the stress exceeds 7 ksi. Then

$$\sigma_3 = \frac{150 - R}{2} \times \frac{1}{3}$$

$$\epsilon_3 = (\sigma_3 - 3.5)/5 \times 10^3$$

$$= \left(\frac{150 - R}{6} - 3.5\right)\Big/5 \times 10^3$$

$$= \frac{129 - R}{30 \times 10^3}$$

$$e_3 = 60\epsilon_3 = 20 \times 10^{-4}(129 - R)$$

(c) *Compatibility*

Since GJH remains straight

$$e_2 \quad \frac{e_1 + e_3}{2}$$

$$2 \times 10 \times 10^{-4} R = 7.5 \times 10^{-4}(150 - R) + 20 \times 10^{-4}(129 - R)$$

and
$$R = 78 \text{ kips}$$

Hence
$$N_1 = 36 \text{ kips}; \quad N_2 = 78 \text{ kips}; \quad N_3 = 36 \text{ kips}$$

Since the area of bar 3 is 3 sq in, the stress in this bar is 12 ksi and therefore the assumption that $\sigma_3 > 7$ was correct.

In many problems the difference between the number of bar forces and the number of joint equilibrium equations will be greater than one. In such problems a corresponding number of unknown forces must be selected. An equal number of compatibility conditions will provide simultaneous equations from which the unknowns can be determined. These equations will be linear if the force–deformation relations of the bars are linear. (They need not be *proportional*.) When the force–deformation relations are non-linear the simultaneous equations similarly will be non-linear.

In this book only bar systems with one redundancy are studied.

9.7 Statically Indeterminate Trusses—Displacement Method

In the displacement method of analysis, the joint displacements are first determined. The bar deformations are first expressed in terms of joint displacements, which are regarded as the unknowns. From the deformations we obtain expressions for the bar forces, and we can then write joint equilibrium equations. For any given joint the number of equilibrium equations will necessarily be the same as the number of degrees of freedom of that joint, hence we shall have the right number of equations from which to solve for the joint displacements.

In the case of a plane pin-jointed truss each joint has two degrees of freedom, that is, it requires two displacement components to specify the new position of the joint. The horizontal and vertical displacement components, u and v, are usually employed. Correspondingly, there are two equilibrium equations for the joint, namely $\sum X = 0$ and $\sum Y = 0$.

The joint displacements having been determined, the bar extensions and forces are easily found.

The problems chosen to illustrate the method are the same as those solved in the previous section. This will make it possible to compare the two methods of solution.

Example 9.14. (The same as example 9.9.) Find the bar forces in the structure of Fig. 9.16. For each bar $A = 0.5$ in² and $E = 20 \times 10^3$ ksi.

Solution. The primary unknown quantities are the displacements of A (u and v) which are both shown as positive in Fig. 9.16. It should be kept in mind that, in reality, these displacements are very small so that the changes in bar inclinations are negligible.

(*a*) *Bar deformations*

Resolve the displacement AA' in the direction of each bar in turn.

$$\left. \begin{array}{l} e_1 = u\cos 0° - v\sin 0° = u \\ e_2 = u\cos 30° - v\sin 30° = 0.866u - 0.5v \\ e_3 = u\cos 60° - v\sin 60° = 0.5u - 0.866v \end{array} \right\} \tag{9.14}$$

These equations are based on the laws of geometry.

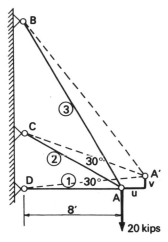

Fig. 9·16

(b) Bar forces

With a Hookean material it is convenient to evaluate first the bar stiffness coefficients, the load required to cause unit extension.

$$k_1 = \frac{EA_1}{L_1} = \frac{20 \times 10^3 \times 0.5}{96} = 104.2 \text{ kips/in}$$

$$k_2 = \frac{EA_2}{L_2} = \frac{20 \times 10^3 \times 0.5}{96/\cos 30°} = 90.2 \text{ kips/in}$$

$$k_3 = \frac{EA_3}{L_3} = \frac{20 \times 10^3 \times 0.5}{96/\cos 60°} = 52.1 \text{ kips/in}$$

Then

$$\left.\begin{array}{l} N_1 = k_1 e_1 \qquad\qquad\qquad\qquad = 104.2u \\ N_2 = k_2 e_2 = 90.2(0.866u - 0.5v) = 78.0u - 45.1v \\ N_3 = k_3 e_3 = 52.1(0.5u - 0.866v) = 26.0u - 45.1v \end{array}\right\} \qquad (9.15)$$

We have now made use of the force–deformation relations for the bars.

(c) Equilibrium

Finally, we use the equations of equilibrium which correspond to the joint displacements u and v.

$$\sum X = 0$$

or

$$N_1 + N_2 \cos 30° + N_3 \cos 60° = 0$$

$$104.2u + (78.0u - 45.1v)0.866 + (26.0u - 45.1v)0.5 = 0$$

$$184.8u - 61.6v = 0 \qquad (9.16)$$

$$N_2 \sin 30° + N_3 \sin 60° - 20 = 0$$

or

$$(78.0u - 45.1v)0.5 + (26.0u - 45.1v)0.866 - 20 = 0$$

$$61.6u - 61.6v - 20 = 0 \tag{9.17}$$

Solution of equations (9.16) and (9.17) yields

$$u = -0.162 \text{ in}; \quad v = -0.487 \text{ in}$$

(d) *Bar forces*

From equations (9.15) we now have the bar forces

$$N_1 = -16.88 \text{ kips}; \quad N_2 = +9.32 \text{ kips}; \quad N_3 = +17.75 \text{ kips}$$

The slight difference between these values, and those obtained by the force method, indicates that the number of significant figures used in the calculations is insufficient to provide reliability in the second decimal place of the answers.

The above procedure is equally applicable to frames in which the material characteristic is non-linear. However, solution of example 9.9 would involve rather extensive algebra. The reason for this is that the stress–strain function of that problem is given in the form $\epsilon = f(\sigma)$ which is convenient for the force method, but would have to be inverted for the displacement method into the form $\sigma = g(\epsilon)$. Since $f(\sigma)$ is a quadratic function, the result is rather unwieldy.

We may note that the inversion introduces no complication in the case of a linear function. For

$$\text{if } \sigma = E\epsilon \quad \text{then} \quad \epsilon = \left(\frac{1}{E}\right)\sigma$$

$$\text{and even if } \sigma = E_1\epsilon + E_2 \quad \text{then} \quad \epsilon = \left(\frac{1}{E_1}\right)\epsilon - \left(\frac{E_2}{E_1}\right)$$

As in the case of the force method of analysis, parallel bar (or compound bar) problems lead to an easier solution than problems in which the bars are inclined to one another.

Example 9.15. (The same as example 9.11.) In the structure of Fig. 9.17 the properties of the bars are:

$$L_1 = 60 \text{ in} \qquad L_2 = 72 \text{ in}$$

$$A_1 = 1.0 \text{ in}^2 \qquad A_2 = 1.2 \text{ in}^2$$

$$E_1 = 30 \times 10^3 \text{ ksi} \quad E_2 = 15 \times 10^3 \text{ ksi}$$

Find the location of the 50 kip load so that the beam GH remains horizontal, and find the force in each bar.

Solution. In this problem the vertical displacement of GH corresponds to joint displacement in the previous example. Since GH remains horizontal there is only one degree of freedom, expressed by the displacement v.

(a) *Bar deformations*

$$e_1 = e_2 = -v$$

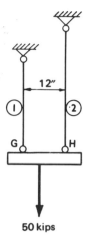

50 kips

Fig. 9·17

(b) *Bar forces*

$$k_1 = \frac{E_1 A_1}{L_1} = 0.5 \times 10^3 \text{ kips/in}; \quad k_2 = \frac{E_2 A_2}{L_2} = 0.25 \times 10^3 \text{ kips/in}$$

$$\left. \begin{array}{l} N_1 = k_1 e_1 = -0.5 \times 10^3 v \\ N_2 = k_2 e_2 = -0.25 \times 10^3 v \end{array} \right\} \tag{9.18}$$

(c) *Equilibrium*

Consider the equilibrium of beam GH

$\Sigma\, Y = 0$

$$N_1 + N_2 - 50 = 0$$

$$-0.75 \times 10^3 v = 50$$

$$v = -66.67 \times 10^{-3} \text{ in}$$

(d) *Bar forces*

From equations (9.18)

$$N_1 = 33.33 \text{ kips} \quad \text{and} \quad N_2 = 16.67 \text{ kips}$$

We might notice that the bar loads are proportioned to the bar stiffness.

The location of the 50 kip load is obtained from the rotational equilibrium of beam GH. This shows that the load must be 4 in from G.

Example 9.16. (The same as example 9.12.) A bar of aluminium, 2 in² in cross-section, and a bar of copper, 1.5 in² in cross-section, are rigidly connected at their ends (Fig. 9.18 (a)). The compound bar is subjected to a compressive force of 30,000 lb applied in such a way that the bars remain straight. The elastic moduli for aluminium and copper are $E_A = 10 \times 10^6$ psi and $E_C = 16 \times 10^6$ psi.

Find the force in each bar.

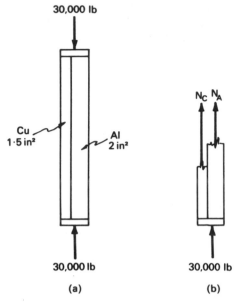

30,000 lb

Cu
1·5 in²

Al
2 in²

30,000 lb

(a)

N_C N_A

30,000 lb

(b)

Fig. 9·18

Solution. Here again the extension, common to both bars, is taken as the main variable.

$$e_A = e_C = e \text{ (say)}$$

$$k_A = \frac{E_A A_A}{L} = \frac{20 \times 10^6}{L} \text{ lb/in}; \quad k_C = \frac{E_C A_C}{L} = \frac{24 \times 10^6}{L} \text{ lb/in}$$

$$N_A = k_A e_A = 20 \times 10^6 e/L; \quad N_C = k_C e_C = 24 \times 10^6 e/L$$

For equilibrium (of the end connecting plate, Fig. 9.18 (b))

$$N_A + N_C + 30,000 = 0$$

$$20 \times 10^6 e/L + 24 \times 10^6 e/L + 30,000 = 0$$

$$e/L = -0.683 \times 10^{-3}$$

Hence

$$N_A = -13,667 \text{ lb} \quad \text{and} \quad N_C = -16,333 \text{ lb}$$

Example 9.17. (The same as example 9.13.) Fig. 9.19 shows a rigid beam GJH supported by three bars, whose cross-sections and spacing are as indicated. A load of 150 kips is attached to the mid-point of the beam. For the stress range of this problem, bars 1 and 2 are Hookean with $E_1 = 20 \times 10^3$ ksi and $E_2 = 12 \times 10^3$ ksi. For bar 3, the stress–strain law is a straight line with a slope of 10×10^3 ksi up to a stress of 7 ksi, and above this stress it is another straight line with half this slope. The equation of the second line is therefore $\sigma = 5 \times 10^3 \epsilon + 3.5$ provided that σ is in ksi.

Find the force in each bar.

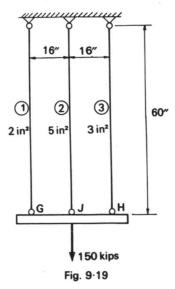

150 kips

Fig. 9·19

Solution. In this example the beam GJH has two degrees of freedom. Its displacement can be defined either by the vertical movement of two points, or by the vertical movement of one point and the angular displacement. Suppose we use v_G and v_H as the variables.
Then

$$e_1 = -v_G$$

$$e_2 = -\frac{v_G + v_H}{2}$$

$$e_3 = -v_H$$

$$N_1 = \frac{E_1 A_1}{L_1} e_1 = \frac{20 \times 10^3 \times 2}{60}(-v_G) = -667v_G$$

$$N_2 = \frac{E_2 A_2}{L_2} e_2 = \frac{12 \times 10^3 \times 5}{60}\left(-\frac{v_G + v_H}{2}\right) = -500v_G - 500v_H$$

For bar 3,

$$\epsilon = \frac{-v_H}{60}$$

$$\sigma = 5 \times 10^3 \left(\frac{-v_H}{60}\right) + 3.5$$

$$N_3 = \sigma_3 A_3 = -250v_H + 10.5$$

For the equilibrium of bar GH, we have
$\Sigma Y = 0$

$$N_1 + N_2 + N_3 = 150$$

$$-667v_G - 500v_G - 500v_H - 250v_H + 10.5 = 150$$

$$-1167v_G - 750v_H = 139.5 \qquad (9.19)$$

$\sum M = 0$ (about J)

$$N_1 = N_3$$

$$-667v_G = -250v_H + 10.5 \qquad (9.20)$$

Solution of equations (9.19) and (9.20) give

$$v_G = -0.054 \text{ in} \quad \text{and} \quad v_H = -0.10 \text{ in}$$

Hence

$$N_1 = 36 \text{ kips}; \quad N_2 = 78 \text{ kips}; \quad \text{and} \quad N_3 = 36 \text{ kips}$$

The force method and the displacement method are not the only possible procedures for the analysis of statically indeterminate structures. For instance, one may select the primary variables taking some from among the forces and some from among the displacements. Such a procedure is sometimes called a *mixed method*. As this clearly implies that there are quite a number of primary variables, we may consider the discussion of such procedures as being beyond the scope of this book.

9.8   Stresses due to Temperature Change and Shrinkage

In a statically indeterminate truss, where by definition the bars are not free to deform independently, a change of temperature will often be accompanied by stresses in the members. Whether or not this occurs will depend upon whether the thermal deformations are mutually compatible. If they are, then stresses will not be induced. The truss of Fig. 9.20 (a) is statically indeterminate, but a uniform temperature rise will result in a general increase in the size of the structure, similar to the effect produced by drawing the truss to a slightly larger scale. The thermal deformations will be compatible and no stresses will be introduced. The truss of Fig. 9.20 (b), on the other hand, is not free to undergo uniform expansion. In this structure the thermal deformations will not be compatible, and stresses will be induced.

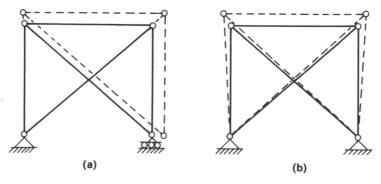

(a)   (b)

Fig. 9·20

The calculation of such *thermal stresses* is based simply on the fact that *total deformations* must be compatible. The solution may be carried out either by the force method or by the displacement method.

Example 9.18. The three bar truss of Fig. 9.21 has the same dimensions as that of example 9.9 (p. 243), i.e., $A = 0.5$ in^2 and $E = 20 \times 10^3$ ksi for each bar. If the coefficient of linear thermal expansion, α, is 15×10^{-6} per °F for each bar, find the bar forces induced by a temperature rise of 60°F.

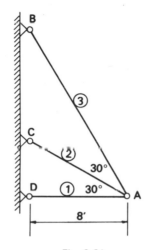

Fig. 9·21

Solution. A force solution following the same steps as that of example 9.8 will be used. For brevity some of the previous results will be used.

(*a*) *Bar forces*

By using the two equilibrium equations (for joint A) we can express all bar forces in terms of one unknown, say the force R in bar 2. Then as in example 9.8 we have (see equation 9.8)

$$\left.\begin{array}{l} N_1 = -0.5774R \\ N_2 = R \\ N_3 = -0.5774R \end{array}\right\} \qquad (9.21)$$

(*b*) *Bar deformations*

The total deformation of any bar is

$$e = \alpha TL + fN$$

where f is the bar flexibility. For all bars the thermal strain is

$$\epsilon_T = \alpha T = 15 \times 10^{-6} \times 60 = 9 \times 10^{-4}$$

Hence, using the flexibility coefficients from p. 244 we have

$$
\left.
\begin{aligned}
e_1 &= \epsilon_T L_1 + f_1 N_1 = 9 \times 10^{-4} \times 96 + 0.0096(-0.5774R) = 0.0864 - 0.0055R \\
e_2 &= \epsilon_T L_2 + f_2 N_2 = 9 \times 10^{-4} \times 111 + 0.0111(R) = 0.0998 - 0.0111R \\
e_3 &= \epsilon_T L_3 + f_3 N_3 = 9 \times 10^{-4} \times 192 + 0.0192(-0.5774R) = 0.1728 - 0.0111R
\end{aligned}
\right\} \quad (9.22)
$$

Note that the only difference between the present problem and example 9.8 is that in these deformation expressions the constant terms represent the deformation due to temperature change instead of that due to external loading. The remainder of the solution follows as before.

(c) *Compatibility*

The discontinuity at C (see Fig. 9.10 (b)) is

$$CC' = -0.5774e_1 + e_2 - 0.5774e_3$$

When the values of e_1, e_2 and e_3 are substituted from equations (9.22) we obtain

$$-0.0497 + 0.0207R = 0$$

$$\therefore \quad R = 2.4 \text{ kips}$$

(d) *Bar forces*

From equations (9.21) we then have

$$N_1 = -1.39 \text{ kips}; \quad N_2 = +2.4 \text{ kips}; \quad N_3 = -1.39 \text{ kips}$$

If desired, the displacements of A could be computed as in example 9.8.

Example 9.19. Solve example 9.18 by the displacement method.

Solution. This time reference is made to the corresponding external load problem (example 9.14) on p. 252.

(a) *Bar deformations*

As for example 9.14, we express the bar deformations in terms of the displacements u, v at A.

$$
\left.
\begin{aligned}
e_1 &= u \\
e_2 &= 0.866u - 0.5v \\
e_3 &= 0.5u - 0.866v
\end{aligned}
\right\} \quad (9.23)
$$

(b) *Bar forces*

The total deformation, e, of each bar is the sum of e_S due to stress and e_T due to temperature.

$$e = e_S + e_T$$

Thus

$$e_S = e - e_T = e - \epsilon_T L$$

The bar force is obtained by multiplying e_S alone by the bar stiffness coefficient k (see p. $\overline{253}$).

$$
\left.
\begin{aligned}
N_1 &= k_1 e_{1S} = k_1(e_1 - \epsilon_T L_1) = 104.2(u - 0.0864) \\
N_2 &= k_2 e_{2S} = k_2(e_2 - \epsilon_T L_2) = 90.2(0.886u - 0.5v - 0.0998) \\
N_3 &= k_3 e_{3S} = k_3(e_3 - \epsilon_T L_3) = 52.1(0.5u - 0.866v - 0.1728)
\end{aligned}
\right\} \quad (9.24)
$$

(c) *Equilibrium*

We have two equations of equilibrium of joint A.

$$\sum X = 0$$
$$N_1 + N_2 \cos 30° + N_3 \cos 60° = 0$$
$$\sum Y = 0$$
$$N_2 \sin 30° + N_3 \sin 60° = 0$$

When the values of N are substituted from equations (9.24) these give

$$184.8u - 61.6v - 21.3 = 0 \atop 61.6u - 61.6v - 12.3 = 0$$

(9.25)

The constant terms depend on the temperature, whereas previously (equations 9.16, 9.17 on p. 253) they depended on the external loading. The other terms are the same as before. Solution of equations (9.25) gives

$$u = +0.073 \text{ in} \quad \text{and} \quad v = -0.127 \text{ in}$$

(*d*) *Bar forces*

From equation (9.24) we now have

$$N_1 = -1.39 \text{ kips}; \quad N_2 = 2.4 \text{ kips}; \quad N_3 = -1.37 \text{ kips}$$

Compound bars, being statically indeterminate structures, will also usually be put under stress by any change of temperature. If two bars side by side (Fig. 9.22) are constrained to remain the same length, then, if they have different coefficients of thermal expansion, stresses will be induced by temperature change.

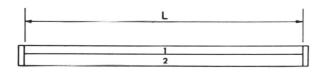

Fig. 9·22

The total deformations must be equal, therefore

$$f_1 N_1 + \alpha_1 TL = f_2 N_2 + \alpha_2 TL$$

and for equilibrium

$$N_1 + N_2 = 0$$

These two equations are easily solved for N_1 and N_2.

The bars may be end to end, in which case the appropriate compatibility condition must be used.

Example 9.20. Three bars are placed end to end (Fig. 9.23) and overall longitudinal deformation is prevented by rigid end supports. The lengths and areas of the bars are given in Fig. 9.23. The coefficients of linear thermal expansion are $\alpha_1 = 6 \times 10^{-6}$, $\alpha_2 = 12 \times 10^{-6}$ and $\alpha_3 = 16 \times 10^{-6}$ per °F. For the stress range of this problem $E_1 = 15 \times 10^6$, $E_2 = 30 \times 10^6$ and $E_3 = 10 \times 10^6$ psi.

Find the stress induced in each bar by a fall in temperature of 50°F. Find also the movement of points B and C.

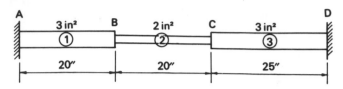

Fig. 9·23

Solution. In this problem, equilibrium of the junctions B and C requires that

$$N_1 = N_2 = N_3 = N \text{ (say)}$$

The bar flexibility coefficients are

$$f_1 = \frac{L_1}{E_1 A_1} = 0.444 \times 10^{-6} \text{ in/lb}; \quad f_2 = 0.333 \times 10^{-6} \text{ in/lb}; \quad f_3 = 0.833 \times 10^{-6} \text{ in/lb}$$

The total deformations are

$$\left.\begin{array}{l} e_1 = f_1 N_1 + \alpha_1 TL_1 = 0.444 \times 10^{-6} N + 6 \times 10^{-6}(-50)\,20 \\ e_2 = f_2 N_2 + \alpha_2 TL_2 = 0.333 \times 10^{-6} N + 12 \times 10^{-6}(-50)\,20 \\ e_3 = f_3 N_3 + \alpha_3 TL_3 = 0.833 \times 10^{-6} N + 16 \times 10^{-6}(-50)\,25 \end{array}\right\} \quad (9.26)$$

For compatibility the sum of the three extensions must be zero.

$$e_1 + e_2 + e_3 = 0$$

Substitution from equations (9.26) gives

$$1.611N + 38{,}000 = 0$$

$$\therefore \quad N = 23{,}586 \text{ lb}$$

i.e.,

$$N_1 = N_2 = N_3 = 23{,}586 \text{ lb}$$

The stresses are

$$\sigma_1 = \frac{N_1}{A_1} = 7862 \text{ psi}$$

$$\sigma_2 = \frac{N_2}{A_2} = 11{,}793 \text{ psi}$$

$$\sigma_3 = \frac{N_3}{A_3} = 7862 \text{ psi}$$

To find the movements of B and C we require the deformations of bars 1 and 3. From equation (9.26)

$$e_1 = 10^{-6}(10{,}480 - 6000) = +0.00448 \text{ in}$$

so that point B moves $+0.0045$ in.
    Again

$$e_3 = 10^{-6}(19{,}660 - 20{,}000) = -0.00034 \text{ in}$$

so that point C moves $+0.00034$ in.

Shrinkage strains, like thermal strains, will also induce stresses if for some reason independent deformation is prevented. A practical example of this occurs in reinforced concrete. For the present discussion this may be regarded as a compound bar of steel and concrete.

Example 9.21. A block of concrete 3 ft long has a steel bar embedded longitudinally in it (Fig. 9.24). The area of the bar is 1 in² and the net area of concrete is 20 in². At the time of casting the concrete there is no stress in the steel. If the concrete were free to contract it would suffer a strain of -2×10^{-4} over a period of four weeks. Assuming that the steel and concrete remain bonded together, find the stress in each material at the end of this period. Take the elastic moduli of concrete and steel as $E_c = 2 \times 10^6$ psi and $E_s = 30 \times 10^6$ psi respectively.

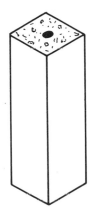

Fig. 9·24

Solution. Let the steel and concrete stresses be σ_s and σ_c.
For equilibrium, the total bar force is zero since there is no external load.

$$N_c + N_s = 0$$

$$20\sigma_c + 1\sigma_s = 0 \qquad (9.27)$$

The bars deform by the same amount, and since they are the same length, the strains are equal

$$\epsilon_c = \epsilon_s$$

The concrete strain consists of the free shrinkage strain (-2×10^{-4}) together with the strain caused by the stress σ_c.

$$\left(\frac{\sigma_c}{2 \times 10^6}\right) - 2 \times 10^{-4} = \frac{\sigma_s}{30 \times 10^6} \qquad (9.28)$$

From equations (9.27) and (9.28) we find that

$$\sigma_s = -3429 \text{ psi} \quad \text{and} \quad \sigma_c = +171 \text{ psi}$$

9.9 Residual Stresses

The methods of analysis developed earlier in this chapter are equally applicable whether the force–deformation relation for the bars is linear or non-linear. In regard to the latter situation, a further matter should be considered. For materials with a non-linear stress–strain relation it is usual for the unloading curve to differ from the loading curve. What effect will this factor have on a bar assembly which is first loaded and then unloaded?

The effect will be different in the case of statically determinate and statically indeterminate trusses. In the former, the bar forces are always related to the external loads by the laws of equilibrium. Hence if the external loads are removed, the bar forces will return to zero. Since each bar may retain some residual deformation, the frame itself will not revert to its original geometry. In the latter type of frame, where bar deformations are not independent, it is most probable that the bars will retain a residual stress as well as a residual deformation upon unloading.

In either type of truss the final state of stress and deformation can be calculated by the superposition of two problems (a) the application of the loads, during which the loading curve is used in the analysis, and (b) the application of the negative of the loads, during which the unloading curve is used. This is illustrated by two examples. In order to save space use will be made of previous calculations.

Example 9.22. The material from which the truss of Fig. 9.25 is constructed has a stress–strain equation

$$\epsilon = 6 \times 10^{-5}(\pm 0.04\sigma^2 + \sigma)$$

while the stresses are increasing, the $+$ and $-$ signs referring to tension and compression respectively.

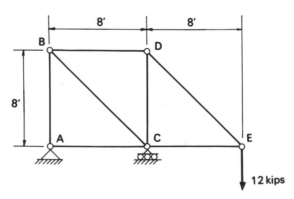

Fig. 9·25

When the stress is decreasing, the stress–strain graph is linear, with a slope of 20×10^3 ksi. All bar dimensions are the same as in examples 9.1 and 9.2.

Find the residual displacements of joint E if the load of 12 kips is first applied and then removed.

Solution. The effect of applying the load of 12 kips to this particular frame was computed in example 9.2 (p. 234). It was found there that the displacements of joint E are

$$u_E = -0.138 \text{ in} \quad \text{and} \quad v_E = -1.456 \text{ in}$$

Unloading the structure is the same as adding a further load of 12 kips upwards which will cancel the previous load. During this second stage the material behaves as if it were elastic with a modulus of 20×10^3 ksi.

The analysis of this problem was carried out in example 9.1 (p. 232). Reversing the sign of those results, we see that an upward force of 12 kips would produce displacements

$$u_E = +0.072 \text{ in} \quad \text{and} \quad v_E = +0.696 \text{ in}$$

The final, or residual displacements of E will therefore be

$$u_E = -0.138 + 0.072 = -0.066 \text{ in}$$

$$v_E = -1.456 + 0.696 = -0.760 \text{ in}$$

The residual displacements of other joints could be found in the same way. Since the truss is statically determinate the bar forces after unloading will be zero.

Example 9.23. The truss of Fig. 9.26 is constructed from the same material as that of Fig. 9.25. During loading $\epsilon = 6 \times 10^{-5} (\pm 0.04\sigma^2 + \sigma)$. The unloading graph is a straight line of slope 20×10^3 ksi. All bar dimensions are the same as in examples 9.9 and 9.10. Find the residual bar forces and joint displacements when the 20 kip load is first applied and then removed.

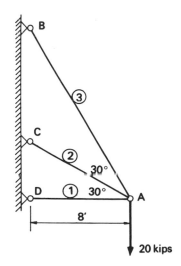

Fig. 9·26

Solution. The application of the 20 kip load constitutes exactly the same problem as example 9.10. Hence the first phase of our present problem will result in the following bar forces and joint displacements:

$$N_1 = -17.60 \text{ kips}; \quad N_2 = +10.48 \text{ kips}; \quad N_3 = +17.05 \text{ kips}$$

$$u_A = -0.49 \text{ in}; \quad v_A = -1.36 \text{ in}$$

We can unload the structure by applying an upward load to cancel the 20 kip load. During this phase the material behaves with a constant tangent modulus of 20×10^3 ksi. The effects can therefore be obtained by reversing the signs of the results of example 9.9.

$$N_1 = +16.91 \text{ kips}; \quad N_2 = -9.30 \text{ kips}; \quad N_3 = -17.74 \text{ kips}$$

$$u_A = +0.16 \text{ in}; \quad v_A = +0.49 \text{ in}$$

The residual forces and displacements will be the sum of the two previous sets. That is

$$N_1 = -0.69 \text{ kips}; \quad N_2 = +1.18 \text{ kips}; \quad N_3 = -0.69 \text{ kips}$$

$$u_A = -0.33 \text{ in}; \quad v_A = -0.87 \text{ in}$$

It might be noted that the residual bar forces, although not zero, satisfy equilibrium at joint A.

9.10   Extension of Procedures

Throughout this chapter, attention has been confined to assemblages of bars which sustain axial force only, and consequently undergo only axial deformation. However, very frequently the bars also sustain bending deformation, and possibly torsion deformation as well. Solution of such problems is carried out on the principles outlined above. It is only necessary to compute the deformation due to all causes and to make sure that compatibility is satisfied.

For instance, the reactions at the supports of the structure of Fig. 9.27 (a) cannot be found by statics. It is statically indeterminate. This is clear because

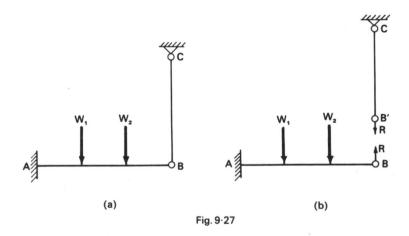

(a)   (b)

Fig. 9·27

deformation of AB must ·be accompanied by deformation of BC. We may regard the force in bar BC as an unknown force R (Fig. 9.27 (b)). It is only necessary to equate the vertical displacement of B to that of B'. The displacement of B is due to bending of AB, while displacement of B' is due to axial deformation of $B'C$. Each displacement is found in terms of R, which can be determined when the displacements are equated.

A similar procedure can be adopted for the structure of Fig. 9.28 (a). The calculation of the deflection of C, due to the loading and to R, Fig. 9.28 (b), presents no difficulty. When this is equated to the deflection of the spring due to R, then R can be determined.

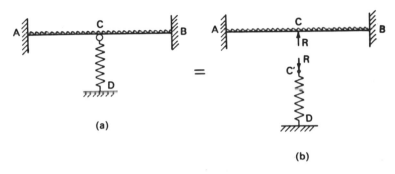

(a)

(b)

Fig. 9·28

When the bars of a frame are joined together by pin joints, relative rotation can occur between the ends of the joined members. Compatibility requires only that the displacements u and v (and w in three-dimensional frames) shall be the same for the ends of adjoining members. However, in many practical frames the joints are rigid, such as welded joints in steelwork, or monolithic joints in concrete construction. This rigidity implies that rotations of adjoining member ends shall also be equal.

Once again no new principles are involved. It is only necessary to calculate the *rotations* of the ends of the members as well as their u and v displacements, and then to ensure that compatibility is preserved in respect of rotation as well as translation.

It is therefore clear that, provided we can calculate the deformation (of all types) of a bar due to external forces, to temperature change and to any other cause, then we are in a position to analyse any structure which is composed of such bars.

Problems

9.1   In the bar assembly of Fig. P9.1, each of the bars is 5 ft long and has a cross-section of 0.2 in². Find the extension of each bar and the vertical displacement of the joints A and D due to the 5 kip load. Take $E = 30 \times 10^3$ ksi.

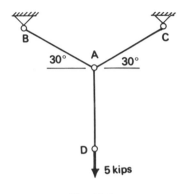

Fig. P9·1

9.2   In Fig. P9.2 the bar AB is 10 ft long and 0.5 in² in cross-section, and the bar AC is 6 ft long and 0.3 in² in cross-section. Both bars are of steel ($E = 30 \times 10^3$ ksi). Find the deformation of each bar due to the 12 kip load and the vertical and horizontal displacement of the joint A.

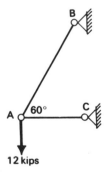

Fig. P9·2

9.3   In Fig. P9.3 the bar AC is 10 ft long and B is vertically below A. The cross-section of AC is 0.2 in² and that of BC is 0.5 in². Find the displacements of joint C if $E = 20 \times 10^3$ ksi.

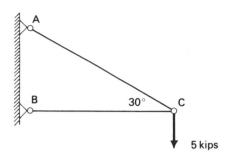

Fig. P9·3

9.4   In problem 9.3, find the displacements of C by means of the theorem of virtual forces.

9.5   In the structure of Fig. P9.5 the bars have the same dimensions as those in problem 9.3. However, the bars are made of a material whose stress–strain relation is $\epsilon = 8 \times 10^{-5}(10^{-1}\sigma^2 + \sigma)$ in tension and $\epsilon = 8 \times 10^{-5}(-10^{-1}\sigma^2 + \sigma)$ in compression, where σ is in ksi. Find the displacements of the joint C caused by the 5 kip load.

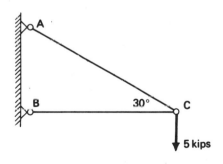

Fig. P9·5

9.6   Fig. P9.6 shows a pin-jointed frame supported by a hinge joint at A and on rollers at B. Dimensions of the bars are as shown. Find the bar forces by statics. Find the bar deformations and hence calculate by geometry the displacements of the joints D, C and E. Take $E = 30 \times 10^3$ ksi.

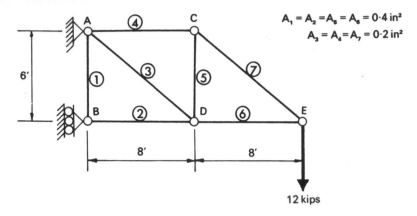

Fig. P9·6

9.7   In the frame of problem 9.6, find the force in the bar AC (a) by the method of virtual displacements and (b) by the method of sections.

9.8   In the frame of problem 9.6, compute the bar deformations by any convenient method. Hence, find the vertical displacements of joints D and E by the method of virtual forces.

9.9   A frame has the same overall dimensions and the same bar sizes as those shown in Fig. P9.6. The present frame is made from a material which has the stress–strain relation shown in Fig. P9.9. Up to a stress of 30 ksi the tangent modulus is 15×10^3 ksi and thereafter the tangent modulus is 5×10^3 ksi. The relation is the same both for tension and compression.

   A downward vertical load of 6 kips is applied at E. Having found the bar deformations, use the method of virtual forces to compute the vertical displacements at D and E and the horizontal displacement at E.

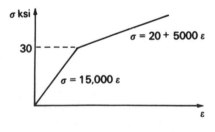

Fig. P9·9

9.10 The frame of Fig. P9.10 is made from material having the stress–strain characteristics indicated in Fig. P9.9. Find the horizontal and vertical displacements of joints B and C.

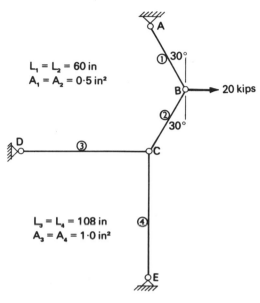

$L_1 = L_2 = 60$ in
$A_1 = A_2 = 0.5$ in²

20 kips

$L_3 = L_4 = 108$ in
$A_3 = A_4 = 1.0$ in²

Fig. P9·10

9.11 A pin-jointed frame has the configuration shown in Fig. P9.1. Each bar is 5 ft long and has a cross-section of 0.2 in². The stress–strain law of the material is

$$\sigma = 24 \times 10^3 \,\epsilon - 24 \times 10^5 \,\epsilon^2$$

where σ is in ksi. Find the displacement of joints A and D (a) when the load at D is 5 kips and (b) when the load at D is 10 kips.

9.12 In the truss of Fig. P9.12 the member DE undergoes an extension of 0.08 in due to temperature rise. The other members are not affected. Use the method of virtual forces to compute the vertical displacement of point C. The panels are of equal length.

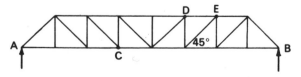

Fig. P9·12

9.13   In the structure of Fig. P9.13, the bar AB extends by 0.12 in and the bar BC contracts by 0.2 in for some reason. Find the component of displacement of joint B in the direction n.

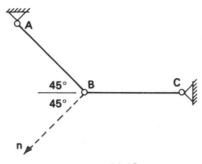

Fig. P9·13

9.14   The bars in the assembly of Fig. P9.14 undergo the following deformations: $e_1 = 0$, $e_2 = -0.3$ in, $e_3 = -0.15$ in. It is required to calculate the vertical displacement of point A.

(*a*) Choose a suitable virtual force system (it is suggested that one bar force be taken as zero).

(*b*) Adopt a different virtual force system and show that the same result is obtained for the displacement.

(*c*) Does the result depend upon the cause of the bar deformations?

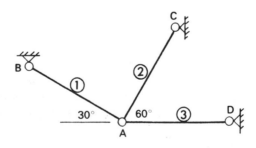

Fig. P9·14

9.15   The rigid bar ABC (Fig. P9.15) is right-angled at B. Due to an accidental blow the bar becomes bent at point D, the angular discontinuity introduced being $1°$. Use the method of virtual forces to find the displacement of point A.

(*Note:* In this problem work will be done by the virtual *bending moment* at *D* when the deformation occurs.)

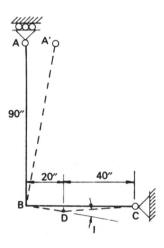

Fig. P9·15

9.16 The statically determinate truss of Fig. P9.16 is loaded as shown. Use the method of virtual displacements to find the force induced in the member *CD*.

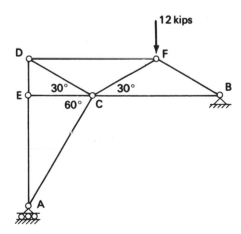

Fig. P9·16

9.17 In the truss of Fig. P9.17 the panels are equal and the angles are 45°. If the force in member CD due to the load at E is $+8$ kips, use the method of virtual displacements to find W.

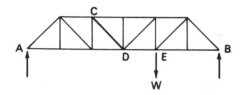

Fig. P9·17

9.18 In the structure of Fig. P9.14, the bar forces are known to be $N_1 = 0$, $N_2 = -3.0$ kips, $N_3 = -1.5$ kips. Find the vertical component of the external force at A by the method of virtual displacements. Compare the calculations with those of problem 9.14.

9.19 In Fig. P9.19 the bars AB, AC and AD are each 5 ft long and 0.2 in² in cross-section. The bar AE is 4 ft long and 0.1 in² in cross-section. All bars are of steel ($E = 30 \times 10^3$ ksi). Find the force in each bar and the displacement of joint A. Use the force method.

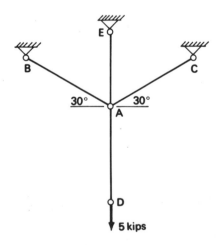

Fig. P9·19

9.20   Solve problem 9.19 by the displacement method.

9.21   Find the forces in the bars of the frame shown in Fig. P9.21. Use the force method. Take $E = 20 \times 10^3$ ksi for all bars. Also calculate the displacements of joint A.

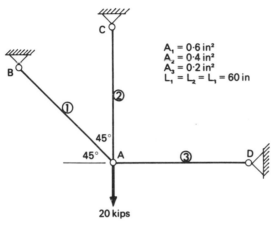

$A_1 = 0.6$ in²
$A_2 = 0.4$ in²
$A_3 = 0.2$ in²
$L_1 = L_2 = L_1 = 60$ in

20 kips

Fig. P9·21

9.22   Solve problem 9.21 by the displacement method.

9.23   A frame has the same dimensions as that shown in Fig. P9.21. Bars AB and AD are made of a linearly elastic material with $E = 20 \times 10^3$ ksi. Bar AC is made of material for which $E = 18 \times 10^3$ up to a stress of 15 ksi. Above this stress $\sigma = 10 + 6 \times 10^3 \epsilon$ (the stress in bar AC is above 15 ksi in this problem). The frame carries the same load as in problem 9.21. Find the force in each bar and the displacements of joint A. Compare the results with those of problem 9.21.

9.24   All bars of the frame shown in Fig. P9.24 have a cross-section of 0.5 in² and are made of a material whose stress–strain characteristic is

$$\epsilon = 8 \times 10^{-5} \left(\frac{\sigma^2}{10} + \sigma \right) \text{ in tension}$$

and

$$\epsilon = 8 \times 10^{-5} \left(-\frac{\sigma^2}{10} + \sigma \right) \text{ in compression}$$

Find the forces in all the bars and the displacements of joint B.

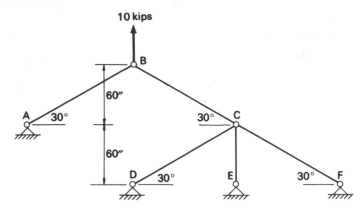

Fig. P9·24

9.25 In the frame of Fig. P9.25 the bars are all the same length. The cross-section of the two lower bars DE and CE is twice that of the two upper bars AE and BE. Find the force in each bar. Take $E = 15 \times 10^3$ ksi.

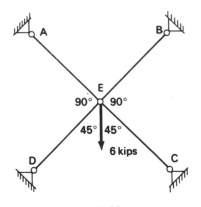

Fig. P9·25

9.26 The frame of the previous problem is made of material for which the stress–strain relation is the same as that given in problem 9.24. Find the force in each bar.

9.27 In Fig. P9.27, the cross-beam BD is supported from two bars AB and CD of equal length. AB has a cross-section of 0.8 in² and its stress–strain equation is $\sigma = 4000 + 15 \times 10^6 \epsilon$ (for the stress range of this problem). CD has a cross-section of 1.5 in² and obeys Hooke's Law with $E = 12 \times 10^6$ psi.

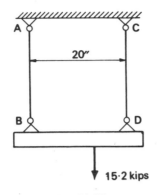

Fig. P9·27

A load of 15.2 kips is suspended from the cross-beam at such a position that BD remains horizontal. Find the stress in each bar and the location of the load.

9.28 The beam BDF (Fig. P9.28) is supported by three rods of equal length. AB is made of copper ($E = 16 \times 10^6$ psi) and has a cross-section of 0.5 in². CD is made of zinc ($E = 12 \times 10^6$ psi) and has a cross-section of 0.4 in². EF is made of aluminium ($E = 10 \times 10^6$ psi) and has a cross-section of 1.0 in². (a) Where must a load of 3000 lb be suspended so that the extension of EF is twice that of AB? (b) What is the stress in each bar?

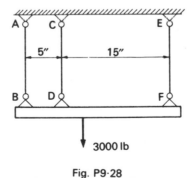

Fig. P9·28

9.29 A solid rod of tungsten ($E = 52 \times 10^6$ psi) has a diameter of 0.25 in. It is embedded in a porcelain rod ($E = 24 \times 10^6$ psi) the outside diameter of which is 1 in. A compressive force of 6000 lb is applied to the compound bar. Assuming that there is no slip between the two materials, find the stress in each.

9.30   A steel bolt, having a length of 48 in and a cross-section of 0.5 in² is placed inside a copper tube the net cross-section of which (excluding the hole) is 1.5 in². After the nut touches the end of the tube it is given half a turn. Find the forces induced in the bolt and in the copper if the bolt thread has a pitch of $\frac{1}{8}$ in. Find also the total energy stored in the system. $E_{copper} = 16 \times 10^6$ psi and $E_{steel} = 30 \times 10^6$ psi.

9.31   Three bars of a nickel alloy ($E = 22 \times 10^6$ psi), a magnesium alloy ($E = 6.5 \times 10^6$ psi) and a titanium alloy ($E = 17 \times 10^6$ psi) are of equal length and have cross-sections of 0.4, 0.6 and 0.8 in² respectively. They are placed side by side and rigidly joined at the ends so that under load they deform equally. If the whole assembly is subjected to a pull of 20,000 lb, find the force and the stress in each bar.

9.32   A reinforced concrete column is 12 in square. In each corner is embedded a 1 in diameter bar of mild steel ($E = 30 \times 10^6$ psi) (Fig. P9.32). The column carries a load of 350 kips. (*a*) Find the stresses in the steel and concrete just after the load is applied. Take E for concrete as 5×10^6 psi.

The creep characteristic of the concrete is such that under constant stress the strain at the end of one year is three times the elastic strain at the instant of loading. In the present problem, if the variation of concrete stress during the year is ignored, the creep will be equivalent to a reduction of E to one-third of its original value. (*b*) On this assumption, find the steel and concrete stresses one year after the load is applied.

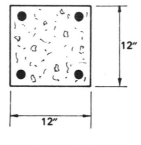

Fig. P9·32

9.33   A mild steel bar of $\frac{1}{2}$ in diameter fits snugly inside a high tensile steel tube of 1 in outside diameter. Both have a Young's modulus of 30×10^6 psi. For the mild steel, the stress–strain graph is straight up to a stress of 40,000 psi after which it is horizontal; the unloading curve is a sloping line parallel to the initial elastic portion. For the high tensile steel the limit of proportionality is 80,000 psi after which the material undergoes strain-hardening.

(a) If the bar is subjected to a tensile force of 45,000 lb, find the stress in each material.

(b) Also find the stress in each material after the force is removed.

9.34 A cast iron cylinder 5 ft long has an internal diameter of 10 in and a thickness of 1 in. The ends are closed by rigid covers secured by a central bolt and nuts. The nuts are screwed up until the tension in the bolt is 40 kips. Internal fluid pressure is now applied, which causes a thrust on each cover of 40 kips. What is the final tension in the bolt? The effective length of the bolt is 60 in and its area of cross-section is 5 in². E for the bolt $= 30 \times 10^6$ psi and E for cast iron is 15×10^6 psi.

9.35 A compound bar is formed of two materials A and B. Material A has a Young's modulus of 30×10^3 ksi and a yield stress of 40 ksi. The area of bar A is 0.6 in². Material B has a Young's modulus of 20×10^3 ksi up to a stress of 16 ksi. Above this stress, the tangent modulus is constant and equal to 10×10^3 ksi. The area of bar B is 0.8 in².

(a) A tensile force of 48 kips is applied to the compound bar. Find the force and stress in each material.

(b) If the external force is removed, find the residual force and stress in each bar. During unloading each material has a linear stress–strain relation, with $E_A = 30 \times 10^3$ ksi and $E_B = 20 \times 10^3$ ksi.

9.36 A composite bar consists, of three bars side by side and fixed together at the ends. The outer bars each have a cross-section of 1.2 in² and are made from material A which has an elastic modulus of 20×10^6 psi and a yield point of 60,000 psi. The centre bar has a cross-section of 1.0 in² and is made from material B which has an elastic modulus of 12×10^6 psi and a yield point of 24,000 psi. For each material the stress–strain curve is straight up to the yield stress and horizontal thereafter.

Find the stress in each material when the composite bar supports a tensile force of (a) 75,000 lb and (b) 150,000 lb.

9.37 Bars of two different materials are firmly fixed together side by side. Each bar is 6 in long and has a cross-section of 2 sq in. The material of bar A has a stress–strain curve whose equation is

$$\sigma = 15 \times 10^6 \epsilon$$

That is to say, it has a constant E of 15×10^6.

The material of bar B has a stress–strain curve consisting of two straight lines,

$$\sigma = 20 \times 10^6 \, \epsilon \qquad \text{when } \sigma < 6000 \text{ psi}$$

and

$$\sigma = 3000 + 10 \times 10^6 \, \epsilon \quad \text{when } \sigma > 6000 \text{ psi}$$

The two lines intersect at a point corresponding to a stress of 6000 psi.

(a) Find the stress in each material if the composite bar is subjected to an axial force of 14,000 lb tension (in this case the stress in each bar is less than 6000 psi).

(b) Find the stress in each material if the composite bar is subjected to an axial force of 56,000 lb tension (in this case the stress in each bar is greater than 6000 psi).

9.38   The frame shown in Fig. P9.16 (without the 12 kip load) experiences a rise in temperature of 60°F. If the bar ED is 10 ft long and the coefficient of linear expansion is 6×10^{-6} per °F, find the deformation of each bar and hence the vertical and horizontal displacements of point D.

9.39   Every bar in the frame of Fig. P9.39 has an area of $2 \, \text{in}^2$ and $E = 10 \times 10^6$ psi. Find the force induced in the bars by a temperature rise of 100°F. Take $\alpha = 8 \times 10^{-6}$ per °F.

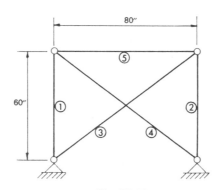

Fig. P9·39

9.40   The three bars AB, AC and AD in Fig. P9.40 are identical. They are made of bronze ($E = 15 \times 10^6$ psi and $\alpha = 10.2 \times 10^{-6}$ per °F) and are pinned at A, B, C and D. If the points B, C and D cannot move, find the stress in each bar when the temperature is raised 100°F.

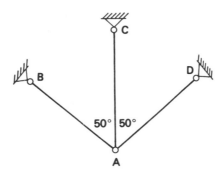

Fig. P9·40

9.41   A bar of zinc $\frac{1}{2}$ in diameter just fits inside a hollow steel tube of $\frac{3}{4}$ in diameter. The assembly is heated 50°C. Find the longitudinal stress in each material if there is no relative movement between the two.

$$E_s = 30 \times 10^6 \text{ psi} \qquad E_z = 15 \times 10^6 \text{ psi}$$

$$\alpha_s = 12 \times 10^{-6} \text{ per °C} \qquad \alpha_z = 40 \times 10^{-6} \text{ per °C.}$$

9.42   Two bars AB and BC of circular cross-section, are joined end on at B and are built into immovable supports at A and C. AB has a length of 18 in and a diameter of 1 in while BC has a length of 30 in and a diameter of $\frac{1}{2}$ in. AB is steel ($E = 30 \times 10^6$ psi and $\alpha = 12 \times 10^{-6}$ per °C) and BC is copper ($E = 15 \times 10^6$ psi and $\alpha - 15 \times 10^{-6}$ per °C).

Find the force in each bar and the movement of the point B when the temperature drops by 60°C.

9.43   A bar of material A is 8 in long and is joined end on to a bar of material B which is 12 in long. The whole bar is compressed to a stress of 7500 psi and then constrained between rigid supports 20 in apart. Each bar has a cross-section of 0.5 sq in.

The temperature of the bars is raised 80°C and it is required to find the stress in each bar after the rise in temperature.

The coefficients of expansion are $\alpha_A = 7 \times 10^{-6}$ and $\alpha_B = 12 \times 10^{-6}$ per °C. The stress–strain curve of A is a straight line with a slope of 20×10^6 psi. In its elastic range, B has a modulus of 10×10^6 psi. It yields at a stress of 15,000 psi after which the stress–strain curve is horizontal.

What is the movement of the junction of the two bars during the heating process?

9.44 A concrete beam has four longitudinal bars embedded in it and arranged symmetrically in the cross-section. The net area of the concrete cross-section is 100 sq in and the area of each bar is 1 sq in.

(*a*) What stresses will be set up in the steel and concrete due to the drying shrinkage of the concrete if the free shrinkage strain is 0.0004? Take $E_s = 30 \times 10^6$ psi and $E_c = 4 \times 10^6$ psi.

(*b*) What end forces must be applied to the ends of the beam in order to restore the steel stress to zero? It is assumed that the steel and concrete strains are equal throughout.

9.45 The three bars in Fig. P9.45 each have a length L and a cross-section of 0.4 in². They are made of the same material, which is an elastic-plastic material with a yield stress of 60 ksi.

(*a*) Find the force in each bar when the 44 kip load is applied.

(*b*) Find the residual bar forces when the load is removed. The unloading curve for the material is parallel to the initial portion of the stress–strain graph.

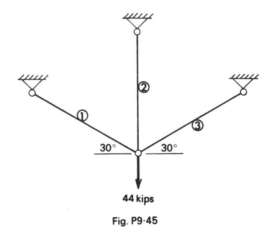

44 kips

Fig. P9·45

9.46 In the assembly shown in Fig. P9.14, each bar is 100 in long and has a cross-section of 1.5 in². The material has a stress–strain characteristic as shown in Fig. P9.9, the unloading curve being a straight line with a slope of 15,000 ksi.

A vertical load of 80 kips is applied at A and then removed. Find the bar forces (*a*) after the application of the load and (*b*) after its removal.

9.47 A beam AB (Fig. P9.47) is simply supported at A, and at B it is suspended from a bar BC. For the beam, $EI = 2400$ kip-ft² and for the bar $EA = 6000$ kips. The beam supports a uniformly distributed load of 1 kip/ft.

Find the vertical displacement of B and of the centre of the beam.

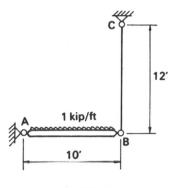

Fig. P9·47

9.48   The structure considered is similar to that of problem 9.47, but the beam is direction-fixed at *A*. The loading is 1 kip/ft as before.

Find the tension in the bar *BC* and the bending moment at the support *A*.

9.49   The two bars *AB* and *BC* (Fig. P9.49) are rigidly connected at *B*. For both members $EI = 2000$ kip-ft². Find the displacements of joints *B* and *C*. Neglect the axial deformation of *BC*.

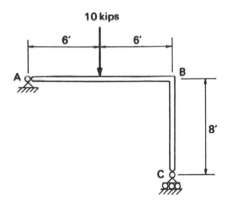

Fig. P9·49

9.50   A beam *AB* ($EI = 4000$ kip-ft²) is direction-fixed at *A* and is supported at *B* on a spring, the stiffness of which is 4 kip/ft. The beam is 24 ft long and supports a load of 8 kips at the mid-point, *C*.

Find the displacement at *B*, and the bending moment at *A* and *C*.

9.51 A beam ACB ($EI = 150,000$ kip-ft^2) is simply supported at A and B, and is supported at the mid-point, C, by a spring of stiffness 75 kip/ft. The beam is 40 ft long and carries a uniformly distributed load of 1 kip/ft.

Find the force in the spring and the bending moment at the centre of the beam. Sketch the bending moment diagram for the beam.

9.52 Two beams AB and CD (Fig. P9.52) are each 24 ft long and simply supported. They are identical in size and they are joined at their mid-point E. Beam AB carries a uniformly distributed load of 0.5 kip/ft.

Find the bending moment at the centre of each beam.

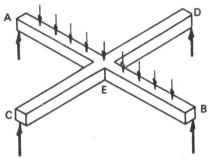

Fig. P9·52

CHAPTER 10

Instability

Any physical state or condition which is not easily changed is said to be *stable*. The usual test for stability is to cause a small disturbance of the existing state of the given system; then if the system resumes its original state when the cause of the disturbance is removed, we know that it is stable.

In the present context we are concerned with the stability of the state of equilibrium. A structure whose function is to withstand or resist loads, must do so with small and definite deformation. In previous chapters we have discussed the problem of finding a configuration of the loaded structure which satisfies the conditions of equilibrium, compatibility and the stress–strain relations of the materials. For the structure to be satisfactory it is necessary to test whether the shape so determined is stable, unless this fact is known from experience.

10.1 Stable, Unstable and Neutral Equilibrium

Suppose that a certain body or system of bodies is in an equilibrium position. We then give it a very small displacement in any direction. If, when the disturbing force is removed the body returns to the original position, then this position is stable. If it moves further away, the original position is unstable. It may neither return nor move further away. Then the original position is said to be one of neutral equilibrium.

A common example is that of a cone (Fig. 10.1), which, if stood on its base, is stable, but which if balanced on its point is unstable. If laid on its side on a flat surface its position is one of neutral equilibrium, since if it is slightly disturbed it merely stays in its new position.

Another illustration is that of a ball rolling on a track (Fig. 10.2). When the ball is at any position where the tangent to the track is horizontal, it is in equilibrium. If it is in a trough, C, and we disturb it slightly, it will return— the position is stable. If it is at the top of a hill, A, and we disturb it slightly,

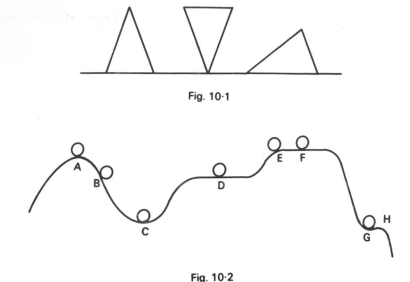

Fig. 10·1

Fig. 10·2

it will not return—it is unstable. If pushed to the left it will fall off the track altogether, while if pushed to the right it will eventually find a stable equilibrium position, *C*. Locations such as *D* or *F* are positions of neutral equilibrium.

Several points should be noted. The disturbance that we cause can be as small as we please, provided it is not zero. We push ball *G* to the right. If the displacement is sufficiently small it will return to *G* so this position is theoretically stable. However, if we push it over hump *H* it will not return. If an engineering structure behaved in this way it is doubtful if it could be considered satisfactory. We might describe it as "stable but not very stable". A theoretical state of unstable equilibrium cannot be realized in practice. If the cone of Fig. 10.1 has a truly sharp point we cannot stand it on its point in reality. If the point is flattened slightly we may be able to stand it up. It is then like the ball at *G*—theoretically stable, but not stable enough for practical purposes.

Moreover, the displacement can be made in any direction we please. If the system fails to recover from *any* such displacement it is not stable. The type of disturbance by which to test stability may therefore call for some intuitive judgement. In many cases this causes no difficulty. In Fig. 10.2, ball *E* is clearly unstable since it will fail to return if pushed to the left. If the track in this diagram is three-dimensional we should have to consider the possibility of displacements normal to the plane of the paper. In this event, even a position such as *C* might be unstable if this were a saddle-shaped depression.

Note that we are testing the nature of the *equilibrium* positions. Positions such as *B* are not equilibrium locations and are of no interest.

10.2 Pin-jointed Frames

Pin-jointed frames provide simple examples of engineering structures whose equilibrium configurations can be stable, unstable or neutral.

The frame of Fig. 10.3 (a) is symmetrical, and under the load system shown it is in equilibrium. Intuitively we recognize this as an unstable configuration. If we give the frame a slight displacement and then consider the member forces, we find that equilibrium cannot be satisfied at all joints, and, furthermore, the resultant force tends to increase the displacement.

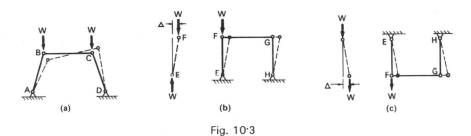

Fig. 10·3

If we specify that the pins at *A*, *B*, *C* and *D* permit joint articulation only in the plane of the frame, then we could orient the frame *A* in a horizontal plane and apply vertical loads at joints *B* and *C*. In this case the frame is in neutral equilibrium, for if it is slightly displaced, the new configuration is also one of equilibrium.

The frame of Fig. 10.3 (b) is also in equilibrium under vertical loads. If we displace *F* by a small amount *Δ* towards the right, no horizontal reaction is induced at *H*. Consequently, the force in bar *FG* is still zero. The forces on bar *EF* therefore constitute a couple, *WΔ*, which tends to rotate the bar further from its original position. This bar suffers a *deforming force* (the couple *WΔ*) tending to increase the displacement, but there is no counteracting *restoring force*. Hence the frame will collapse.

It is clear that it will collapse to the position shown in Fig. 10.3 (c). If we examine the stability of this configuration by giving it a small displacement, *Δ*, towards the right, we find as before that no force is induced in bar *FG*. Bar *EF* is now subjected to a *restoring* couple *WΔ* and no *deforming* couple. Consequently, the frame reverts to its original position. If we displace the frame to the left, we find that, again, the restoring force predominates. We conclude that the configuration is stable, a fact which was intuitively obvious. Notice that to prove the *instability* of frame *B* it was necessary to find only one disturbance from which it did not recover. To prove the *stability* of frame *C* it was necessary to check that it would recover from all disturbances. (Strictly speaking we have not tried all disturbances, but those which induce bar forces will always exhibit stability provided the stresses are not too high.)

Frames of the type just considered are called mechanisms. In such frames there are certain changes of geometry which do not produce changes in the bar forces. Mechanisms are required for systems in motion, such as machine parts, but are unsatisfactory as structures to resist loads. A frame of this type will offer no resistance to loads in general, so that for most load systems the question of equilibrium does not arise. It will support special load systems, and with these the equilibrium may be either stable or unstable.

It is interesting to examine the total energy in the frames of Figs 10.3 (b) and (c), i.e., the sum of the internal strain energy and the gravitational potential energy of the load *W*. In each case, the change in the strain energy due to the very small displacement is negligibly small. The main variation is the change of potential energy of *W*. In the case of the frame of Fig. 10.3 (b) this energy decreases when the frame is displaced, the decrease being equal to *W* times the downward displacement of *F*. The configuration shown is one of maximum energy. In the case of the frame of Fig. 10.3 (c) the potential energy of *W* increases whichever way the frame is displaced, and this represents a configuration of minimum energy.

A position of minimum energy is always stable. This is understandable because for such a system to move, energy must be supplied to it from elsewhere. A position of maximum energy, however, is always unstable. A system in a state of maximum potential energy can move simply by giving up some of its energy. Frequently, a consideration of the total potential energy of a system is a convenient method of investigating stability.

10.3　Critical Loads

The frame of Fig. 10.3 (b) is in unstable equilibrium irrespective of the value of *W*. This is because the frame offers no resistance to change of geometry, and hence when a small displacement is made, no restoring force whatever is brought into play. In a load resisting structure, deformation will produce a restoring force system, and then the question is whether the restoring forces exceed the deforming forces or not. If they do, the configuration is stable, if not, it is unstable. Usually the structure proves to be stable for low loads, but exhibits instability when the load reaches a critical value.

The two bar frame, or toggle, shown in Fig. 10.4 provides a convenient illustration. For a small value of *W* the configuration shown is stable, but when *W* reaches a certain value the structure will "snap through". That is to

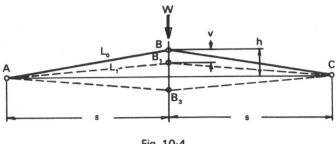

Fig. 10·4

say it will change suddenly from a position B_1 to a position such as B_3. We can call this value of W the *critical load*.

To find the critical load we first determine the relationship between W and the vertical deflection, v, of the pin B. It will be assumed that the stresses in the bars remain within the proportional range, and that no other form of instability occurs before the snap through. It is also assumed that displacement normal to the plane of the paper is prevented.

In terms of the horizontal projection, s, which is constant, the bar lengths are

$$L_0 = (s^2 + h^2)^{1/2}$$

and

$$L_1 = \{s^2 + (h - v)^2\}^{1/2}$$

When the deflection of B is v, the decrease in length, e, of AB is

$$e = L_0 - L_1$$

$$\text{compressive strain in } AB = \epsilon = \frac{e}{L_0} = 1 - \frac{L_1}{L_0}$$

$$\text{compressive stress in } AB = \sigma = E\left(1 - \frac{L_1}{L_0}\right)$$

$$\text{force in } AB = N = EA\left(1 - \frac{L_1}{L_0}\right)$$

$$\text{vertical component of } N = N\left(\frac{h - v}{L_1}\right) = AE(h - v)\left(\frac{1}{L_1} - \frac{1}{L_0}\right)$$

$$W = 2EA(h - v)\,[\{s^2 + (h - v)^2\}^{-1/2} - \{s^2 + h^2\}^{-1/2}] \qquad (10.1)$$

Note that it was necessary to base the equilibrium on the geometry of the deformed shape, and not on the original geometry, since the distance $(h - v)$ is significantly different from h.

Equation (10.1) corresponds to the graph shown in Fig. 10.5 (a). Every point on this graph represents an equilibrium configuration. Any point on the part 01 of the graph represents a stable configuration. Consider the point K which corresponds to load W_1. If we increase v by Δ (i.e., push the bar downward), the bar forces increase and produce an upward component on pin B equal to W_2. This component is the restoring force, while the load W_1 is the

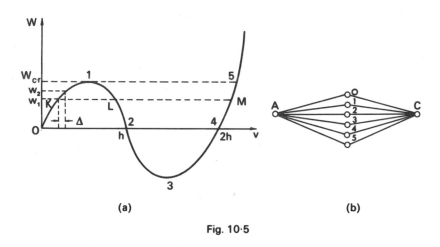

Fig. 10·5

deforming force. Since $W_2 > W_1$, the pin B will return to its original position corresponding to point K on the graph. Similarly, if the pin B is displaced upwards by Δ the component of the bar forces would decrease. In this case, this component is the deforming force, while the load W_1 is the restoring force, so that, once again, B would return to the undisturbed position. The stability of the system is therefore a direct consequence of the positive slope of the $W-v$ graph. For the time being we are ignoring other possible forms of deformation.

In a similar manner it can be shown that where the slope of the graph is negative, the points correspond to unstable configurations. Hence the critical load, W_{cr}, is the load at the turning point 1 of the graph. This can easily be found from equation (10.1). With the load equal to W_{cr} (point 1), if the deflection is slightly decreased the system will recover, but if the deflection is increased it will not recover. Thus the condition is unstable. Point 1 represents the first unstable position, and point 3 the last. All positions in between are unstable.

Fig. 10.5 (b) shows the geometry of the structure corresponding to the various points on the graph. At point 2, $v = h$ and the bars are in a straight line. They have considerable forces in them but these forces have no vertical component, hence $W = 0$. Position 4 is the unstressed position similar to position O. Theoretically, any position between 1 and 3 can be maintained in equilibrium by a suitable value of W, but, as explained previously, such unstable equilibrium states cannot be realized in practice. For an equilibrium position such as L (Fig. 10.5 (a)), the slightest disturbance would cause the geometry to change either to that at K, or to that at M.

If the load W is increased uniformly from zero, the structure will deform until it reaches the position 1 ($W = W_{cr}$) when it will snap through to position 5, in which both bars are in tension. For higher loads, the structure always takes up a stable shape unless the *stresses* reach a critical value.

Let us examine what happens to the total energy of any system when a small disturbance occurs. We assume that the load deformation graph is like the one in Fig. 10.6, and, in fact, the system could be the one previously discussed. For simplicity we can also assume that the load W is a gravitational load acting downward, and that v is a downward deflection. The internal energy in the structure, if it is elastic, is represented by the area under the W–v graph. The loss of potential energy of the load W for any deflection v is Wv.

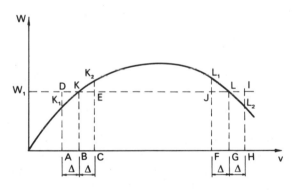

Fig. 10·6

Consider first the equilibrium position K with load W_1. If we displace the system upward by Δ to K_1 the load W_1 increases its potential energy by $W_1 \Delta$ (rectangle $ABKD$) and the internal energy decreases by the area $ABKK_1$. There is, therefore, a net increase of energy in the whole system equal to the area $KK_1 D$. Now if we displace the system downward by Δ to K_2, the load W_1 loses potential energy $W_1 \Delta$ (rectangle $BCEK$) and the internal energy increases by the area $BCK_2 K$. Again, therefore, there is a net increase of energy in the whole system equal to $KK_2 E$. In each case energy must be supplied to enable the system to move, so we conclude that the structure would not make the movement unaided. Furthermore, since the system is conservative, the stored energy will be greater in the disturbed position, and this excess energy will restore the structure to the position K upon removal of the disturbing agent. The position K is one of minimum energy for the load W_1 and is a position of stable equilibrium.

The situation is quite different if we consider the equilibrium position L with the same load W_1. An upward displacement of Δ to L_1 results in an increase of potential energy of W_1 equal to $W_1\Delta$ ($FGLJ$), and a decrease in the internal energy equal to the area $FGLL_1$. There is, thus, a net decrease of energy equal to LJL_1. Similarly, a downward displacement of Δ results in a decrease of energy equal to LIL_2. The position L is one of maximum energy for the load W_1 and is a position of unstable equilibrium.

If, as in earlier chapters, we consider the slope of the $W - v$ graph to be a stiffness coefficient k, then a displacement increment of Δ is accompanied by a force increment of $k\Delta$. The change in the total potential energy of the system is $\frac{1}{2}k\Delta^2$. This value corresponds to the little triangles $KK_1 D$, $KK_2 E$, $LL_1 J$ or $LL_2 I$ discussed in the previous paragraphs. The term $\frac{1}{2}k\Delta^2$ will be positive or negative according to the sign of k, which again shows that the structure is stable where the slope of the graph is positive, and unstable where the slope is negative. The stiffness, k, represents the resistance of the structure to change of configuration. The equilibrium will thus be stable if k is positive, neutral if k is zero and unstable if k is negative.

It might be noted that if the system is not conservative, for instance, if the stresses are inelastic, the change of internal stored energy may be less than the energy supplied to the structure, that is, less than the corresponding area under the graph. In this event the structure may not return from the disturbance of K to K_2, and K could represent a position of neutral equilibrium. On the other hand, the position L is certain to be unstable, since a disturbance of L to L_2 is certain to result in a decrease of energy at least equal to LIL_2.

As an example of inelastic instability we might consider a simply supported beam carrying a central load W (Fig. 10.7), and made of material which has a stress–strain curve such as that of Fig. 10.8 (a). In the first instance we think of conditions in the elastic range. The internal bending stresses are tending to restore the beam to the straight position. This tendency is balanced by W.

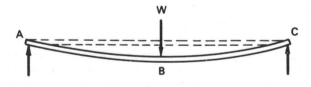

Fig. 10·7

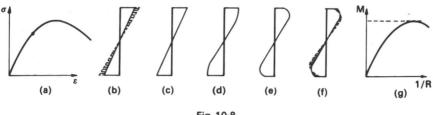

Fig. 10·8

If we push the beam down slightly the internal stresses increase while the external bending moment due to the original loads remains the same. Hence, when the disturbing force is removed the beam returns to its former position. Conversely, if we push the beam upward the stresses decrease and again the beam reverts to the equilibrium position. The equilibrium is stable.

It is seen that the recovery of the beam from a downward disturbance depends on the fact that this displacement is accompanied by an increase in the internal bending moment. When the shape of the cross-section is known, the bending moment formed by the stress increments (shaded) is easily computed, and this is the restoring bending moment. Suppose the material of the beam has a stress–strain law as in Fig. 10.8 (a). Then, as loading progresses, the stress-distribution will pass through the phases as shown in Fig. 10.8 (c), (d) and (e). If now a downward disturbance is made, the stress at the outer fibres will decrease. The stress variation is shown shaded in Fig. 10.8 (f). A stage will be reached, depending on the shape of cross-section, at which the incremental bending moment will be negative. In other words, a greater deflection will produce a smaller resisting moment. The beam is then unstable.

For successive values of the curvature (ρ) we could compute the internal couple and obtain the graph of Fig. 10.8 (g). The maximum value M_{cr} marks the onset of instability. From this, the maximum load capacity of the beam, for a given load system, can be calculated.

A beam made of a yielding, and strain hardening material (Fig. 10.9 (a)),

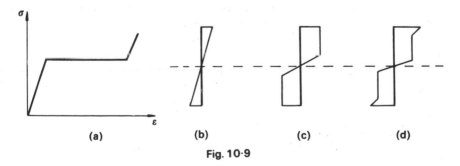

Fig. 10·9

would pass through a phase of practically neutral equilibrium (Fig. 10.9 (c)). At increased values of deflection it may reach a state of stable equilibrium again (Fig. 10.9 (d)) where increased deflection once more produces an increment of stresses.

10.4 Bifurcation Problems

A somewhat different situation arises when we consider the two-bar structure of Fig. 10.10 (a). Joints A, B and C are in a straight line, and the external force

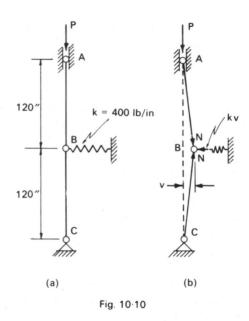

(a)   (b)

Fig. 10·10

P acts along this line, which we assume to be vertical. If it were not for the spring the structure would simply be a mechanism and would be unstable for any value of P. The spring imparts a certain degree of stability, and for moderate values of P it causes the bars to return to the vertical position after a small disturbance. For simplicity it will be assumed that displacement normal to the plane of the paper is prevented.

To test the stability we displace joint B by a small distance v (Fig. 10.10 (b)) and consider the equilibrium of joint B. The bar forces N ($= P$) each have a horizontal component which tends to displace B further. Thus, since v is very small,

$$\text{the total deforming force} = 2P \times \frac{v}{120} \text{ lb}$$

We note that the deforming force is proportional to P.

A restoring force is exerted by the spring, and since the spring stiffness is 400 lb/in,

$$\text{the total restoring force} = kv = 400v \text{ lb}$$

We note that the restoring force is independent of P.

Since the deforming force increases with P while the restoring force does not, there will be a critical value of P ($= P_{cr}$) at which the structure is in equilibrium in the deformed state. This will occur when the restoring force is equal to the deforming force. Then

$$2P_{cr}\frac{v}{120} = 400v$$

and

$$P_{cr} = 24{,}000 \text{ lb}$$

At this load the structure is in neutral equilibrium. For higher loads it is unstable.

We may plot a graph of P against v for all equilibrium positions (Fig 10 11). The graph comprises all points along the P axis, which correspond to the straight configuration, and also points along FGH. We do not draw FGH very long because we have assumed that v is small when computing P_{cr}. The points along OG correspond to positions of stable equilibrium, those along FGH to neutral equilibrium, and those above G to unstable equilibrium.

Only at the value of $P = 24{,}000$ lb is it possible to find equilibrium positions other than the straight one. This value is a "critical load".

In this problem, the points on the graph of Fig. 10.11 which represent equilibrium configurations lie on two distinct lines. The graph forks at the

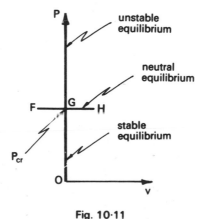

Fig. 10·11

critical load. Because of this, problems of this type are sometimes called *bifurcation problems*.

In the foregoing problem, there is only one kind of configuration other than the straight one, and the structure is said to have one degree of freedom. If we take a similar structure with two hinges (Fig. 10.12 (b)), two different configurations are possible, and we shall find that two critical loads occur.

To make a change we introduce hinge springs which resist any angular change between the adjacent bars. If a more physical device is desired, one may imagine the mechanism of Fig. 10.12 (a). The stiffness of the hinge spring is given as the internal bending moment it will exert per unit angular deformation.

To test the stability of the structure we give it a small deformation. The structure now has two degrees of freedom, and it is necessary to define the

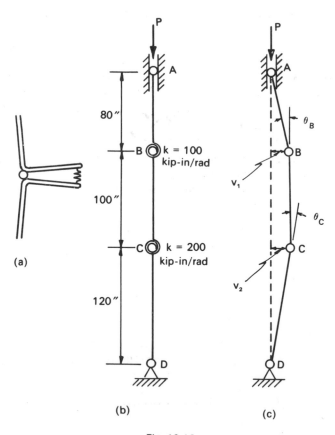

Fig. 10·12

displacement of B ($= v_1$) and the displacement of C ($= v_2$) in order to specify the deformed configuration, which is shown in Fig. 10.12 (c).

There is now an angle change at B and at C. The inclinations of AB, BC and CD to the line AD are respectively $v_1/80$, $(v_2 - v_1)/100$ and $-v_2/120$. Consequently, the angle changes are

$$\theta_A = \frac{v_1}{80} - \frac{v_2 - v_1}{100} = \frac{9}{400} v_1 - \frac{1}{100} v_2$$

and

$$\theta_B = \frac{v_2 - v_1}{100} + \frac{v_2}{120} = -\frac{1}{100} v_1 + \frac{11}{600} v_2$$

At B, the moment of the external forces, which tends to increase the angle change, is Pv_1. The spring, on the other hand, exerts a moment of $k_B \theta_B = 1000 \theta_B$ which is tending to close the angle. The joint B will be in equilibrium if

$$1000 \theta_B = Pv_1$$

Similarly, the deforming moment at C is Pv_2, while the restoring moment is $2000 \theta_C$, and equilibrium requires that

$$2000 \theta_C = Pv_2$$

When the values of θ_B and θ_C are substituted, the two equilibrium conditions become

$$100 \left(\frac{\cdot 9}{400} v_1 - \frac{1}{100} v_2 \right) = Pv_1$$

$$200 \left(-\frac{1}{100} v_1 + \frac{11}{600} v_2 \right) = Pv_2$$

or, after re-arrangement

$$\left(\frac{9}{4} - P \right) v_1 \qquad\qquad - v_2 = 0$$

$$-v_1 + \left(\frac{11}{6} - \frac{P}{2} \right) v_2 = 0 \qquad (10.2)$$

Now for most values of P these two homogeneous equations are only satisfied by the values $v_1 = v_2 = 0$, which correspond to the undeformed structure. However, at values of P which make the determinant of coefficients zero, the equations can be satisfied by non-zero values of v_1 and v_2. This signifies that at these special values of P, equilibrium in a deformed shape is possible. Equating the determinant to zero we have

$$\begin{vmatrix} \left(\dfrac{9}{4}-P\right) & -1 \\ -1 & \left(\dfrac{11}{6}-\dfrac{P}{2}\right) \end{vmatrix} = 0$$

or

$$12P^2 - 71P + 75 = 0 \qquad (10.3)$$

Equation (10.3) is the *characteristic equation* of this structure. Solution of this gives

$$P = 1.38 \text{ or } 4.54 \text{ kips}$$

Notice that we now have two critical values of P because the structure has two degrees of freedom. Each value of P_{cr} corresponds to a particular type of deformation, i.e., to a particular ratio of v_1 to v_2. These critical values, which are a feature of a set of homogeneous equations, are called *characteristic values* or *eigen values*.

The deformed shape, or *characteristic mode* of deformation, which corresponds to each of the critical values of P, can be determined by the substitution of P into equations (10.2).

(a) The first characteristic value, $P_{cr} = 1.38$, is substituted into equations (10.2)

$$\left(\dfrac{9}{4} - 1.38\right)v_1 \qquad\qquad - v_2 = 0$$

$$-v_1 + \left(\dfrac{11}{6} - 0.69\right)v_2 = 0$$

Each of these equations yields the relation

$$v_2 = 0.87v_1$$

which shows that the equations are satisfied by these values when $P = 1.38$. The structure is in neutral equilibrium so the actual value of v_1 is immaterial provided it is small. The displacement v_2 must be $0.87v_1$ (Fig. 10.13 (a)).

(b) The second characteristic value, $P_{cr} = 4.54$, is substituted into equations (10.2).

$$\left(\dfrac{9}{4} - 4.54\right)v_1 \qquad\qquad - v_2 = 0$$

$$-v_1 + \left(\dfrac{11}{6} - 2.27\right)v_2 = 0$$

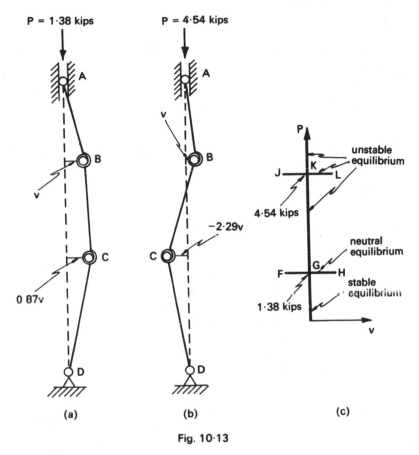

Fig. 10·13

Each of these equations yields the relation

$$v_2 = -2.29v_1$$

The negative sign indicates that the equilibrium configuration which corresponds to $P = 4.54$ requires the displacements v_1 and v_2 to be in opposite directions (Fig. 10.13 (b)).

It should be noted that each critical load or characteristic value is associated with a particular shape of deformation or *characteristic shape*. For instance, when $P = 4.54$, the structure will be in equilibrium only in the shape of Fig. 10.13 (b). It will not be in equilibrium if deformed according to Fig. 10.13 (a). Moreover the shape of Fig. 10.13 (b), although satisfying equilibrium, is not a stable configuration.

Fig. 10.13 (c) shows a graph of lateral displacement of joint B (or joint C) corresponding to equilibrium positions of the structure. All points on the P axis represent equilibrium, and so, also, do points on the branches FGH and JKL. All points above G, however, represent *unstable* equilibrium.

Bifurcation of the equilibrium graph will always occur when the problem gives rise to a set of homogeneous equations. A bifurcation occurs at each characteristic value.

It will be appreciated that in a bar with additional hinges there will be another equilibrium equation for each additional hinge, the characteristic equation will increase by one degree, and there will be one more characteristic value.

The structures studied in this section were assumed to possess a mathematical perfection which could not be realized in practice. In both problems the bars were assumed to lie in a perfectly straight line with the external loads applied exactly along this line. The slightest deviation from either of these conditions would take the problem out of the class of bifurcation problems and the characteristic values would not occur.

The true bifurcation problem is a mathematical abstraction, but the lowest critical load provided by this idealization does represent an upper bound to the capacity of the similar, but slightly imperfect, practical structure. This question will be referred to again in the next section.

10.5   Elastic Instability of Bars in Compression

We consider now a single bar subjected to compressive forces whose line of action does not coincide with the bar axis but has a small eccentricity, e, therefrom. If the bar is fairly slender, its lateral deflection will have an appreciable influence on the value of the bending moment at a given point along the bar. In other words, the bending moment must be calculated from the geometry of the deformed bar, not from the original geometry.

Fig. 10.14 (a) shows such a member AB, where the eccentricity of the load is e_A at A and e_B at B. If y (and, consequently, the deflection v) is taken positive upward, then the bending moment at a typical point is $-P(e + v)$, where v is at present unknown. From Chapter 4 we know that if the stresses are in the proportional range, and if we make the small deflection approximation $(1/R = d^2v/dx^2)$, the internal bending stresses form a couple equal to $EI(d^2v/dx^2)$, therefore we have

$$\text{the external moment } M_{\text{ext}} = -P(e + v)$$

and

$$\text{the internal moment } M_{\text{int}} = +EI\frac{d^2v}{dx^2}$$

(10.4)

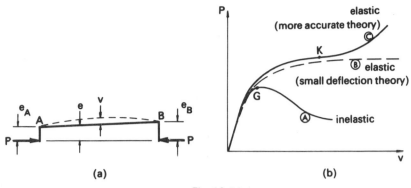

Fig. 10·14

Since $M_{ext} = M_{int}$ for equilibrium, we have two equations from which to find M and v which, as we have seen, are interdependent in this problem. Either variable can be eliminated. Since we require v at present, we eliminate M, and obtain

$$-EI\frac{d^2 v}{dx^2} = P(e + v)$$

or

$$\frac{d^2 v}{dx^2} + \frac{P}{EI}v = -\frac{P}{EI}e \qquad (10.5)$$

where e is a known function of x. If we let

$$\lambda = \sqrt{P/EI}$$

equation (10.5) becomes

$$\frac{d^2 v}{dx^2} + \lambda^2 v = -\lambda^2 e \qquad (10.6)$$

Suppose that e is constant; the solution of equation (10.6) is then

$$v = A \cos \lambda x + B \sin \lambda x - e$$

where A and B are to be found from the boundary conditions of the member. Taking the origin midway along the beam, we have

$$v = 0 \text{ when } x = -l/2 \qquad \therefore \quad 0 = A \cos \lambda l/2 - B \sin \lambda l/2 - e$$
$$v = 0 \text{ when } x = +l/2 \qquad \therefore \quad 0 = A \cos \lambda l/2 + B \sin \lambda l/2 - e$$

from which

$$A = e \sec \frac{\lambda l}{2} \quad \text{and} \quad B = 0$$

For any load P, the deflection curve is therefore

$$v = e \sec \frac{l}{2}\sqrt{\frac{P}{EI}} \cos x \sqrt{\frac{P}{EI}} - e$$

or

$$(v + e) = e \sec \frac{l}{2}\sqrt{\frac{P}{EI}} \cos x \sqrt{\frac{P}{EI}} \qquad (10.7)$$

At the centre, $x = 0$, the relation between v_0 and P is

$$(v_0 + e) = e \sec \frac{l}{2}\sqrt{\frac{P}{EI}} \qquad (10.8)$$

This expression plots as the dotted curve B in Fig. 10.14 (b). The tangent to this curve becomes horizontal when

$$\sqrt{\frac{P}{EI}} = \frac{\pi}{l} \quad \text{and} \quad v = \infty$$

However, this no longer represents our problem, since the solution is only valid for small values of v.

If, for slightly larger deflections, the correct expression for $1/R$ is used, i.e.,

$$\frac{1}{R} = \frac{d^2 v/dx^2}{[1 + (dv/dx)^2]^{3/2}}$$

it is found that the P–v_0 graph is as indicated by curve C in Fig. 10.14 (b). We see that at no stage does our member become unstable. Nevertheless, there is a certain value of P at which the rate of deflection with respect to load is quite large, i.e., dP/dv_0 is small, though not zero. For many practical purposes the member would be considered useless beyond this value of P.

If the member is fairly stiff, the stresses at the position of maximum bending may pass beyond the proportional range before K is reached. When this occurs the deflection begins to increase more rapidly as a result of inelastic straining, and the load–deflection graph then follows a curve such as curve A (Fig. 10.14 (b)).

With inelastic behaviour there is a definite value of the load (point G) at which instability occurs. But when the material remains elastic, true instability does not occur, and the location of point K, which is the limit of practical usefulness, is somewhat vague. In this regard it is helpful to consider the curves for different values of e, including the limiting case where $e = 0$ (Fig. 10.15).

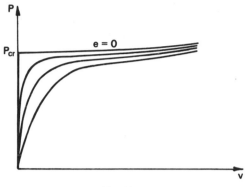

Fig. 10·15

We find that as value of e is made smaller, the curvature of the P–v graph becomes sharper. The graphs approach the heavy line (Fig. 10.15) in the limit as $e \to 0$. If we put $e = 0$ in equation (10.8) we see that $v = 0$ for most values of P, and this is reasonable since, with a centrally loaded bar, the straight position is clearly an equilibrium configuration. An exception may occur when

$$\sqrt{\frac{P}{EI}} = \frac{\pi}{l} \quad \text{because then} \quad \frac{l}{2}\sqrt{\frac{P}{EI}} = \frac{\pi}{2}$$

the secant term becomes infinite, and the right-hand side is therefore in-determinate. This value of P is called the critical load, P_{cr}.

The case of zero eccentricity is unique in several respects. Whereas the other graphs represent bars which can be realized in practice, the case of $e = 0$ is a theoretical concept since it is impossible to make a perfectly straight bar and to load it exactly along its axis. Moreover, for other values of e, the graph is a single smooth curve, but with $e = 0$ the graph is discontinuous, and has more than one branch. That is to say, for some loads it is possible to find more than one position of equilibrium. We see that the limiting case, with $e = 0$, falls into the category of bifurcation problems. The present problem is closely associated with those studied in section 10.4.

The mathematically straight (weightless) bar, loaded as shown in Fig. 10.16, was first examined by the mathematician Leonard Euler, and this problem in

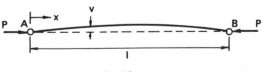

Fig. 10·16

therefore referred to as the Euler column problem. It is obvious that for all values of P the stresses might become excessive and the bar might fail by crushing. This will be ignored for the time being. The question of interest is—"Is there any equilibrium shape other than the straight one?"

Suppose that there is such a shape, and suppose it is such that at x from the end of the bar the deflection is v. Every element must be in equilibrium, and hence for any typical element the bending moment, $-Pv$, due to the end forces must be equal to the couple formed by the internal stresses. If we consider only small values of v, this couple is $EI(d^2v/dx^2)$. We must have

$$EI\frac{d^2v}{dx^2} = -Pv \qquad (10.9)$$

This is like equation (10.5) with $e = 0$. Re-arranging, as in the previous example, we obtain the equation

$$\frac{d^2v}{dx^2} + \lambda^2 v = 0 \qquad (10.10)$$

where $\lambda = \sqrt{P/EI}$. The solution of equation (10.10) is

$$v = A\cos\lambda x + B\sin\lambda x \qquad (10.11)$$

Suppose that x is measured from one end of the bar. The boundary condition $v = 0$ when $x = 0$ shows that $A = 0$. This signifies that a cosine curve will not satisfy the end conditions. The boundary condition $v = 0$ when $x = l$ yields

$$0 = B\sin\lambda l \qquad (10.12)$$

This is true either if $B = 0$ or if $\sin\lambda l = 0$. If we accept the first, we have lost both terms of equation (10.11). This signifies that $v = 0$ (the straight position) is an equilibrium configuration, which we already know.

If we accept that $\sin\lambda l = 0$, we find that this is possible only for certain values of λ, namely

$$\lambda = 0, \frac{\pi}{l}, \frac{2\pi}{l} \cdots \frac{n\pi}{l}$$

where n is any integer either positive, negative or zero. These isolated values of λ for which equation (10.10) can have a solution other than zero are the characteristic values or eigen values of the equation.

Since λ is a function of P, each characteristic value of λ corresponds to a particular value of P

$$\lambda = \sqrt{\frac{P}{EI}}$$

$$\therefore \quad P = EI\lambda^2$$

and the characteristic values of P are

$$P = 0, \; EI\frac{\pi^2}{l^2}, \; EI\frac{2^2\pi^2}{l^2} \ldots EI\frac{n^2\pi^2}{l^2}$$

The equilibrium shape corresponding to any characteristic value of P is called the characteristic function. For

$$P = EI\frac{n^2\pi^2}{l^2}, \quad \lambda = \frac{n\pi}{l}$$

and the deflected shape is, from equation (10.8)

$$v = B\sin n\pi\frac{x}{l} \tag{10.13}$$

This is a sine wave having a maximum amplitude B, and a node point at $x = 0$ and at $x = l$. It has n half waves in the length of the rod. This means that at a load $P = EI(n^2\pi^2/l^2)$ the bar is in equilibrium either if it is straight, or if it is bent into n half sine waves, the amplitude being arbitrary. The amplitude can only be small because this result is derived from small deflection theory. If the critical value of P corresponding to $n = 0, 1, 2\ldots$ are denoted by P_0, $P_1, P_2\ldots$ the graph of deflection (at mid-wave) against load is as shown in Fig. 10.17. Every point on the P axis is an equilibrium position. Each branch

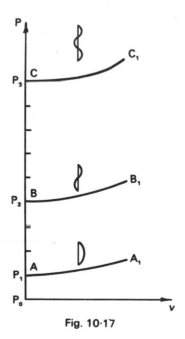

Fig. 10·17

corresponds to a specific shape, i.e., to a specific characteristic function. The above calculations show the branches to be horizontal, but when v ceases to be small, more accurate theory indicates that the branches curve upward.

The significance of the branch at $P = 0$ can be deduced by the reader. It should be noted that the half wavelength is l/n and, in the foregoing work, the functions have not been restricted to the range $0 < x < l$.

We have now to determine whether any of these equilibrium positions are unstable. Considering, first, the straight configuration, we test it by deforming it slightly into a half sine wave $v = v_0 \sin \pi x/l$. The restoring forces are the internal moments set up by the bending stresses. These depend only on the deformation we have imposed, and are independent of P. The forces tending to increase the deformation are the bending moments Pv, which clearly depend on P. We know that the two sets are equal when $P = P_1 (= \pi^2 EI/l^2)$. Hence

for $P < P_1$ the restoring forces > deforming forces

for $P = P_1$ the restoring forces = deforming forces

for $P > P_1$ the restoring forces < deforming forces

Thus the straight position is stable for $P < P_1$ and unstable for $P > P_1$. For $P = P_1$ the equilibrium is neutral. The investigation of the stability of positions represented by the branch lines can be undertaken in the usual way, that is, by causing a slight disturbance and testing the relative magnitude of the restoring and deforming forces. Since the small deflection theory of bending is inadequate for this test, the algebra will not be carried out. We should find that points on the first branch, AA_1 (Fig. 10.17), represent positions of stable equilibrium (or neutral equilibrium close to the P axis), while points on all the other branches correspond to positions of unstable equilibrium.

The portion OAA_1 of this graph exhibits properties similar to those of the previous examples. Where the slope of the graph is positive, the equilibrium is stable, and where the slope is zero, the equilibrium is neutral. The other branches appear to violate these rules since the equilibrium is unstable although the slope is nowhere negative. However, it must be noted that each branch of the graph is only valid for one *shape* of deformation. Thus the branch BB_1 represents equilibrium positions provided the rod is bent into two half waves with a node at B (Fig. 10.18). If we make a disturbance which retains the node at B, then the bar will recover, and the positive slope of the graph signifies stability in this very limited sense. However, for real stability we must be free to make *any* disturbance, and if we displace the mid-point, B, the configuration is found to be unstable.

Fig. 10·18

The reader will probably have noticed the similarity between the problem just studied, and the strut with two hinges examined in section 10.4 (see Fig. 10.12). The hinge springs of the previous problem are similar in function to the internal bending stresses in the present case, namely, they tend to straighten the bar when it is bent. The two-hinged strut had two degrees of freedom, yielded two simultaneous equilibrium equations (10.2), which in turn gave rise to two characteristic values. The continuous Euler strut has an infinite number of degrees of freedom. It yields the differential equation (10.10) which in turn gives rise to an infinite number of characteristic values. In both cases, each characteristic value is associated with a particular deformed shape, and if the two shapes of Fig. 10.13 are examined, their general resemblance to a half sine wave, and full sine wave, respectively, will be noticed

The study of this rather abstract problem, namely the axially-compressed perfectly straight rod, has provided us with the curve which represents the limiting case, as $e \to 0$, of the family of graphs of Fig. 10.15. It yields, also, the value of the critical load at which finite deflections occur for no load increase. This critical load, P_{cr} $(= \pi^2 EI/L^2)$ is called the Euler buckling load. Real compression members, or struts, which are nearly straight and nearly axially loaded, will follow one of the graphs close to this limit line. If such a strut is tested in compression, it will exhibit very little deformation until the load is approximately equal to P_{cr}, and then the lateral displacement will increase very rapidly for practically no load increase. For further increases of load, the lateral deflection takes place less rapidly as indicated by the upward trend of the graph. At the critical load, the strut is said to "buckle", and although it is not truly an instability phenomenon it has a similar appearance.

10.6   Other Characteristic Value Problems

Many other structural problems find a solution by means of bifurcation theory. In many cases a critical load occurs as a result of compressive stresses in some region of the structure. We have examined a strut which is pinned at each end. Struts with other end conditions can be investigated in a similar way. They obey a differential equation having the same form as equation (10.5), but the boundary conditions are different and this results in different values for the critical load for the different problems (Fig. 10.19).

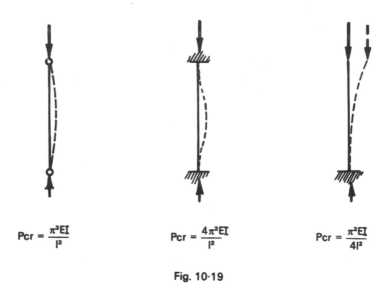

$$P_{cr} = \frac{\pi^2 EI}{l^2} \qquad P_{cr} = \frac{4\pi^2 EI}{l^2} \qquad P_{cr} = \frac{\pi^2 EI}{4l^2}$$

Fig. 10·19

A beam, whose depth is considerably greater than its width, may fail by lateral instability. Fig. 10.20 (a) shows a deep cantilever. Fig. 10.20 (b) shows a deep beam simply supported at the ends and also provided with lateral supports at the ends, so that the end cross-sections cannot twist. If we give either of these beams a small deformation such that the cross-sections are laterally displaced, then for loads above a critical value, W_{cr}, the beam will not recover from this displacement. If this is so, we shall have found an equilibrium position other than the vertical one.

To set up the necessary equation for calculating W_{cr} we note that for a given cross-section in the deformed position of the beam, the load has a transverse component which causes lateral bending, and it also has a moment about the centroid of the cross-section which causes torsion. Each of these effects tends to increase the deformation, and these deforming forces vary from

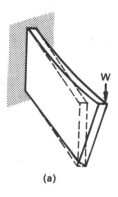

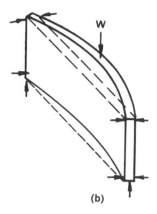

Fig. 10·20

one cross-section to another. On the other hand, the deformation has induced stresses which are tending to restore the beam to the vertical. The lateral curve in the centroidal axis involves lateral bending stresses. The variation of twist from one section to another involves torsion stresses. At W_{cr} it is possible to find a deformed position such that restoring and deforming forces are balanced both as regards lateral bending and torsion. The algebra is more complicated than for the Euler column and will not be discussed in detail.

A thin plate in shear (Fig. 10.21 (a)) can buckle into waves having crests parallel to the lines of principal tension. The buckling is essentially caused by

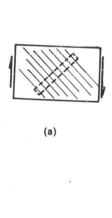

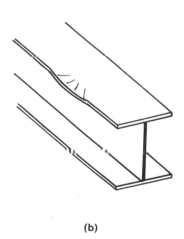

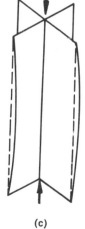

Fig. 10· 21

the principal compressive stresses. An element of the plate, such as the dotted one shown, is behaving somewhat like an Euler column, although the boundary conditions are more complex. The compressive flange of a plate girder (Fig. 10.21 (b)), if made too thin, will buckle at the unsupported edges. Very similar is the behaviour of a column built up of thin plates as shown in Fig. 10.21 (c). Here, again, the unsupported edges of the plates will tend to buckle. We might consider the buckling of an individual fin of this column, and thereby compute a critical load. However, if we assume that all four buckle together, producing a bodily rotation of the mid-height cross-section, we find that the restoring forces are then less than before, and we arrive at a smaller critical load. It will be the *smallest critical load* that we can find which will control the strength of the member.

10.7   Primary and Secondary Deformation

The instability phenomena which have so far been investigated do not seem to be entirely similar in regard to the characteristics which they exhibit. Although the same basic criterion of instability has been applied to each of the foregoing problems, the behaviour pattern of the various structures appears to vary considerably. We can distinguish three main types:

1. The structure is inherently unstable at all loads.
2. The structure possesses a load–deformation graph which is a smooth curve, and instability commences at a deformation corresponding to a maximum in this curve.
3. The structure possesses a load–deformation graph with a number of branches, only one of which represents stable configurations. The branch points correspond to eigen values, and the deformation is zero for loads below the first eigen value.

Why do these problems differ in this way? The first group are usually not called *structures*, they are mechanisms. That is to say, they can undergo certain types of deformation without stresses being induced. In structures of groups 2 and 3, on the other hand, any deformation is accompanied by a change of stress.

When a structure such as this is loaded, it deforms, and although we frequently consider only one component of its deformation, in reality it probably suffers several different components simultaneously. Take the simple beam of section 10.3 as an example. Under vertical load it obviously deflects vertically. This we might call the *primary deformation*. However, unless it is perfectly straight, with the principal axis of its cross-sections perfectly vertical, and the load perfectly centred on this axis, and so on, it will, to some extent, bend laterally and twist. These we shall call *secondary deformations*. In the idealized problem studied, secondary deformations do not occur, but they will occur in any real problem.

We then see that in section 10.3 we studied the variation of primary deformation with load. For every load there is a finite primary deformation, and this results in smooth graphs such as those of Figs 10.5 and 10.8 (g). A maximum in these graphs represents instability. A strut with a load deliberately off-line (Fig. 10.14) will clearly bend, and the lateral deflection is a primary deformation. The graph of Fig. 10.14 (b) is of the same type as before.

If the strut is axially loaded, the only primary deformation is axial shortening. Lateral deflection is a secondary deformation, and in an idealized example it will not occur. However, it is necessary in testing for stability to consider a disturbance in the direction of any secondary deformation. Such a test gives rise to a characteristic value problem. The centrally loaded beam of Fig. 10.20 (b) is tested for stability by applying a *lateral* disturbance. If pursued in detail, this test would give rise to a homogeneous differential equation. This equation has characteristic values, and for a load corresponding to the first of these values, a lateral deflection can be maintained, *even though the beam were perfectly constructed*. It may be, of course, that the beam would fail in vertical deflection (primary deformation) before this load was reached.

Characteristic value problems result from the investigation of an idealized situation. However, they are of great importance since the real situations can approach this ideal closely, as we have seen in the example of the Euler strut (Fig. 10.11).

In the investigation of the toggle (Fig. 10.4), only the primary deformation— axial shortening of the bars—was taken into account. From this a certain critical load was calculated at which instability occurred. However, instability due to a secondary mode might have occurred at a lower load. In particular, if the bars of the toggle were slender they might have suffered Euler buckling before the snap through load was reached (see problem 10.10 at the end of the chapter).

For any given load system there is usually one specific mode of primary deformation. Frequently there are a number of possible modes of secondary deformation. In the examples investigated in this chapter, the secondary deformation mode which would lead to the smallest critical load was selected by intuition. Where the selection of the "right" type of deformation is not obvious, it becomes necessary to investigate every possible mode for instability, and thus to evaluate the lowest of the critical loads.

10.8   Inelastic Instability of Struts

Neither primary nor secondary instability is confined to the proportional range of stress, although, as a rule, the calculations become more difficult when the stress–strain relation is non-linear. Previously, in section 10.3, we discussed qualitatively the instability of a beam in bending, the instability being that of vertical deflection. As an instance of instability of a secondary deformation, when the stresses are in the non-linear range, we take the inelastic buckling of a straight axially loaded column. In effect, it is the same as the Euler column problem, except that the stress at the critical load is beyond the proportional limit.

As before, to test for instability we introduce a small lateral displacement into the loaded column, and discuss whether or not it will recover. Specifically, the stress existing in the column just prior to the test does not enter the calculation. It is the *incremental stresses* caused by the bending that are of interest. Suppose the axial load produces a stress σ_1, which is below the proportional limit (Fig. 10.22 (a)). The small curvature introduced by our lateral displacement will cause a small increase of the strains on one side, and a small decrease of those on the other. The change of stress at any point is E times the change of strain; the incremental stresses are shown shaded in Fig. 10.22 (a) and they form a couple equal to $EI(d^2v/dx^2)$.

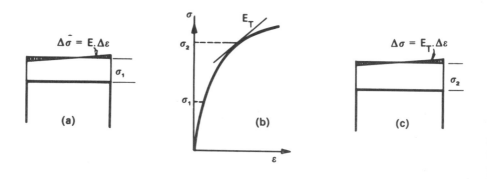

Fig. 10·22

When the load produces a stress σ_2 which is above the proportional limit (Fig. 10.22 (c)), the incremental strains introduced by the test displacement are exactly the same as before. The stresses, however, are not. Since the stress varies only slightly from σ_2 we can take it that the change stress at any point is E_T times the change of strain, where E_T is the tangent modulus at the stress σ_2 (Fig. 10.22 (b)). The incremental stresses (Fig. 10.22 (c)) will thus form a couple equal to $E_T I(d^2v/dx^2)$. The investigation then follows on exactly the same lines as in the Euler problem (see equation 10.9). Clearly, this will lead to a critical value of P given by

$$P_{cr} = \frac{\pi^2 E_T I}{l^2} \tag{10.14}$$

This is known as the tangent modulus formula, and it was first proposed by Engesser in 1889. In this form it does not give a specific value for P_{cr}, since E_T is itself a function of P. Hence we might re-arrange the equation as

$$\frac{P_{cr}}{E_T} = \frac{\pi^2 I}{l^2} \tag{10.15}$$

This gives a specific value for P_{cr}/E_T, and from this P_{cr} can easily be found by a suitable re-plotting of the stress–strain graph, or the relevant part of it. For instance, we might plot σ against σ/E_T.

The *tangent modulus* theory of Engesser assumes that in the region of σ_2 a strain increase $\Delta\epsilon$ is accompanied by a stress increase $E_T \cdot \Delta\epsilon$, while a strain decrease $\Delta\epsilon$ is accompanied by a stress decrease $E_T \cdot \Delta\epsilon$. The latter is not true. We know that on unloading, most materials follow a stress–strain graph parallel to that of the proportional range (Fig. 10.23 (c)). Von Kármán showed how this could be taken into account. Because of the difference in the effective modulus on the two sides of the neutral axis, the theory is often called the *double-modulus* theory. According to this theory, the incremental stress distribution will be as shown in Fig. 10.23 (a).

As usual, it is assumed that the strains vary linearly across the section, with zero bending strain along the line *BB*, which is the neutral axis of bending. Since the stresses on the "tensile" side of *BB* are E times the strains, while

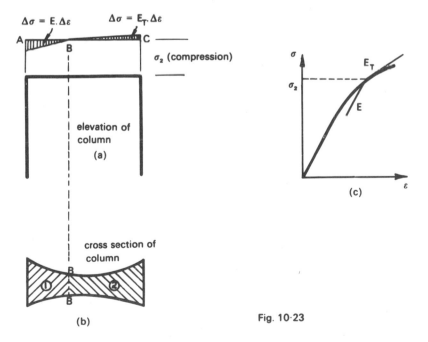

Fig. 10·23

on the "compressive" side they are E_T times the strains, the stress variation is not linear. The neutral axis BB, therefore, does not pass through the centroid, G, but is so located that the total tensile and compressive forces are equal. (Throughout the argument, only the incremental stresses are being considered.) This means that BB divides the cross-section into two parts (1) and (2) (Fig. 10.23 (b)), such that

$$EQ_1 = E_T Q_2$$

where Q_1 and Q_2 are the first moments of the two parts about BB. The location of BB will depend upon the shape of the cross-section, as well as upon the relative values of E and E_T.

Having located BB for a particular value of E_T, the relation between the column curvature (caused by the test displacement), and the moment resultant of the incremental stresses, can easily be found by the methods of Chapter 4. We find that

$$M = \frac{EI_1}{R} + \frac{E_T I_2}{R}$$

where I_1 and I_2 are the second moments of the two parts of the cross-section about BB. Then, if we assume that $1/R = d^2v/dx^2$, we can write

$$M = \frac{d^2 v}{dx^2}(EI_1 + E_T I_2) \tag{10.16}$$

From this point we can proceed as in the classical Euler problem, using $(EI_1 + E_T I_2)$ in place of the simple EI. We shall thus obtain a critical value of P given by

$$P_{cr} = \frac{\pi^2(EI_1 + E_T I_2)}{l^2} \tag{10.17}$$

The calculation of P_{cr} from this equation is by no means direct, since for each assumed value of E_T the neutral axis BB has to be located, and the values of I_1 and I_2 re-calculated.

10.9   Instability of other Physical Conditions

We have been concerned with the stability of only one type of physical condition, namely *equilibrium*. It is quite possible for other conditions to exhibit the instability characteristics, although these phenomena are not relevant to the present book.

A system may be in a state of motion to which it will not return after a small disturbance. For instance, a stretched string can be made to vibrate as a sine wave with a node point at the centre. If this condition is disturbed by giving the slightest displacement to the mid-point the motion will instantly change to a vibration in the fundamental mode. The condition of motion in the first mode is an example of *dynamic instability.*

A shaft rotating about its longitudinal axis can be shown to be in an unstable state of motion at certain speeds of revolution.

Physically, these problems are not related to those studies in the present chapter. There is a similarity in the mathematics of analysis because of the characteristic of instability. No matter what physical condition is being examined, it is unstable if it will not recover from a small disturbance.

Problems

In the following problems it is assumed that displacement normal to the plane of the paper is prevented.

10.1 The rigid bar AB (Fig. P10.1) is pinned at A and supported at B by a horizontal spring. In the unloaded position the bar is at 60° to the horizontal. If the vertical force W at B is gradually increased, what angle will the bar make with the horizontal when it becomes unstable? The spring is sufficiently long that it can be assumed to remain horizontal as B moves.

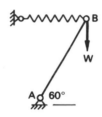

Fig. P10·1

10.2 A homogeneous beam has a rectangular cross-section 3 in wide and 10 in deep. It is simply supported over a span of 12 ft and supports a central load W. The stress–strain curve for the material is as shown in Fig. P10.2 (b), and the ascending and descending portions of the curve have the same slope.

At what value of W will instability due to bending occur? Lateral instability is prevented. Figure P10.2 is on page 316.

10.3 In the structure of Fig. P10.3, the hinges A, B and C lie in a straight line. The bars AB and BC are each of length l and may be taken as rigid. Hinge A is free to move vertically, and B is supported laterally by a spring of stiffness k. Find the value of P at which the structure becomes unstable. The bars may be considered as rigid. Figure P10.3 is on page 316.

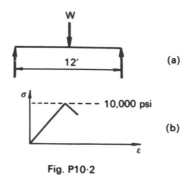

Fig. P10·2

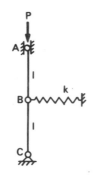

Fig. P10·3

10.4   In the rigid bar structure of Fig. P10.4, the hinge A is free to move vertically, while hinges B and C are each supported laterally by springs of stiffness k. The bars each have a length l ($= L/3$). Indicate the two possible modes of instability and find the value of P at which each occurs.

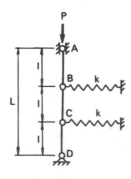

Fig. P10·4

10.5 The structure of Fig. P10.5 is similar to that of the previous problem except that the springs at *B* and *C* have stiffnesses of *k* and 2*k* respectively. What are the two eigen values of *P*?

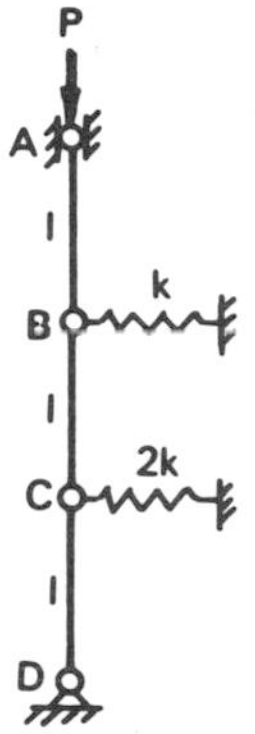

Fig. P10·5

10.6 In the rigid bar structure of Fig. P10.6 bars *AB* and *CD* are vertical. Horizontal displacement is opposed by a spring of stiffness 500 lb/in. At what value of the vertical load *W* will the structure become unstable?

10.7 The structure of Fig. P10.7 is similar to that of problem 10.6 except that, owing to an error of construction, the member *AB* is initially out of vertical by an amount *e*.

If *u* is the horizontal displacement of joint *B* due to loading, plot a graph of *W* against *u* for (*a*) *e* = 0.1 in; (*b*) *e* = 0.5 in; and (*c*) *e* = 1.0 in. Assume that the spring stiffness remains constant irrespective of how great *u* may become. Neglect the vertical displacement of *B* and plot the graphs up to *u* = 10 in.

For the case of *e* = 1.0 in, plot the graph for higher values of *u* but take account of the vertical movement of *B*. Does the structure become unstable?

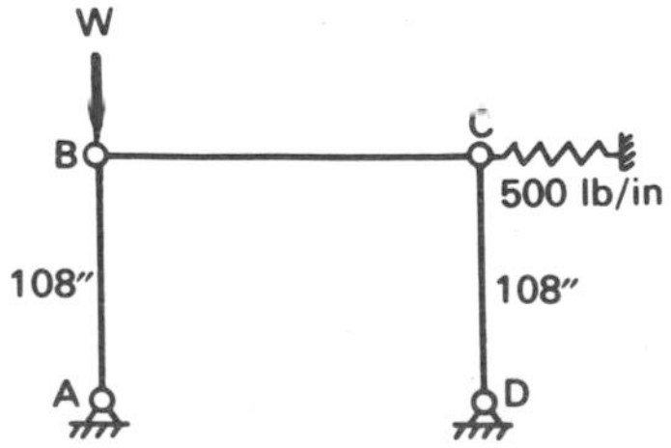

Fig. P10·6

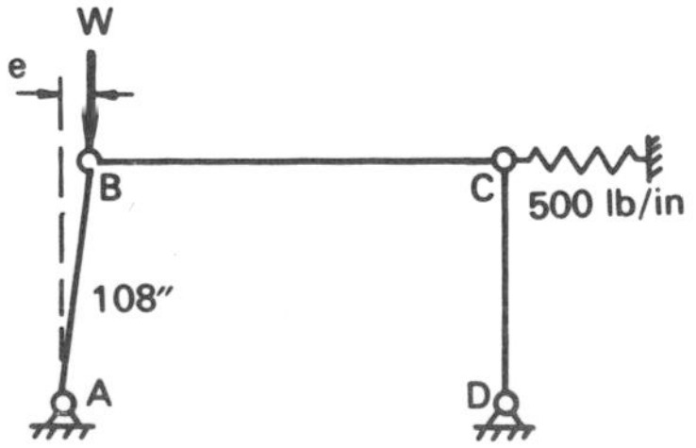

Fig. P10·7

10.8   The structure of Fig. P10.8 is similar to that of the previous problem except that a further hinge is introduced at E. Vertical movement of this hinge is opposed by a spring of stiffness k_E. Find the critical value of W (*a*) when $e = 0$ and $k_E = 20$ lb/in; (*b*) when $e = 0.2$ in and $k_E = 20$ lb/in; and (*c*) when $e = 0.2$ in and $k_E = 40$ lb/in.

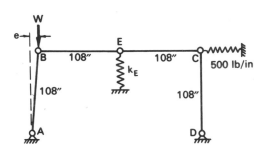

Fig. P10·8

10.9   In the rigid bar structure of Fig. P10.9, bars AB and CD are vertical. CD is 90 in long, and for this member $E = 10 \times 10^6$ psi and $I = 0.4$ in^4. Find the critical value of W for this structure.

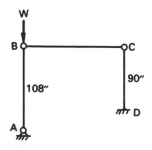

Fig. P10·9

10.10   The two rigid bars AB and BC (Fig. P10.10) are initially at 45° to the horizontal. A is pinned and movement of C is restricted by a horizontal spring of stiffness 200 lb/in. Calculate the load P at which the structure becomes unstable.

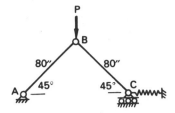

Fig. P10·10

10.11 The column ABC (Fig. P10.11) is hinged at A and B and fixed at C. The part AB may be taken as rigid. In terms of E and I for the portion BC, calculate the lowest critical load for the column.

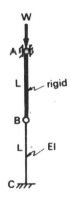

Fig. P10·11

10.12 A column AB is direction-fixed at both ends, but A is free to move vertically. Calculate the lowest critical load if the value of EI is constant.

10.13 A column of length L is made up of rigid bars AB, BC, CD and DE (Fig. P10.13) connected by spring hinges. Stability is maintained by the hinge springs which exert an internal moment tending to oppose any angular deformation. The stiffness of each spring is 400,000 lb-in per radian.

Find the load W at which instability occurs if $L = 150$ in. Assume that the critical deformation mode is such that the displacements of B and D are equal.

Sketch the deformation modes corresponding to the two values of W_{cr} which you obtain.

What other deformation mode is possible?

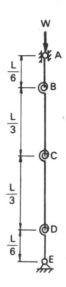

Fig. P10·13

10.14   The instability load of a pin-ended column of length L and stiffness EI can be computed approximately by imagining the column to be divided into three equal segments. The flexural stiffness of each segment is concentrated at a spring hinge at the centre of the segment, the spring stiffness being $EI/(L/3)$. When the hinges are connected by rigid arms the system reduces to that of Fig. P10.13 where $k = 3EI/L$.

Find the percentage error introduced by such an approximation.

Develop a similar model based on division of the column into four equal segments. What is the percentage error obtained by using this model?

CHAPTER 11

Dynamic Loading

The analysis of structures which are in a state of accelerated motion forms an extensive field of study, and it is intended here to give only a brief introduction. Attention will be confined to bars in which the deformation is proportional to the load—often referred to as "elastic" systems. Even here a rigorous treatment may be complex, but many practical problems are capable of approximate solution by quite simple methods.

The method of approach is similar to that used in earlier chapters, but where equilibrium equations were previously used, equations of motion now apply. Distribution of mass becomes important for the first time. In reality, mass is distributed in a known manner throughout the bar or structure. However, the calculations are often simplified if the mass is assumed to be localized at a number of "mass points". This gives us a *lumped system* in place of the actual *distributed system*.

11.1 Elastic Systems—Free Vibration

As a first example we will consider the vibration in an axial direction of a bar to the end of which is attached a body of mass m (Fig. 11.1 (a)). For simplicity, we ignore the mass of the bar, which, as far as the motion is concerned, then merely acts the part of a weightless spring. It is possibly easier to imagine the behaviour if we actually picture the bar as a spring with the same axial stiffness (Fig. 11.1 (b)). If the body is given a displacement in the x direction, and then released, it will oscillate with simple harmonic motion.

The stiffness of the spring must be equal to the axial stiffness of the bar.

$$k = EA/L \tag{11.1}$$

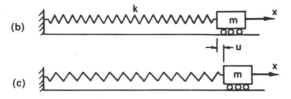

Fig. 11·1

When the displacement of the body is $+u$ (Fig. 11.1 (c)) the tension in the spring is ku and the force exerted on the body is $-ku$. Then, by Newton's second law,

$$m\frac{d^2u}{dt^2} = -ku$$

Differentiation with respect to t will be denoted by dots in the usual way. Thus

$$m\ddot{u} = -ku \tag{11.2}$$

If we put $p^2 = k/m$ then

$$\ddot{u} + p^2 u = 0 \tag{11.3}$$

The solution of this equation is

$$u = A\cos pt + B\sin pt \tag{11.4}$$

The values of the two constants A and B depend upon the amplitude of the vibration and the position from which we decide to measure time. Suppose that a is the amplitude and that we measure t from the instant when u has its maximum value ($= a$). Then, when $t = 0$, $u = a$ and $\dot{u} = 0$. With these initial conditions $A = a$ and $B = 0$, and the motion is described by the equation

$$u = a\cos pt \tag{11.5}$$

The quantity p is called the *circular frequency*, and is measured in radians per second. It is a function of the spring (bar) stiffness and the mass of the oscillating body. Sometimes it is expressed in terms of the *weight* of the body.

$$p = \sqrt{\frac{k}{m}} \quad \text{or} \quad p = \sqrt{\frac{kg}{W}}$$

The period, or time for one cycle of oscillation, is given by

$$T = \frac{2\pi}{p} \text{ sec} \tag{11.6}$$

and the frequency, n, is given by

$$n = \frac{p}{2\pi} \text{ cycles per sec} \tag{11.7}$$

The quantities T and n are called the natural period and frequency of vibration of the system. They are independent of the amplitude of vibration.

The axial force in the bar at any instant is given by

$$N = ku \tag{11.8}$$

and the maximum value is

$$N_{max} = ka \tag{11.9}$$

The axial force will be tensile for $u = +a$ and compressive for $u = -a$.

In the absence of friction and any other forces which might tend to retard the motion, the oscillations will continue indefinitely. The system is then said to be *undamped*. In such a case the sum of the potential and kinetic energies remains constant, i.e., the system is conservative.

The potential energy (P.E.) consists of the strain energy in the bar. When the extension is u, the strain energy is $\frac{1}{2}ku^2$ (see Chapter 3). In Fig. 11.2 (a) the axial force is plotted against displacement. The strain energy is equal to

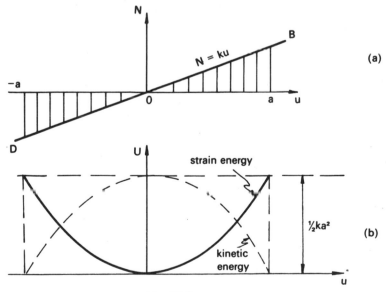

Fig. 11·2

the area under this graph. It is zero for $u = 0$ and positive for any other value of u. The strain energy is plotted against displacement in Fig. 11.2 (b).

The kinetic energy is confined to the body since the bar is assumed to have no mass. Consequently, the kinetic energy (K.E.) is $\frac{1}{2}m(\dot{u})^2$ when the extension is u. The kinetic energy is also plotted against displacement in Fig. 11.2 (b). We can see that a gradual interchange takes place between strain energy in the bar when $u = \pm a$, and kinetic energy in the body when $u = 0$. The energy equation

$$\text{K.E.} + \text{P.E.} = \text{constant}$$

or

$$\tfrac{1}{2}m(\dot{u})^2 + \tfrac{1}{2}ku^2 = \text{constant} \tag{11.10}$$

contains the same information as the equation of motion, equation (11.2). If we differentiate equation (11.10) with respect to u, we obtain

$$m\ddot{u} + ku = 0$$

which is identical with equation (11.2).

We note from Fig. 11.2 (b) that the equilibrium position ($u = 0$) corresponds to the position of minimum potential energy.

The analysis of a similar bar suspended vertically (Fig. 11.3 (a)) follows the same lines. The main difference is that the equilibrium position is not now one of zero bar deformation. In the equilibrium position (Fig. 11.3 (b)) the deformation is $mg/k \ (= W/k)$. This is often called the static deformation and denoted by u_{st}.

$$u_{st} = \frac{mg}{k} = \frac{W}{k} \tag{11.11}$$

If the body is now displaced (Fig. 11.3 (c)), oscillation will take place about the equilibrium position. If the amplitude is a, the maximum and minimum deformations will be $(u_{st} + a)$ and $(u_{st} - a)$.

At any displacement u, the resultant (downward) force on the body is $W - N$ or $mg - ku$.

The equation of motion is thus

$$m\ddot{u} = mg - ku \tag{11.12}$$

If we define $\bar{u}$ as the displacement measured from the equilibrium position (Fig. 11.3 (c)), we have

$$\bar{u} = u - u_{st} = u - mg/k$$

so that

$$\ddot{\bar{u}} = \ddot{u}$$

Equation (11.12) can then be expressed in the form

$$m\ddot{\bar{u}} = -k\bar{u} \tag{11.13}$$

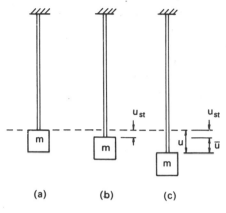

Fig. 11·3

This is similar to equation (11.2) with $\bar{u}$ instead of u. Consequently, the motion will be described by the equation

$$\bar{u} = a \cos pt \qquad (11.14)$$

where a is the amplitude. The natural period and frequency are the same as before.

$$T = \frac{2\pi}{p} \text{ sec} \quad \text{and} \quad n = \frac{p}{2\pi} \text{ cycles/sec}$$

where

$$p = \sqrt{\frac{k}{m}} = \sqrt{\frac{kg}{W}} = \sqrt{\frac{g}{u_{st}}}$$

The maximum and minimum values of the axial force are

$$N_{max} = k(u_{st} + a) = W + ka$$

and

$$N_{min} = k(u_{st} - a) = W - ka$$

As before, if the system is undamped, the sum of the potential and kinetic energies is constant. As the position of the body changes, then, in addition to the change in the strain energy and kinetic energy there is also a change in the potential energy by virtue of the change in height of the body. The latter will be called the gravitational energy (G.E.). Its value will depend on the choice of datum, which is arbitrary. The potential energy of the system now comprises the strain energy in the bar and the gravitational energy of the body. If the position of the body is denoted by h, measured *downward* from some datum, then

$$\text{P.E.} = -mgh + \tfrac{1}{2}ku^2 \qquad (11.15)$$

The energy equation is

$$\tfrac{1}{2}m(\dot{u})^2 + (-mgh + \tfrac{1}{2}ku^2) = 0 \qquad (11.16)$$

Differentiation with respect to u results in the equation of motion,

$$m\ddot{u} - mg + ku = 0 \tag{11.17}$$

since $dh/du = 1$. This equation is the same as equation (11.12).

In Fig. 11.4 (a), the line DOB is a graph of axial force against displacement. The axis of u is drawn vertically to correspond with the direction of displacement. The kinetic energy is a maximum at the equilibrium position and zero at the ends of the travel (Fig. 11.4 (b)). The strain energy varies parabolically with u, and is zero at zero deformation (full line in Fig. 11.4 (b)). The gravitational energy varies with u. Three different graphs are shown which correspond to datum levels chosen at D_1, D_2 and D_3. If we add the gravitational and strain energies we obtain the potential energy of the system (Fig. 11.4 (c)). Although we have different graphs depending upon the choice of datum, the potential energy is always a minimum at the equilibrium position.

The amplitude of vibration will depend upon the magnitude of the disturbance which initiates the vibration. Vibrations are commonly set up by bodies falling upon the structure. We may imagine that a body of weight W falls onto a stop at the lower end of a suspended bar (Fig. 11.5), and then remains attached to the bar as it vibrates. In this case, the energy which initiates vibration can easily be calculated.

Suppose that the body, of weight W, falls through a height h before reaching the stop (Fig. 11.5). The extension of the bar is u_{max} when the body finally comes to rest at the lowest point of the travel. At this instant there is no kinetic

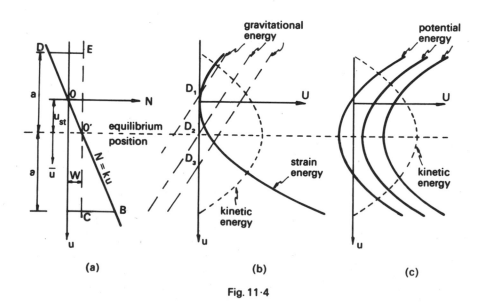

(a)   (b)   (c)

Fig. 11·4

Fig. 11·5

energy in the system so we may equate the loss of gravitational energy in the body to the gain of strain energy in the bar

$$\text{Loss of gravitational energy} = W(h + u_{\max})$$

$$\text{Gain of strain energy} = \tfrac{1}{2}ku_{\max}^2$$

where k is the stiffness of the bar. Hence

$$\tfrac{1}{2}ku_{\max}^2 = Wh + Wu_{\max}$$

or

$$u_{\max} = \frac{W}{k} \pm \sqrt{\left(\frac{W}{k}\right)^2 + 2h\frac{W}{k}} \tag{11.18}$$

The two solutions correspond to the fact that the system has zero kinetic energy both at the lowest and the highest point of the travel. We note that

$$\frac{W}{k} = u_{\text{st}}$$

the deflection under a static load W.

Equation (11.18) could be written as

$$\left.\begin{aligned}u_{\max}\\u_{\min}\end{aligned}\right\} = u_{\text{st}} \pm a \tag{11.19}$$

where a is the amplitude. The amplitude is given by

$$a = \sqrt{\left(\frac{W}{k}\right)^2 + 2h\frac{W}{k}} \tag{11.20}$$

The maximum deflection might also be expressed in terms of the static deflection and the distance of fall. Putting $W/k = u_{st}$ in equation (11.18), we have

$$u_{max} = u_{st} + \sqrt{u_{st}^2 + 2hu_{st}}$$

$$= u_{st}\left[1 + \sqrt{1 + \frac{2h}{u_{st}}}\right] \tag{11.21}$$

If the extension of the bar is small compared with the distance through which the weight falls prior to making contact with the bar, then the loss of gravitational energy can be taken as Wh approximately. Then we should have

$$\tfrac{1}{2}ku_{max}^2 \approx Wh$$

or

$$u_{max} \approx \sqrt{\frac{2hW}{k}} \tag{11.22}$$

If we put $h = 0$ in equation (11.18) we obtain

$$u_{max} = \frac{W}{k} + \sqrt{\left(\frac{W}{k}\right)^2} = 2\frac{W}{k} = 2u_{st}$$

We must distinguish between the condition represented by putting $h = 0$ and the equilibrium condition studied earlier in the book. If we want to preserve equilibrium at all times, the external load must be *applied gradually* so that it "keeps pace" with the extension of the bar. If the load is applied all at once then, instantaneously, the bar, having no deformation, and therefore no internal force, is not in equilibrium with the external load. In this latter case the load is said to be *suddenly applied*, and it is this condition which is represented by putting $h = 0$ in equation (11.18). It can be seen that a suddenly applied load will cause an extension twice that which would be produced by the same load applied gradually.

Example 11.1. A steel bar ($E = 30 \times 10^6$ psi) is 100 in long and has a cross-sectional area of 2 in². A body weighing 5000 lb is fixed to the lower end (Fig. 11.6). At the upper end of the bar is a stop which is designed to support the assembly on a platform. When the stop is 0.02 in above the platform the assembly is allowed to fall. It is assumed that when the stop reaches the platform it is restrained from lifting, so that the bar is thereafter capable of being put into compression.

(a) Find the maximum stress induced in the bar.

(b) Find also the amplitude, frequency and period of the oscillations which are set up. Assume that there is no damping, and neglect the mass of the bar.

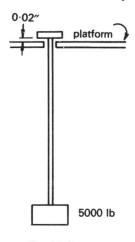

Fig. 11·6

Solution. (a) The maximum stress in the bar occurs when the deformation is a maximum, i.e., at the lowest point of the travel.

$$k = \frac{EA}{L} = \frac{30 \times 10^6 \times 2}{100} = 6 \times 10^5 \text{ lb/in}$$

The maximum extension is given by equation (11.18)

$$u_{max} = \frac{W}{k} + \sqrt{\left(\frac{W}{k}\right)^2 + 2h\frac{W}{k}}$$

$$u_{st} = \frac{W}{k} = \frac{5000}{6 \times 10^5} = \frac{1}{120} \text{ in}$$

$$\therefore \quad u_{max} = \frac{1}{120} + \sqrt{\left(\frac{1}{120}\right)^2 + \left(2 \times 0.02 \times \frac{1}{120}\right)}$$

$$= 0.0083 + 0.0201$$

$$= 0.0284 \text{ in}$$

Note that the static deflection is 0.0083 in and the amplitude of vibration is 0.0201 in. Also, since the amplitude exceeds the static deflection, the bar will at times be in compression.

$$N_{max} = ku_{max} = 6 \times 10^5 \times 0.0284$$

$$\sigma_{max} = \frac{N}{A} = \frac{6 \times 10^5 \times 0.0284}{2} = 8520 \text{ psi}$$

(b) As noted above

$$a = 0.0201 \text{ in}$$

To find T and n, we first calculate the circular frequency, p.

$$p = \sqrt{\frac{kg}{W}} = \sqrt{\frac{6 \times 10^5 \times 386}{5000}} = 215 \text{ radians/sec}$$

(Note that g is expressed in inch and second units.)
  Then the frequency,

$$n = \frac{p}{2\pi} = 34.3 \text{ cycles/sec}$$

and the period

$$T = \frac{1}{n} = 0.029 \text{ sec/cycle}$$

The system which has been examined so far is said to have one degree of freedom since the specification of only one quantity (the extension of the bar in this case) is required to define the configuration of the system at any one instant. The analysis can be applied directly to any elastic system with one degree of freedom provided the relevant stiffness coefficient is known. For instance, we can apply the results to the oscillation of a (weightless) beam supporting a body at any point (Fig. 11.7). The body can be situated at any point on the beam, and k is computed as the force required, at the point where the body is situated, to produce unit deflection at this same point.

To be consistent with the notation of earlier chapters, displacement normal to the bar (i.e., in the y direction) will be denoted by v.

Example 11.2. A cantilever is 6 ft long. It has a rectangular section 6 in wide and 1 in deep, and its elastic modulus, E, is 4×10^6 psi. A body weighing 10 lb is dropped from a height of 2 in onto the beam at a point 5 ft from the support (Fig. 11.8). It is assumed that it remains attached to the beam as the latter vibrates.

Find the maximum stress induced in the beam, and the amplitude and frequency of the vibration which is set up. Neglect the weight of the beam.

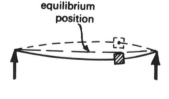

Fig. 11·7

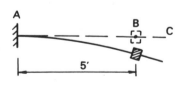

Fig. 11·8

Solution. To find the flexural stiffness of the beam we need to know the moment of inertia of the cross-section.

$$I = \frac{6 \times 1^3}{12} = \tfrac{1}{2} \text{ in}^4$$

The cantilever is effectively 5 ft long since the portion beyond the load does not enter the problem. The bending stiffness of a cantilever in regard to a vertical end force (see Chapter 4) is

$$k = \frac{3EI}{L^3}$$

This is the force required at B (Fig. 11.8) to produce a deflection of 1 in at B.

$$\therefore \quad k = \frac{3 \times 4 \times 10^6 \times 0.5}{60^3} = 27.8 \text{ lb/in}$$

Let v_{max} = the maximum deflection of the beam at B during the oscillations. Then, equating the energy lost by the falling weight to the strain energy stored in the beam, we obtain

$$10(2 + v_{max}) - \tfrac{1}{2}kv_{max}^2$$

$$= \tfrac{1}{2} \times 27.8 v_{max}^2$$

whence

$$v_{max} = 0.36 \pm 1.25$$

$$\therefore \quad v_{max} = 1.61 \text{ in}$$

The maximum beam stress clearly occurs at the support, and one way to find this stress is to find the equivalent static load which would produce a deflection of 1.61 in. This is

$$W_{st} = kv = 27.8 \times 1.61 = 44.8 \text{ lb}$$

With a load of 44.8 lb at B, the bending moment at A will be

$$M = 44.8 \times 60 \text{ lb-in}$$

and

$$\sigma = \frac{My}{I} = \frac{44.8 \times 60 \times 0.5}{0.5} = 2690 \text{ psi}$$

It is noted that the stress is 4.48 times the stress which would be produced by a static load of 10 lb.

The amplitude is $v_{max} - v_{st}$. The calculation of the static deflection is simple but unnecessary, since in the solution of the quadratic equation for v_{max} (i.e., $v_{max} = 0.36 \pm 1.25$) the first term represents the static deflection, and the second is the amplitude of vibration. From this we see that the amplitude is 1.25 in.

To find the frequency we first calculate p.

$$p = \frac{kg}{W} = \frac{27.8 \times 386}{10} = 32.7 \text{ radians/sec}$$

Then

$$n = \frac{p}{2\pi} = 5.2 \text{ cycles/sec}$$

In problems of this kind it is often easier to find the flexibility than the stiffness. For instance, in example 11.2, we might apply a unit force at 5 ft from the support of the cantilever, and calculate the deflection at this point as $L^3/3EI$. This is the value of the relevant flexibility coefficient, f, and by inversion we then have $k = 3EI/L^3$.

If a static load rests upon a structure, the sum of the vertical reactions is equal to the load. However, if the system is in motion, the sum of the reactions will, for part of the time, be greater than the supported load. It may be considerably greater.

Fig. 11·9

Consider the system of Fig. 11.9 in which the structure is replaced by an equivalent spring. Under static conditions the reaction, R, is equal to W, the weight of the supported body. If the body is dropped from a height onto the spring, the reaction (which is always equal to the axial force in the spring) will depend upon the spring stiffness. We can calculate the maximum value by the method of example 11.1.

From equation (11.18) we have an expression for the maximum bar extension.

$$u_{\max} = \frac{W}{k} + \sqrt{\left(\frac{W}{k}\right)^2 + 2h\frac{W}{k}}$$

Then

$$R_{\max} = N_{\max} = ku_{\max} = W + \sqrt{W^2 + 2hWk}$$

$$= W\left[1 + \sqrt{1 + \frac{2hk}{W}}\right]$$

or

$$R_{\max} = W\left[1 + \sqrt{1 + \frac{2h}{u_{\text{st}}}}\right] \tag{11.23}$$

If the weight, W, is suddenly applied (i.e., $h = 0$) then R is equal to $2W$. If it is dropped from a height, the value of R will depend upon k. It is possible to arrange for R to have any value in excess of $2W$ by designing a spring of suitable stiffness.

Example 11.3. A body of 100 lb weight is dropped from a height of 1 in onto the end of a spring. Find the spring stiffness such that the maximum support reaction is 500 lb.

Solution.

$$R_{max} = W\left[1 + \sqrt{1 + \frac{2h}{u_{st}}}\right]$$

$$500 = 100\left[1 + \sqrt{1 + \frac{2}{u_{st}}}\right]$$

$$\therefore \quad u_{st} = \frac{2}{15} \text{ in}$$

i.e.,

$$\frac{W}{k} = \frac{2}{15}$$

and

$$k = \frac{15W}{2} = 750 \text{ lb/in}$$

For any value of R_{max} greater than $2W$ it will always be possible to determine a finite value of k. Whether or not a suitable spring or structure can be designed is a matter which is outside the scope of the present discussion.

11.2   Elastic Systems—Forced Vibration

In the previous section we considered an elastic system which, once set in motion, is acted upon by a constant force (the weight of the oscillating body), or by zero force where the system is oscillating horizontally. The resulting motion is the free vibration of the system. The frequency of oscillation, which is called the *natural frequency* of the system, depends upon the elastic properties of the structure, characterized by the stiffness, and upon the mass of the vibrating system.

However, if the system is subjected to a fluctuating external force of a continuing nature this will modify the motion. Vibration of this type is called a *forced vibration*. It can arise from the effects of out-of-balance forces in rotating machinery or from other causes. The external force is called a *disturbing force*.

We shall, as before, consider a very simple system, namely, an elastic system with one degree of freedom. It will be characterized as a body of mass m acted upon by a weightless spring (Fig. 11.10). The disturbing force is represented by the function $F\sin\bar{p}t$. This function is referred to as the *forcing* function. (The circular frequency of the disturbing force is $\bar{p}$, while that of the free vibration is p, which is equal to $\sqrt{k/m}$.)

Fig. 11·10

When the extension of the spring is u, the total force acting on the body in the x direction is $-ku + F\sin \bar{p}t$. The equation of motion of the body is therefore

$$m\ddot{u} = -ku + F\sin \bar{p}t$$

or

$$\ddot{u} + \frac{k}{m}u = \frac{F}{m}\sin \bar{p}t \tag{11.24}$$

If we put $p^2 = k/m$ and $\phi^2 = F/m$ then

$$\ddot{u} + p^2 u = \phi^2 \sin \bar{p}t \tag{11.25}$$

The complementary solution (with the right-hand side equal to zero) will be the same as that for free vibration (equation 11.4). The particular solution, which must be introduced to satisfy the particular forcing function, is of the form

$$u = C\sin \bar{p}t$$

and the value of C must be chosen to satisfy equation (11.25). Substituting into equation (11.25), we obtain

$$-\bar{p}^2 C\sin \bar{p}t + p^2 C\sin \bar{p}t = \phi^2 \sin \bar{p}t$$

whence

$$C = \frac{\phi^2}{p^2 - \bar{p}^2} = \frac{\phi^2}{p^2[1 - (\bar{p}/p)^2]}$$

The complete solution is

$$u = A\cos pt + B\sin pt + C\sin \bar{p}t \tag{11.26}$$

If we assume that the system is at rest prior to the application of the external force, and measure t from the instant of its application, then we have that when $t = 0$, $u = 0$ and $\dot{u} = 0$. These initial conditions give

$$A = 0 \quad \text{and} \quad B = -\frac{\bar{p}}{p}C = \frac{-\bar{p}\phi^2}{p^3[1 - (\bar{p}/p)^2]}$$

Then

$$u = \frac{-\bar{p}\phi^2}{p^3[1 - (\bar{p}/p)^2]}\sin pt + \frac{\phi^2}{p^2[1 - (\bar{p}/p)^2]}\sin \bar{p}t \tag{11.27}$$

A graph of this equation would consist of the sum of two sine waves of different amplitude and frequency. When the frequencies $\bar{p}$ and p are appreciably different, the resulting graph is as shown in Fig. 11.11 (a). When $\bar{p}$ and p are close together, the resultant wave is similar to that of Fig. 11.11 (b) which exhibits the phenomenon known as "beats". In either case, the maximum amplitude occurs when the two terms are in phase, when it is equal to the sum of the separate amplitudes.

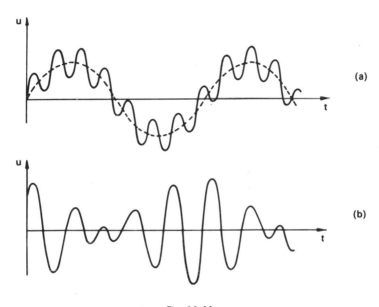

Fig. 11·11

It can be seen from equation (11.27) that the amplitude of each component wave depends, not only upon the *magnitude* of the disturbing force (represented by ϕ), but also upon the circular frequency both of the disturbing force, $\bar{p}$, and of the natural vibration, p. In particular, when these frequencies are equal, the amplitude of the oscillation theoretically becomes infinite, and resonance occurs. Thus the condition for resonance is

$$\bar{p}^2 = p^2 = k/m \qquad (11.28)$$

In practice, the system always experiences a certain amount of damping, and the infinite amplitude at resonance is never achieved. Even so, the amplitude may become very large. When the frequency of the disturbing force and the weight of the moving body are given, then resonance can be avoided only by changing the stiffness of the supporting structure. This, in turn, alters the natural frequency.

11.3 Systems with Two Degrees of Freedom

A system whose configuration, at any instant, can be defined by a single quantity, is said to have one degree of freedom. A single mass moving along a given path is such a system. If the mass moves in a horizontal line as considered previously, then the horizontal displacement at a given instant completely specifies the state of the system.

If the mass were free to move in a plane, two quantities would be required to define its position. Alternatively, if we have two masses, each capable of independent movement and each following a specified path, this system will also have two degrees of freedom, since each mass will require the specification of one quantity in order to define the configuration of the whole system.

The investigation of any system with two degrees of freedom leads to two simultaneous equations—one equation of motion corresponding to each degree of freedom. To illustrate the method, a two-mass system (Fig. 11.12) will be discussed.

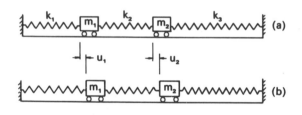

Fig. 11·12

Fig. 11.12 (a) shows the system in its equilibrium position, and Fig. 11.12 (b) shows the two masses displaced from this position by u_1 and u_2 respectively. The deformation, e_1, of spring 1 is u_1; the deformation, e_2, of spring 2 is $(u_2 - u_1)$; and the deformation, e_3, of spring 3 is $-u_2$. If the spring stiffnesses are k_1, k_2 and k_3, then the spring tensions are $k_1 u_1$, $k_2(u_2 - u_1)$ and $-k_3 u_2$ respectively. The equations of motions of the masses are

$$m_1 \ddot{u}_1 = -k_1 u_1 + k_2(u_2 - u_1) = -(k_1 + k_2)u_1 + k_2 u_2$$

and

$$m_2 \ddot{u}_2 = -k_2(u_2 - u_1) - k_3 u_2 = +k_2 u_1 - (k_2 + k_3)u_2$$

If we give mass m_1 a unit displacement while mass m_2 is undisplaced (Fig. 11.13), then to maintain the masses in these positions it will be necessary to exert on mass m_1 a force $(k_1 + k_2)$, and on mass m_2 a force $-k_2$. It is convenient to denote these forces by K_{11} and K_{21}. Symbol K_{11} (read as K one-one not K eleven) signifies the force required to hold m_1, when m_1 is given unit displacement. Symbol K_{21} is the force required to hold m_2, when m_1 is given a unit displacement. These holding forces are equal and opposite to the forces exerted by the springs.

Fig. 11·13

If mass m_2 is given unit displacement, while m_1 is undisplaced, then to hold the masses in position we shall need a force of $-k_2$ (denoted by K_{12}) on m_1, and a force of $(k_2 + k_3)$ (denoted by K_{22}) on m_2.

With this notation the equations of motion can be written in the form

$$m_1 \ddot{u}_1 = -K_{11} u_1 - K_{12} u_2$$

$$m_2 \ddot{u}_2 = -K_{21} u_1 - K_{22} u_2$$

or, alternatively,

$$m_1 \ddot{u}_1 + K_{11} u_1 + K_{12} u_2 = 0$$

$$m_2 \ddot{u}_2 + K_{21} u_1 + K_{22} u_2 = 0 \tag{11.29}$$

It will be shown that a system with two degrees of freedom possesses two natural modes of vibration, and that any other manner of oscillation can be regarded as a combination of these natural modes.

The system is said to be oscillating in a natural mode if, at every instant, the ratio of the displacement of one mass to the displacement of the other is constant. Suppose that

$$u_2 = ru_1$$

where r is a constant. If r is positive, then at any instant the masses are displaced in the same direction. If r is negative, then the masses always have unlike displacements. Clearly, the two masses have zero displacement at the same instant. If the displacement of mass m_1 is given by

$$u_1 = a_1 \cos(pt + \alpha) \tag{11.30}$$

then the displacement of m_2 will be given by

$$u_2 = ra_1 \cos(pt + \alpha)$$

$$= a_2 \cos(pt + \alpha) \tag{11.31}$$

The amplitudes of the two motions are a_1 and a_2; also $a_2 = ra_1$. We see that for natural modes of vibration, the circular frequency p of the motion of each mass is the same.

When the expressions for u_1 and u_2 (given by equations 11.30 and 11.31) are substituted into the equations of motion (equations 11.28), we obtain

$$-a_1 p^2 m_1 \cos(pt + \alpha) + K_{11} a_1 \cos(pt + \alpha) + K_{12} a_2 \cos(pt + \alpha) = 0$$

$$-a_2 p^2 m_2 \cos(pt + \alpha) + K_{21} a_1 \cos(pt + \alpha) + K_{22} a_2 \cos(pt + \alpha) = 0$$

After eliminating the common factor $\cos(pt + \alpha)$ this gives

$$(K_{11} - m_1 p^2) a_1 + K_{12} a_2 = 0$$
$$K_{21} a_1 + (K_{22} - m_2 p^2) a_2 = 0 \tag{11.32}$$

These are two equations in terms of the amplitudes a_1 and a_2. However, since the equations are homogeneous (all terms are linear and there are no constant terms) they will yield a non-trivial solution only if the determinant of coefficients is zero. In this determinant the K values are simply functions of the spring stiffnesses, the values of the masses are given, but the value of p is unknown. Thus, only specific values of p will permit a solution other than $a_1 = a_2 = 0$, and these are the values of p which cause the determinant of coefficients of equation (11.23) to be zero. These values are the *characteristic values* or *eigen values* of equations (11.32). The reader might note the very close similarity between the present problem and the instability problem with two degrees of freedom studied in section 10.4.

from which

$$\begin{vmatrix} (K_{11} - m_1 p^2) & K_{12} \\ K_{21} & (K_{22} - m_2 p^2) \end{vmatrix} = 0$$

$$K_{11} K_{22} - K_{11} m_2 p^2 - K_{22} m_1 p^2 + m_1 m_2 p^4 - K_{21} K_{12} = 0$$

or

$$p^4 - \left(\frac{K_{11}}{m_1} + \frac{K_{22}}{m_2}\right) p^2 + \frac{K_{11} K_{22} - K_{21} K_{12}}{m_1 m_2} = 0 \tag{11.33}$$

This quadratic equation in p^2 can be solved to give two values of p^2 and hence two values of $|p|$. Each is a circular frequency corresponding to one of the *natural frequencies* of the system ($n = p/2\pi$). When either of the values of p is substituted into equations (11.32), these equations can be solved, not for absolute values of a_1 and a_2, but for a value of the ratio a_2/a_1. The natural mode of vibration is characterized by the frequency and by the relative amplitudes of the two masses. The absolute values of the amplitudes will depend upon the magnitude of the initial disturbance.

Example 11.4. A two-mass system is illustrated in Fig. 11.14 (a). The weights and spring stiffnesses are as shown. Find the two natural frequencies of vibration of the system and the relative amplitudes in each mode of vibration.

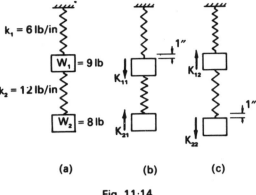

Fig. 11·14

Solution. Suppose that body 1 is displaced 1 in from its equilibrium position (Fig. 11.14 (b)), while body 2 is restrained against displacement. Spring 1 is then stretched by 1 in, while spring 2 is compressed by 1 in. The forces required to maintain the system in this deformed state are (see Fig. 11.14 (b))

$$K_{11} = 6 + 12 = +18 \text{ lb} \quad \text{and} \quad K_{21} = -12 \text{ lb}$$

A similar operation is now performed with body 2 displaced 1 in (Fig. 11.14 (c)). Now spring 2 is stretched while spring 1 is undeformed. The forces required to maintain this state of deformation are

$$K_{12} = -12 \text{ lb} \quad \text{and} \quad K_{22} = +12 \text{ lb}$$

We are given that

$$m_1 = \frac{W_1}{g} = \frac{9}{386} \text{ slugs}$$

and

$$m_2 = \frac{W_2}{g} = \frac{8}{386} \text{ slugs}$$

Hence the equations (11.32) can be written explicitly

$$\left(18 - \frac{9}{386}p^2\right)a_1 \qquad\qquad - 12a_2 = 0$$

$$-12a_1 + \left(12 - \frac{8}{386}p^2\right)a_2 = 0 \tag{11.34}$$

Equating the determinant of coefficients to zero we have

$$\begin{vmatrix} \left(18 - \dfrac{9}{386}p^2\right) & -12 \\[2ex] -12 & \left(12 - \dfrac{8}{386}p^2\right) \end{vmatrix} = 0$$

or

$$p^4 - (3.5 \times 386)p^2 + 386^2 = 0$$

Hence

$$p^2 = 121 \text{ or } 1230$$

If the two values of $|p|$ are denoted by p_1 and p_2 then

$$p_1 = 11.0 \text{ radians/sec}$$

$$p_2 = 35.1 \text{ radians/sec}$$

These are the circular frequencies associated with the two natural modes of vibration. The natural frequencies are therefore

$$n_1 = \frac{11.0}{2\pi} = 1.75 \text{ cycles/sec}$$

$$n_2 = \frac{35.1}{2\pi} = 5.59 \text{ cycles/sec}$$

The natural periods are $T_1 = 0.572$ sec and $T_2 = 0.179$ sec.

(a) First natural mode

The natural mode of vibration with the lowest frequency is called the fundamental mode. In this case, for the fundamental mode, $n = 1.75$ and $p^2 = 121$. If this value of p^2 is substituted into equations (11.34) the ratio of amplitudes can be found.

$$15.18a_1 - 12a_2 = 0$$

$$-12a_1 + 9.45a_2 = 0$$

Each of these equations is satisfied by the ratio $a_2/a_1 = 1.26$. Thus, when the system is vibrating in the fundamental mode, the frequency of vibration is 1.75 cycles/sec and at every instant the displacement of mass 2 is $+1.26$ times the displacement of mass 1, that is to say is 1.26 times as great and in the same direction.

(b) Second natural mode

With this system with only two degrees of freedom, there is only one natural mode of vibration higher than the fundamental mode. For the higher mode, $n_2 = 5.59$ cycles/sec and $p^2 = 1230$. When this value of p^2 is substituted into equations (11.34) we obtain

$$-10.67a_1 - 12a_2 = 0$$

$$-12a_1 - 13.49a_2 = 0$$

Each of these equations is satisfied by the ratio $a_2/a_1 = -0.89$. Thus for this mode of vibration the frequency is 5.59 cycles/sec and at every instant the displacement of mass 2 is -0.89 times the displacement of mass 1, that is to say 0.89 times as great and in the opposite direction.

As mentioned before, the system can oscillate in a manner other than one of the natural modes. However, the motion must be such that it can be regarded as the sum of two natural modes of vibration. Such a motion is called a mixed-mode vibration.

The two-mass system of example 11.2 can be used for illustration. The upper part of Fig. 11.15 illustrates the combination of two natural mode vibrations for the mass m_1. The amplitude of the fundamental mode (Fig.

11.15 (a)) has been arbitrarily taken as '1 in, while that of the higher mode (Fig. 11.15 (b)) has been taken as 0.2 in. The combination of these two natural mode waves is shown in Fig. 11.15 (c), which represents a mixed-mode oscillation of the upper mass m_1.

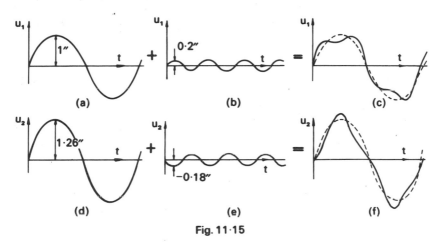

Fig. 11·15

The corresponding motion of the lower mass m_2 is completely determined. Fig. 11.15 (d) shows the fundamental mode component. The amplitude is 1.26 in since it must be 1.26 times the amplitude of the upper mass. Similarly, the amplitude of the higher mode component (Fig. 11.15 (e)) is 0.18 in, i.e., −0.89 times that of the upper mass. The combination is shown in Fig. 11.15 (f).

An infinite number of mixed-mode oscillations are possible since the amplitudes of the two natural modes are arbitrary. Fig. 11.16 shows a different combination of two natural mode oscillations. Here the amplitude of the higher mode component is larger than before, and the amplitude of the fundamental component is smaller. In addition, the initial conditions are different.

The combination of two natural mode vibrations can be looked upon as a possible solution of the differential equations of motion (equations 11.29). These equations are satisfied by the displacement functions corresponding to the first natural mode, namely

$$u_1^{(1)} = a_1^{(1)} \cos{(p^{(1)} t + \alpha^{(1)})}$$
$$u_2^{(1)} = r^{(1)} a_1^{(1)} \cos{(p^{(1)} t + \alpha^{(1)})}$$

(11.35)

They are also satisfied by the functions

$$u_1^{(2)} = a_1^{(2)} \cos{(p^{(2)} t + \alpha^{(2)})}$$
$$u_2^{(2)} = r^{(2)} a_1^{(2)} \cos{(p^{(2)} t + \alpha^{(2)})}$$

(11.36)

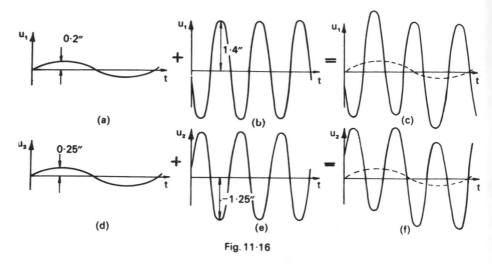

Fig. 11·16

where the symbols refer specifically to the second natural mode. The differential equations will also be satisfied by any linear combination of these displacement functions, that is, by functions of the form

$$u_1 = cu_1^{(1)} + du_1^{(2)}$$
$$u_2 = cu_2^{(1)} + du_2^{(2)}$$

$$(11.37)$$

If the values of the two natural frequencies are rational numbers, then any mixed-mode vibration will be periodic—there will come a time when the configuration of the system will be the same as it was at $t = 0$, and thereafter the motion will be repeated. If the natural frequencies are irrational, such a periodicity will not occur.

So far, the behaviour of a system with two degrees of freedom has been illustrated by means of two masses and a series of springs. However, the method of analysis can be applied equally well to any system where the springs are replaced by elastic structures. Equations (11.32) can be considered as typical of such systems. The coefficients K_{11} and K_{12} are the forces which must be exerted on bodies 1 and 2, respectively, in order to maintain a configuration where the displacements of the bodies are unity and zero respectively. Similarly, K_{21} and K_{22} are the forces required to maintain displacements of zero and unity. We now consider two bodies attached at different points to a cantilever whose mass is neglected.

Example 11.5. A cantilever 12 ft long supports two bodies (Fig. 11.17 (a)), one of weight 10 lb at the free end A, and one of weight 6 lb at the point B, 9 ft from the support. If the value of EI for the bar is 120,000 lb-in², find the frequencies and relative amplitudes of each of the two natural modes of vibration.

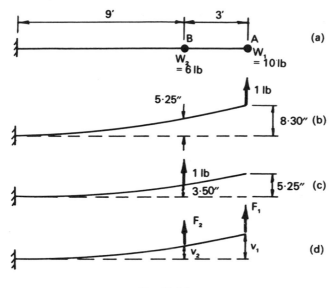

Fig. 11·1/

Solution. The first step in problems of this sort is the calculation of the quantities K_{11}, K_{12}, K_{21} and K_{22} as defined above. Perhaps the easiest way of doing this is first to apply a unit force at each of the points A and B, and to calculate the deflections at A and B for each force.

A force of 1 lb is applied at A (Fig. 11.17 (b)). The deflections at A and B are calculated by the methods of Chapter 4 (the moment-area method is convenient in this problem).

$$v_A = 8.30 \text{ in}; \quad v_B = 5.25 \text{ in}$$

Now a force of 1 lb is applied at B (Fig. 11.17 (c)) and the deflections computed as

$$v_A = 5.25 \text{ in}; \quad v_B = 3.50 \text{ in}$$

The deflections at A and B due to a combined system of forces F_1 and F_2 (Fig. 11.17 (d)) are then

$$v_1 = 8.30F_1 + 5.25F_2$$
$$v_2 = 5.25F_1 + 3.50F_2$$
(11.38)

These equations are solved in order to express F_1 and F_2 in terms of v_1 and v_2.

$$F_1 = 2.35v_1 - 3.52v_2$$
$$F_2 = -3.52v_1 + 5.57v_2$$
(11.39)

The required K values are obtained directly from these equations
With $v_1 = 1$ and $v_2 = 0$

$$K_{11} = 2.35 \quad \text{and} \quad K_{21} = -3.52$$

With $v_1 = 0$ and $v_2 = 1$

$$K_{12} = -3.52 \quad \text{and} \quad K_{22} = 5.57$$

The remainder of the work follows the same form as in example 11.2. The equations for the amplitudes (equations 11.32) are

$$\left(2.35 - \frac{W_1}{g}p^2\right)a_1 \qquad\qquad - 3.52a_2 = 0$$

$$(11.40)$$

$$-3.52a_1 + \left(5.57 - \frac{W_2}{g}p^2\right)a_2 = 0$$

These have solutions only when

$$\begin{vmatrix} \left(2.35 - \dfrac{10}{386}p^2\right) & -3.52 \\ -3.52 & \left(5.57 - \dfrac{6}{386}p^2\right) \end{vmatrix} = 0$$

i.e.,

$$p^4 - 386\left(\frac{2.35}{10} + \frac{5.57}{6}\right) + 386^2\left(\frac{2.35 \times 5.57 - 3.52^2}{10 \times 6}\right) = 0$$

Hence

$$p^2 = 3.86 \text{ or } 445$$

and

$$p = 1.96 \text{ or } 21.1 \text{ radians/sec}$$

(a) First natural mode

By substituting $p^2 = 3.86$ into equation (11.40) we find that $a_2/a_1 = +0.64$. The configuration corresponding to this mode of vibration is therefore as shown in Fig. 11.18 (a).

$$p = 1.96 \text{ radians/sec}$$

$$n_1 = \frac{1.96}{2\pi} = 0.31 \text{ cycles/sec}$$

$$T_1 = \frac{2\pi}{1.96} = 3.2 \text{ sec}$$

(a)

(b)

Fig. 11·18

(b) *Second natural mode*

By substituting $p^2 = 455$ into equation (11.40) we find that $a_2/a_1 = -2.61$. The configuration for this mode is shown in Fig. 11.18 (b).

$$p = 21.1 \text{ radians/sec}$$

$$n_2 = \frac{21.1}{2\pi} = 3.36 \text{ cycles/sec}$$

$$T_2 = \frac{2\pi}{21.1} = 0.30 \text{ sec}$$

It might be noted that, as in the previous example, the bodies are moving in the same direction at every instant in the first mode of vibration, while in the second mode they are always moving in opposite directions.

Each natural mode of vibration is associated with a particular configuration of deformation, which is a *characteristic shape* corresponding to the *characteristic value* of p. The similarity with the corresponding instability problem will easily be seen.

11.4 Systems with Many Degrees of Freedom

It is not proposed to study in detail the motion of systems which have more than two degrees of freedom. However, it might be mentioned that the analysis follows the same lines as that of the last section.

There will be one equation of motion for every degree of freedom. Where the system has n degrees of freedom we shall have n equations similar to the two equations of (11.32), and the form of these n equations might be anticipated from the nomenclature used in equations (11.32). When the determinant of coefficients is equated to zero, an equation of the nth degree in p^2 is obtained instead of the second-degree equation (11.33). Solution provides n possible values of p, and each corresponds to a particular mode of vibration.

11.5 Distributed Systems

The systems studied so far have contained a finite number of masses located at discrete points. The elastic supporting structure has been simulated by springs. In effect, the elastic, or stiffness, characteristic of the structure has been simulated by springs having elasticity but no mass; while the mass of the structure has been concentrated at mass points which possess mass, but no elasticity. As mentioned at the beginning of the chapter, such a system is called a *lumped system*.

In the real structure, both mass and elasticity are distributed throughout the structure. It is really a *distributed system*. Although in many practical structures solution of the true distributed system leads to complex analysis, in simple structures the continuous system can be solved without difficulty.

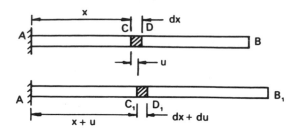

Fig. 11·19

Consider, first, a uniform bar of length L which is vibrating freely in a longitudinal direction. The end A is fixed in position, while the end B oscillates in the x direction (Fig. 11.19). A small element of length between cross-sections C and D, a distance x from A, is regarded in the same way as an isolated mass in the earlier analysis. At a given instant the length of the element is $(dx + du)$ and the deformation is therefore du. Since u is a function both of x and of t, partial derivatives must be used. We have

$$\text{the strain at } x = \epsilon = \frac{\partial u}{\partial x}$$

$$\text{the stress at } x = \sigma = E\frac{\partial u}{\partial x}$$

$$\text{the axial force at } x = N = EA\frac{\partial u}{\partial x}$$

The difference of axial force between C_1 and D_1 is therefore

$$dN = EA\frac{\partial^2 u}{\partial x^2}dx$$

This force dN is the accelerating force and is similar to the spring force exerted on the mass in previous examples. If m is the mass per unit length of the bar, then the mass of the element CD is $m\,dx$. The equation of motion is thus

$$(m\,dx)\frac{\partial^2 u}{\partial t^2} = EA\frac{\partial^2 u}{\partial x^2}dx$$

If a dot is used to denote differentiation with respect to t, while a dash is used to denote differentiation with respect to x, this equation becomes

$$u'' - \frac{m}{EA}\ddot{u} = 0 \qquad (11.41)$$

If the bar is vibrating in one of the natural modes, then the displacement of any point will have the form

$$u = X \cos(pt + \alpha) \tag{11.42}$$

where X is some function of x alone. Then

$$u'' = X'' \cos(pt + \alpha) \quad \text{and} \quad \ddot{u} = -p^2 X \cos(pt + \alpha)$$

With these values, equation (11.41) becomes

$$X'' \cos(pt + \alpha) + \frac{p^2 m}{EA} X \cos(pt + \alpha) = 0$$

or

$$X'' + \lambda^2 X = 0 \tag{11.43}$$

where

$$\lambda = p \sqrt{\frac{m}{EA}} \tag{11.44}$$

From equation (11.43)

$$X = C_1 \sin \lambda x + C_2 \cos \lambda x$$

so that

$$u = (C_1 \sin \lambda x + C_2 \cos \lambda x) \cos(pt + \alpha)$$

Now when $x = 0$, $u = 0$, irrespective of the value of t

$$\therefore \quad C_1 \sin \lambda x + C_2 \cos \lambda x = 0 \quad \text{when } x = 0$$

From this we have that $C_2 = 0$.

We also know that at the free end, **B**, the axial force is zero irrespective of the value of t.

$$N = EA \frac{\partial u}{\partial x} = EAC_1 \lambda \cos \lambda L \cos(pt + \alpha)$$

Hence

$$\cos \lambda L = 0$$

$$\lambda L = \frac{\pi}{2}, \frac{3\pi}{2}, \frac{5\pi}{2}$$

Since

$$\lambda = p \sqrt{\frac{m}{EA}}$$

then

$$p = \frac{\pi}{2L} \sqrt{\frac{EA}{m}}, \frac{3\pi}{2L} \sqrt{\frac{EA}{m}}, \text{ etc}$$

These *characteristic values* of p are the only ones for which equation (11.43) has a solution other than the trivial solution $X = 0$. The determination of these values of p for which the differential equation (11.43) has a solution is thus similar to the step in the earlier problem where we evaluated two values of p for which the two simultaneous equations of (11.32) had a solution (p. 338). In that problem, the system had two degrees of freedom, whereas the present system, being distributed, has an infinite number of degrees of freedom. It is not surprising, therefore, that it has an infinite number of natural modes of vibration.

In the fundamental mode,

$$p_1 = \frac{\pi}{2L} \sqrt{\frac{EA}{m}}$$

the frequency

$$n_1 = \frac{p_1}{2\pi} = \frac{1}{4L} \sqrt{\frac{EA}{m}}$$

$$\lambda_1 = \frac{\pi}{2L}$$

and

$$u = C_1 \sin \frac{\pi}{2L} x \cos(pt + \alpha)$$

The maximum displacement at any instant occurs at the free end, and i^r we denote the amplitude of this point by a, then

$$C_1 = a$$

and

$$u = a \sin \frac{\pi}{2L} x \cos\left(\frac{\pi}{2L} \sqrt{\frac{EA}{m}} t + \alpha\right) \tag{11.45}$$

$$\epsilon = \frac{\partial u}{\partial x} = \frac{\pi}{2L} \frac{a}{} \cos \frac{\pi}{2L} x \cos\left(\frac{\pi}{2L} \sqrt{\frac{EA}{m}} t + \alpha\right) \tag{11.46}$$

Equation (11.45) indicates that at any instant the displacement of the various points along the bar varies sinusoidally with x, being zero at A and a maximum at B. Equation (11.46) indicates that the strain also varies sinusoidally, being zero at B and a maximum at A.

A similar analysis can be carried out for a bar vibrating transversely. A uniform bar will be considered, simply supported at its ends, and having a mass m per unit length. As before, the equation of motion is written for a typical element of length dx and mass $(m dx)$.

In Chapter 4 we saw that the relation between a transverse force w (per unit length) and deflection v is (with EI constant)

$$w = EI \frac{d^4 v}{dx^4}$$

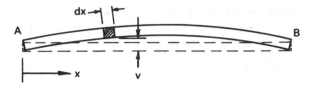

Fig. 11·20

Consequently, with the bar in a bent state (Fig. 11.20) we conclude that the force exerted on the typical element is

$$-w\,dx = -EI\frac{\partial^4 v}{\partial x^4}\,dx$$

The equation of motion of the element is then

$$(m\,dx)\frac{\partial^2 v}{\partial t^2} = -EI\frac{\partial^4 v}{\partial x^4}\,dx$$

or

$$v^{IV} + \frac{m}{EI}\ddot{v} = 0 \tag{11.47}$$

For vibration in a natural mode

$$v = X\cos(pt + \alpha) \tag{11.48}$$

where X is a function of x. Then

$$X^{IV}\cos(pt+\alpha) - \frac{p^2 m}{EI}X\cos(pt+\alpha) = 0$$

or

$$X^{IV} - \lambda^4 X = 0 \tag{11.49}$$

where

$$\lambda^1 = \frac{p^2 m}{EI}$$

The solution of equation (11.48) is

$$X = C_1 \sin\lambda x + C_2\cos\lambda x + C_3\sinh\lambda x + C_4\cosh\lambda x$$

and from (11.48)

$$v = X\cos(pt + \alpha)$$

Now the boundary conditions for this beam are that when $x = 0$, $v = 0$ and $\partial^2 v/\partial x^2 = 0$; and when $x = L$, $v = 0$ and $\partial^2 v/\partial x^2 = 0$. These conditions must apply for all values of t. They lead to the values

$$C_2 = C_3 = C_4 = 0$$

and

$$C_1 \sin \lambda L = 0$$

The last condition is true only if

$$\lambda L = \pi, 2\pi, 3\pi \ldots$$

or

$$\lambda^2 = \frac{\pi^2}{L^2}, \frac{4\pi^2}{L^2}, \frac{9\pi^2}{L^2} \ldots$$

Since

$$\lambda^2 = p \sqrt{\frac{m}{EI}}$$

then

$$p = \frac{\pi^2}{L^2} \sqrt{\frac{EI}{m}}, \frac{4\pi^2}{L^2} \sqrt{\frac{EI}{m}} \text{ etc.}$$

As before, these values of p correspond to the natural modes of vibration. In the fundamental mode,

$$p_1 = \frac{\pi^2}{L^2} \sqrt{\frac{EI}{m}}$$

and

$$n_1 = \frac{\pi}{2L^2} \sqrt{\frac{EI}{m}}$$

If a is the maximum amplitude—the amplitude at the mid-point in the fundamental mode—then the equation for the deflection becomes

$$v = a \sin \frac{\pi x}{L} \cos \left(\frac{4\pi^2}{L^2} \sqrt{\frac{EI}{m}} t + \alpha \right) \qquad (11.50)$$

11.6 Approximations to Distributed Systems

When the analysis of a structure as a distributed system proves unduly difficult, an approximate solution can often be obtained by considering a lumped system. The mass of the structure is assumed to be concentrated at a number of points. The accuracy depends largely on the number of mass points adopted. As the number of such points is increased, the assumed mass distribution approaches more and more closely to the real distribution. When the number of mass points becomes extremely large, arithmetic round-off errors may begin to affect the result adversely. However, as it will be essential to use a computer for the solution of a large number of equations, such considerations are seldom a major consideration in practice.

The use of a lumped system in place of a distributed system is illustrated in problem 11.23 at the end of this chapter.

The general mathematical similarity between the work of Chapter 10 and that of Chapter 11 will no doubt have been noticed. The analysis of a bar in longitudinal vibration (equation 11.43 on p. 347) is very similar to the instability analysis of the same bar in axial compression (equation 10.10 on p. 304). In each case, the distributed system can be approximated by a lumped system in order to simplify the analysis. In the case of instability analysis the elastic strut can be replaced by a series of rigid bars connected by spring-loaded hinges; the elasticity of the bar is, in effect, concentrated at these hinges. In the case of dynamic analysis, the bar can be replaced by a weightless bar to which masses are attached at intervals; the mass of the bar is concentrated at mass points.

Problems

11.1   A body weighing 5000 lb is supported by a vertical brass rod which is fastened at its upper end. The rod is 100 in long, has a cross-section of 0.5 in^2 and $E = 15 \times 10^6$ psi. If the rod vibrates in an axial direction, find the natural frequency and period of the system. Neglect the weight of the rod.

11.2   A horizontal timber beam spans 20 ft and has a cross-section 6 in square. A load of 1000 lb is suspended from the mid-point of the beam. If the system is set in motion in the vertical direction, find the natural frequency and period of the system. Take $E = 2 \times 10^6$ psi and neglect the weight of the beam.

11.3   In problem 11.2, if the 1000 lb were shifted to the quarter point of the beam instead of the mid-point, what would the frequency and period of the system be?

11.4   A steel cantilever ($E = 30 \times 10^6$ psi) is 20 in long, 2 in wide and $\frac{1}{2}$ in deep. It carries a body of weight W at its free end, and is set in motion vertically. Find the natural frequency and period of vibration (*a*) if $W = 100$ lb, (*b*) if $W = 200$ lb, and (*c*) if $W = 300$ lb. Neglect the weight of the cantilever.

11.5   A steel rod, 2 in diameter and 100 in long, is cantilevered vertically. At the free end it carries a disk of radius 40 in and weight 500 lb. The assembly is set into torsional vibration. Find the natural frequency and period. Take $G = 11.6 \times 10^6$ psi and neglect the weight of the rod. (The moment of inertia of a disk is $\frac{1}{2}MR^2$ or $\frac{1}{2}WR^2/g$.)

11.6 As in problem 11.5, a disk of weight 500 lb and radius 40 in is set into torsional vibration at the end of a cantilever 100 in long. In the present problem, the cantilever is a composite rod comprising a copper rod of 1 in diameter inside a steel tube of inner diameter 1 in, and outer diameter 2 in (see problem 6.13). Find the natural frequency and period of vibration if G for steel is 11.6×10^6 psi and G for copper is 5.8×10^6 psi.

11.7 A body weighing 5 kips is supported by a high tensile steel wire of length 700 in. For a length of 400 in the wire has a cross-section of 0.2 in², and for the remainder of the length the cross-section is 0.1 in². Find the natural frequency of the system in longitudinal vibration. Take $E = 30 \times 10^6$ psi.

11.8 A 2 kip load is supported at the mid-point C of a simply supported beam ACB, which has a total span of 24 ft. The part AC has an EI value of 1.5×10^6 lb-ft² while the part CB has an EI value of 4.5×10^6 lb-ft². Find the frequency of vertical oscillation of the system. (*Note*: Find the flexibility coefficient of the system by applying a unit load at the centre of the beam and computing the deflection by the conjugate beam method.)

11.9 A steel bar of 1 in diameter and 5 ft long is put into tension by a load of 5000 lb suddenly applied. If $E = 30 \times 10^6$ psi, determine the maximum stress and elongation produced. Neglect the mass of the bar.

11.10 A bar is $2\frac{1}{4}$ inches in diameter and 45 in long. Determine (*a*) the elongation (*b*) the stress produced in it, by a weight of 400 lb falling 3 in before commencing to stretch the bar. $E = 30 \times 10^6$ psi. Neglect the mass of the bar.

11.11 A short steel column of circular section is 24 in long and tapers from 8 in diameter at the base to 4 in diameter at the top. Find the maximum stress produced in the column if a load of 500 lb drops vertically from a height of 6 in onto the top of the block. Also find the maximum deflection produced. The mass of the column may be neglected. Take $E = 30 \times 10^6$ psi.

11.12 A sliding weight of 400 lb is dropped down a vertical rod which is suspended from the top and is provided with a collar at the bottom. The length of the rod is 12 ft and the area of cross-section is 0.4 in². In order to reduce the shock a helical buffer spring is placed on the collar. The spring stiffness is 1000 lb/in. Taking account of the work done in compressing the spring and in stretching the bar, find the height, measured from the top of the uncompressed spring from which the weight must be dropped in order to produce a maximum stress of 10,000 psi in the rod. Take $E = 30 \times 10^6$ psi.

11.13 A steel cantilever 48 in long, tapers in width from 6 in at the support, to zero at the free end. The depth is constant at $\frac{1}{2}$ in. $E = 30 \times 10^6$ psi.

(*a*) Find the deflection of the free end under a static load of 10 lb situated at the free end.

(*b*) Find the maximum deflection if the same load is dropped from a height of 3 in.

11.14   A suspended steel wire ($E = 30 \times 10^6$ psi) has a cross-sectional area of 0.01 in². A weight of 100 lb is dropped from a height of 6 in onto a stop at the lower end of the wire. What must be the length of the wire so that the axial force induced shall not exceed 300 lb?

11.15   A steel wire of length 2000 in and cross-sectional area 0.02 in² is suspended. At the lower end is a stop onto which a weight of 200 lb falls. To reduce the shock load in the wire a helical spring is interposed between the stop and the falling weight. If the weight falls 6 in before reaching the spring, find the spring stiffness such that the axial force in the wire is limited to 600 lb.

11.16   Two blocks are supported by springs as shown in Fig. P11.16. For each natural mode of vibration, find the frequency, the period, and the ratio of the amplitudes of the two blocks.

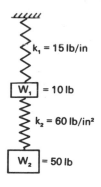

$k_1 = 15$ lb/in

$W_1 = 10$ lb

$k_2 = 60$ lb/in²

$W_2 = 50$ lb

Fig. P11·16

11.17   Two blocks are supported by springs as shown in Fig. P11.17. For each natural mode of vibration, find the frequency, the period, and the ratio of the amplitudes of the two blocks.

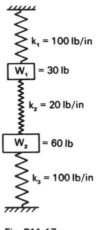

$k_1 = 100$ lb/in

W_1 = 30 lb

$k_2 = 20$ lb/in

W_2 = 60 lb

$k_3 = 100$ lb/in

Fig. P11·17

11.18   A horizontal cantilever 160 in long supports a body of weight 100 lb at the centre, and a body of weight 50 lb at the free end. If the value of EI for the beam is 2×10^6 lb-in^2, find the natural frequency and the ratio of amplitudes in each natural mode of vibration. Neglect the weight of the beam.

11.19   A beam 300 in long is simply supported at its ends. The value of I is 16.67 in^4. The beam supports two pieces of machinery each of 800 lb weight. The machines are situated 100 in from each end of the beam. If the weight of the beam is neglected, find the frequency, the period, and the ratio of the amplitudes of the two machines for each natural mode of vibration. Take $E = 30 \times 10^6$ psi.

11.20   A beam 300 in long is supported at points 200 in apart, and overhangs 100 in. A body of weight 1 kip is attached midway between the supports, and another body of the same weight is attached at the free end of the overhang. The value of EI is 10^5 kip-in^2. Find the frequency, the period, and the ratio of the amplitudes for each natural mode of vibration. Neglect the weight of the beam.

11.21   A steel bar AB is 2 in diameter and 20 ft long (Fig. P11.21). It is completely fixed at each end and has a cross-bar DCE fixed to it transversely at the centre. The cross-bar is 4 ft long, and carries a body of weight 0.5 kip at each extremity. Determine the characteristics of the natural modes of vibration. Neglect the weight of the bars and also the deformation of the cross-bar. Take $E = 30 \times 10^3$ ksi and $G = 12 \times 10^3$ ksi. The deflection due to bending at the centre of a fixed-ended beam carrying a central load W is $WL^3/192EI$.

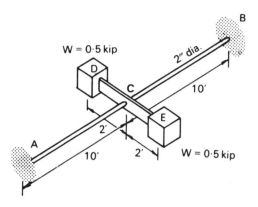

Fig. P11·21

11.22   For the mass and spring system shown in Fig. P11.22 find the frequency and the period of each of the natural modes of vibration.

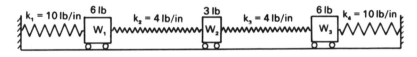

Fig. P11·22

11.23   A beam of length 300 in weighs 2400 lb and is simply supported at its ends. The value of EI is 5×10^8 lb-in^2. Find the frequency of the lowest mode of vibration (*a*) if the mass is considered as uniformly distributed, (*b*) if one-third of the mass is situated at the two one-third points, and (*c*) if one-half of the mass is situated at the mid-point.

APPENDIX A

Statics of Bars

A.1   Internal Force on a Cross-section

If a rigid bar is acted upon by external forces, then, in general, a force must be transmitted across any cross-section of the bar. The rigidity of the bar, in fact, depends upon its ability to transmit this force.

The bar AB (Fig. A.1 (a)) is in equilibrium under the action of any general, three-dimensional system of external forces. Any part of the bar is also in equilibrium. The force E which must be exerted upon the cut end C (Fig. A.1 (b)) is evidently the equilibrant of the external loads on the segment AC, and it can be determined by considering the equilibrium of the free-body AC. The force E' which is the reaction to E is similarly the equilibrant of the external forces acting upon the segment $C'B$ (Fig. A.1 (c)). It may be determined by

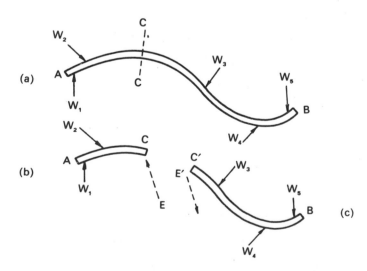

Fig. A·1

considering the equilibrium of the free-body $C'B$. Since the external forces on the whole bar are in equilibrium, E' is equal and opposite to E.

A.2 Stress-resultants

Force E is the resultant of the stress system which acts on the face C. Hence E is called a stress-resultant. It is usually expressed in terms of its components which are likewise known as *stress-resultants*.

The most common method is to specify E by six orthogonal Cartesian components, three forces and three couples. At the cross-section in question, C, the x axis is usually taken in a direction tangential to the axis of the bar and passing through the centroid of the cross-section. The principal axes of the cross-section (see Appendix B) are taken as the y and z axes. This set of co-ordinate axes are called the axes of the member at C.

The signs of the six component stress-resultants are determined in the same way as those for stresses (see Chapter 1). A sense is arbitrarily adopted for the x axis, i.e., for a bar AB, x either runs from A to B, or from B to A. Let us suppose it is chosen in the sense A to B. Then when we isolate the free-bodies AC and $C'B$ (Fig. A.2 (a) and (b)), we observe that x is directed outward from the cut face C, which is consequently termed the face of positive incidence. It is directed inward on face C', which is called the face of negative incidence.

Any component stress-resultant on face C (positive incidence) whose direction agrees in sense with the member axes is said to be positive. The components of E', on face C', are all in the opposite direction to the components of E. Consequently, we may equally well say that on the face of

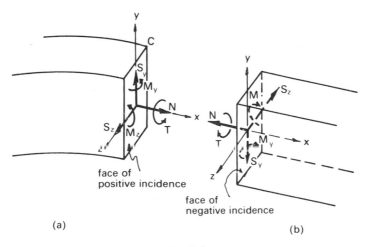

(a)   (b)

Fig. A·2

negative incidence a component stress-resultant is positive if its direction disagrees in sense with the member axes. The positive sense of the stress-resultants is shown in Figs A.2 (a) and (b). The sense for the couples is determined by the right-hand screw rule; they may be represented by the circular arrows or by double-headed arrows.

The pair of force components in the x direction is called the *axial force* and is denoted by N. The transverse force components are called *shear forces* and are denoted by S_y or S_z. The pair of couples acting around the x axis constitute the *twisting moment* (or torque), and is denoted by T. The couples around the y and z axes are called *bending moments*, and are denoted by M_y or M_z.

For a particular beam AB with a given system of external loads, the six component stress-resultants at a given section, C, are calculated by considering the equilibrium either of the free-body AC or the free-body CB. It is advisable to draw the chosen free-body, and to indicate thereon the given external forces and the unknown stress-resultants drawn with the positive sense. The equations of equilibrium are then employed.

Example A.1. The beam of Fig. A.3 (a), rectangular in cross-section, forms a quadrant of a circle lying in the horizontal plane. It is cantilevered from B and carries two loads at A— a vertical load of 5 kips, and a horizontal load of 7 kips directed towards the centre of the circle. Find the stress-resultants at C, midway between A and B.

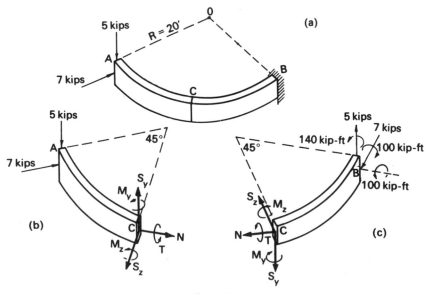

Fig. A·3

Solution. Let the x axis run from A to B. At any section take y upward and z radially outward.

It is most convenient to study free-body AC which is drawn in Fig. A.3 (b). The external loads are shown and the positive directions of the stress-resultants at C. The six equations of equilibrium are now applied. The axes referred to below are the member axes at C.

Parallel to x:	$+7\cos 45° + N$	$= 0$	$\therefore$	$N = -7/\sqrt{2}$ kips
Parallel to y:	$-5 + S_y$	$= 0$	$\therefore$	$S_y = +5$ kips
Parallel to z:	$-7\cos 45° + S_z$	$= 0$	$\therefore$	$S_z = +7/\sqrt{2}$ kips
Moments about x:	$-5 \times 20(1 - \cos 45°) + T = 0$		$\therefore$	$T = +29.3$ kip-ft
Moments about y:	$-7 \times 20\cos 45° + M_y$	$= 0$	$\therefore$	$M_y = +99.0$ kip-ft
Moments about z:	$+5 \times 20\cos 45° + M_z$	$= 0$	$\therefore$	$M_z = -70.7$ kip-ft

Alternatively, we may study the free-body CB which is drawn in Fig. A.3 (c). The external forces comprise the reactions at support B, which have to be calculated. At section C, the stress-resultants are shown with positive sense. Since we are now dealing with the face of negative incidence these are opposite to the directions of the co-ordinate axes. Application of the equilibrium equations will now yield the values of the stress-resultants. They will be the same in magnitude and sign as those determined above.

The values obtained for the stress-resultants will vary according to the position chosen for the origin of co-ordinates within the cross section, and this is to some extent arbitrary. Even the description of the total force on the cross-section CC in terms of six orthogonal components is arbitrary. For instance, instead of the two shear components S_y and S_z, we could define the shear as a single force S acting at θ to the y axis. Two values are still required, of course, to fully describe the shear force. The two bending moment components can also be combined in a similar manner. In some problems this alternative method of describing the shearing and bending produced by the external loads is more convenient.

In general, the stress-resultants vary from section to section along the bar. That is to say, each stress-resultant can be regarded as a function of x, where x is the distance along the bar axis measured from some arbitrary origin. Frequently the origin is taken at one end of the bar, but this is not essential. Sometimes it is convenient to take the origin elsewhere in order to simplify the resulting algebraic functions. In the case of tapered bars it is advisable to measure x from the origin of the taper. This is done in order that the section properties can be expressed as simple functions of x.

A.3   Equilibrium Relations between Load, Shear Force and Bending Moment

Suppose that a bar is subjected to a distributed external load,* the intensity of which varies from point to point. At any particular point, the load can be

---

* In theory, point loads, or concentrated loads, are often introduced for convenience. Since it is impossible in reality to apply finite loads to an infinitesimally small area, all loads are really distributed. The point load is simply an idealization for a load of high intensity over a short length of the bar.

resolved into components w_x, w_y and w_z. Then w_z is the component load per unit length normal to the bar axis in the z direction, and so on.

Equilibrium requires that certain relationships exist between the loading and the stress-resultants. These relationships will be derived here only for a straight segment of the bar. In Fig. A.4, the upper diagram shows the forces

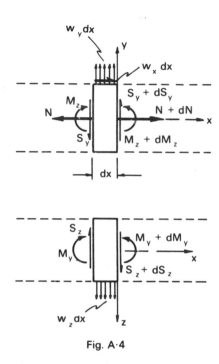

Fig. A·4

acting in the xy plane on an elemental length of the bar, while the lower diagram shows the forces in the xz plane on the same element. The axial force and the w_x load are shown only in the upper diagram.

From Fig. A.4, equilibrium in the x direction gives

$$-N + N + dN + w_x\, dx = 0$$

$$w_x = -\frac{dN}{dx} \qquad\qquad \text{(A.1)}$$

or

$$N = -\int w_x\, dx \qquad\qquad \text{(A.1a)}$$

Equilibrium in the y direction gives

$$-S_y + S_y + dS_y + w_y\,dx = 0$$

$$w_y = -\frac{dS_y}{dx} \qquad (A.2)$$

or

$$S_y = -\int w_y\,dx \qquad (A.2a)$$

Rotational equilibrium around the z axis gives

$$+S_y\,dx - w_y\,dx\frac{dx}{2} - M_z + M_z + dM_z = 0$$

In this equation the moment of the force $w_x\,dx$ has been ignored because, in the simple theory of bars, it is assumed that the dimensions of the cross-section are small. Moreover, surface forces in the x direction rarely occur in practice. Then, neglecting the second order differential term we have

$$S_y = -\frac{dM_z}{dx} \qquad (A.3)$$

or

$$M_z = -\int S_y\,dx \qquad (A.3a)$$

Referring now to the lower diagram of Fig. A.4, equilibrium in the z direction gives

$$-S_z + w_z\,dx + S_z + dS_z = 0$$

$$w_z = -\frac{dS_z}{dx} \qquad (A.4)$$

or

$$S_z = -\int w_z\,dx \qquad (A.4a)$$

Rotational equilibrium gives

$$-S_y\,dx + w_z\,dx\frac{dx}{2} - M_y + M_y + dM_y = 0$$

$$S_z = \frac{dM_y}{dx} \qquad (A.5)$$

or

$$M_y = \int S_z\,dx \qquad (A.5a)$$

It will be sufficient to consider examples where the loading is in the xy plane only, since the relationships for the xz plane are similar to those for the xy plane, although there is a difference in sign between equation (A.3) and (A.5).

Example A.2. For the beam of Fig. A.5 (a), express the bending moment and shear force at any section as functions of x, where x is measured from A.

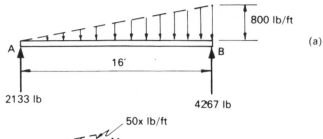

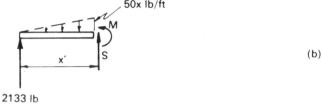

Fig. A.5

Solution. We can use the free-body equilibrium procedure as before. At x ft from A the load intensity is $50x$ lb/ft. From the free-body of Fig. A.5 (b) we see that

$$S = -2133 + \left(50x\frac{x}{2}\right) = -2133 + 25x^2 \text{ lb}$$

$$M = 2133x - 25x^2\frac{x}{3} = 2133x - \frac{25}{3}x^3 \text{ lb-ft}$$

Alternatively, we can employ the relationships of equations (A.2a) and (A.3a) (subscripts are omitted for convenience).

At x ft from A

$$w = -50x \text{ lb/ft}$$

(Downward loads are negative since y is positive upward.)

$$S = -\int w\,dx = +50\frac{x^2}{2} + c_1$$

When $x = 0$, $S = -2133$ $\therefore c_1 = -2133$

$$S = -2133 + 25x^2 \text{ lb}$$

$$M = -\int S\,dx = 2133x - 25\frac{x^3}{3} + c_2$$

When $x = 0$, $M = 0$ $\therefore c_2 = 0$

$$M = 2133x - 25\frac{x^3}{3} \text{ lb-ft}$$

It should be noted that both solutions are based on equilibrium considerations. For the derivation of expressions for S and M in terms of x, the free-body method is usually more convenient, especially if the loading is discontinuous. However, the differential relationships are extremely important in the development of the simple bar theory.

A.4   Graphs of Stress-resultants

The variation of the stress-resultants along a bar is conveniently illustrated by drawing graphs of these functions, the bar axis being used as a base-line. The following examples involve loads in the xy plane only, but similar procedures apply for loads in the xz plane.

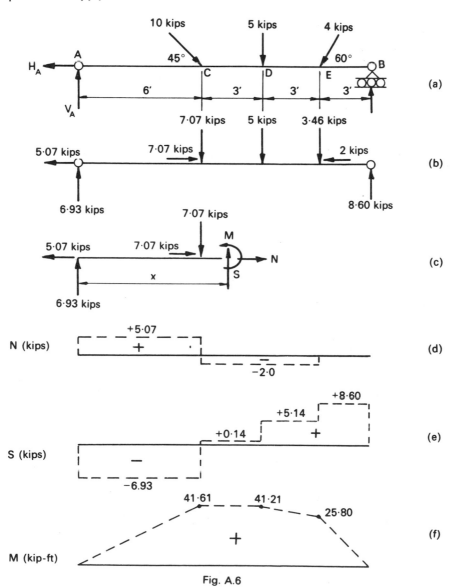

Fig. A.6

Example A.3. Draw graphs of N, S and M for the beam of Fig. A.6 (a).

Solution. It is convenient to resolve the inclined loads into their x and y components (Fig. A.6 (b)).

Since the loading is a discontinuous function of x, so also are the quantities N, S and M. The expressions for these quantities must therefore be derived separately for each segment of the beam. The calculations for segment CD are taken as typical.

Fig. A.6 (c) is a free-body taken to a typical point within segment CD. Equilibrium in the x direction gives

$$N = -2 \text{ kips}$$

Equilibrium in the y direction gives

$$S = +0.14 \text{ kip}$$

Rotational equilibrium gives

$$M = 42.42 - 0.14x$$

Expressions for the other segments are determined in a similar manner, and the graphs are then drawn (Figs A.6 (d), (e) and (f)).

For a straight beam loaded by concentrated forces, the quantities N and S are constant between load points. Where a y load is applied, the shear force S_y changes abruptly by an amount equal to this load. The graph of S for this problem could therefore be drawn by inspection of Fig. A.6 (b) without calculation. Similar remarks apply to the graph of N. Moreover, for a straight beam with point loads, the graph of M consists of a series of straight lines. For this graph it would therefore be sufficient to compute M at the loaded points.

The graphs can be checked to some extent by ensuring that the relationship between S and M is obeyed. Since $S = -dM/dx$, the ordinate of the S graph at any point must be equal to the slope of the M graph at the same point but opposite in sign. For instance, over the segment DE, the slope of the M graph $= (25.80 - 41.21)/3 = -5.14$ kips. The ordinate of the S graph in this region should therefore be $+5.14$ kips and this is so.

Example A.4. Construct graphs of S and M for the beam of Fig. A.7 (a).

Solution. From the loading we can see that the graph of S will have steps of -140 at A, $+100$ at B, $+200$ at C and -160 at D. Between the loads the graph will be horizontal. The graph of Fig. A.7 (b) can therefore be drawn directly.

For the M graph we compute M at B and C.

$$M_B = 140 \times 4 = 560 \text{ kip-ft}; \quad M_C = 160 \times 5 = 800 \text{ kip-ft}$$

When these two values are known the graph of M (Fig. A.7 (c)) can be drawn.

It would appear that in the two previous examples the shear force is indeterminate at sections of the beam where point loads are applied. It should be realized, however, that the point load is an idealization (see footnote to page 359). In fact, the load must be distributed over a finite length of the beam.

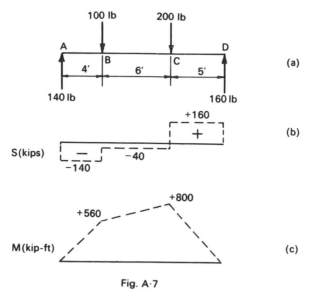

Fig. A·7

Fig. A.8 shows that with the real load, the graph of S has a finite slope and the graph of M is curved. If the load is idealized to a point force, the S graph becomes vertical, and the M graph acquires a sharp change of slope. In practice, the difference is usually not significant.

When the load is distributed over a considerable length of the beam, the graphs are drawn accordingly. We know that $S = -\int w\,dx$ and $M = -\int S\,dx$. Consequently, if the intensity of load is constant, S will be a linear function of

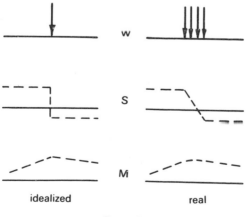

Fig. A·8

x, and M will be parabolic. For a beam loaded as shown in Fig. A.9 (a), the general shape of the S and M graphs is therefore known. The detailed construction of the graphs is simplified if we observe certain characteristics.

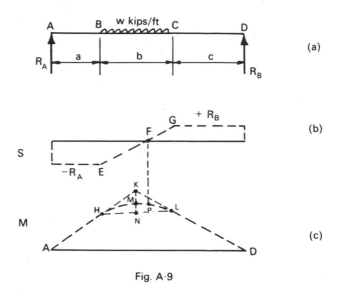

Fig. A·9

Once the reactions have been calculated, the graph of S can be drawn immediately (Fig. A.9 (b)). The slope of EFG is equal to the intensity of load, w, and the difference of ordinates at E and G is equal to wb, the total load on the segment BC.

If the total load were applied at the mid-point of BC, the bending moment graph would follow the line AKD (Fig. A.9 (c)). Point K is then easily found. Since the load is in fact distributed, the graph HKL must be modified to the parabola HML. This parabola is easily drawn if it is realized that M bisects NK, and that the tangent at M is parallel to HL. Because of the differential relationship we know that the turning point, P, of the M graph must lie below the point F on the shear force graph. It is of interest to note that the height NM at the centre of the parabolic segment is equal to $wb^2/8$, where w is the load intensity, and b is the length of the loaded segment.

For a cantilever of length L carrying a uniformly distributed load of intensity $-w$ (i.e. downward), the graphs of S and M are shown in Fig. A.10. At the free end, the value of S is zero, hence the tangent to the M graph is horizontal at this point.

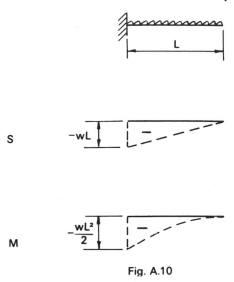

S $-wL$

M $-\dfrac{wL^2}{2}$

Fig. A.10

Provided that all the external forces on a bar are known, including the reactions, then the stress-resultant due to a given force system is the sum of the stress resultants due to individual loads which comprise the system. This often provides a means of obtaining the diagrams of stress-resultants, since the diagrams corresponding to individual loads are usually known. A simple illustration is given in Fig. A.11.

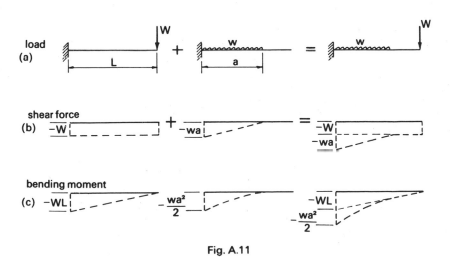

Fig. A.11

Example A.5. Draw graphs of shear force and bending moment for the beam of Fig. A.12 (a) which is subjected to a couple at the point C.

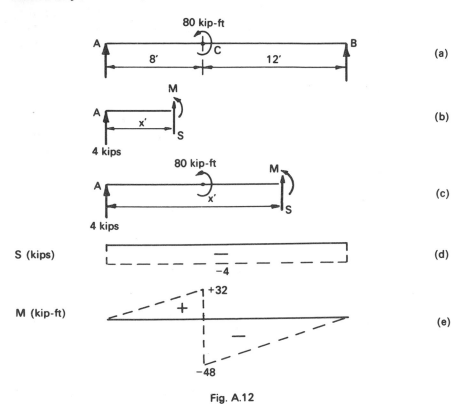

Fig. A.12

Solution. The reactions are calculated.

$$R_A = +4 \text{ kips}; \quad R_B = -4 \text{ kips}$$

For a point within the segment AC, S and M are found by means of the free-body of Fig. A.12 (b).

$$S_{AC} = -4 \text{ kips}; \quad M_{AC} = 4x \text{ kip-ft}$$

For a point within the segment CB, S and M are found by means of the free-body of Fig. A.12 (c).

$$S_{CB} = -4 \text{ kips}; \quad M_{CB} = 4x - 80 \text{ kip-ft}$$

These equations produce the graphs of Fig. A.12 (d) and (e). Note that the slope of the M graph is $+4$ at all points and this is consistent with the S graph, which everywhere has an ordinate of -4.

The reader should replace the couple at C by two unlike parallel forces, say -40 kips at 1 ft to the left of C, and $+40$ kips at 1 ft to the right of C, and draw graphs of S and M for this loading condition.

In the case of bent or curved bars, the usual practice is to draw the bar with its proper shape and then to plot the graphs with ordinates normal to this bar.

In the bent bar of Fig. A.13 (a), the separate segments of the bar are straight and the external loads are all point loads. Therefore, we know that N and S will be constant along a given segment of the bar, and M will vary linearly.

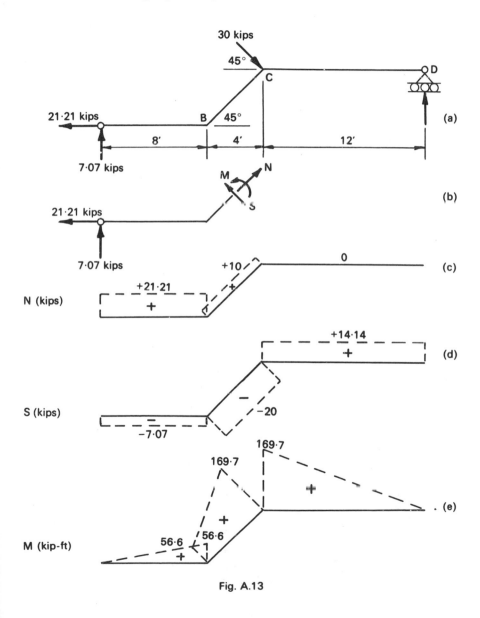

Fig. A.13

Hence, by calculating the main values of these functions, the graphs can be obtained (Fig. A.13 (c), (d) and (e)). For segment BC, a free-body as in Fig. A.13 (b) will be used. Note that N must be drawn tangential to the bar and S normal to the bar *at the cut section*.

Problems

A.1   Find the axial force, shear force and bending moment at the mid-point of the bar of Fig. PA.1.

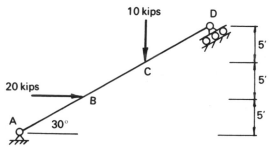

Fig. PA·1

A.2   Fig. PA.2 shows a bent bar $ABCD$ supported on rollers at A and pinned at D. The mid-points of AB, BC and CD are respectively E, F and G. Find the axial force, shear force and bending moment at the points E, F and G.

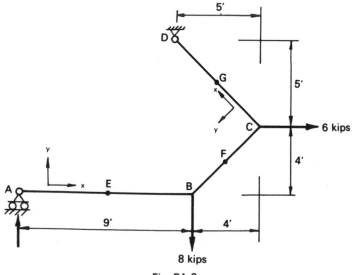

Fig. PA·2

A.3 For the bar of Fig. PA.3 calculate M, S and N at the mid-points of AB and BC.

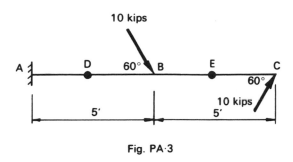

Fig. PA·3

A.4 For the bar shown in Fig. PA.4 write down $M_A M_B M_C M_D$. Find the shear force at A, B, just to the left of C, just above C, just to the left of E and just to the right of E. Find the axial force in the portions AB, BC, and CD.

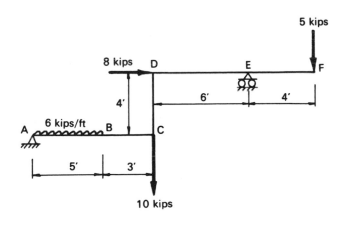

Fig. PA·4

A.5 Draw the graphs of M and S for the bar AD of Fig. PA.5
(a) When it carries the 12 kip load only.
(b) When it carries the 20 kip load only.
(c) When it carries both loads together.

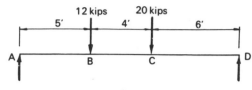

Fig. PA·5

A.6 For the beam of Fig. PA.6, express the bending moment and shear force at any section in terms of its distance from A. Note that separate expressions will be required for AB and BC. Draw the diagrams of M and S for the complete beam.

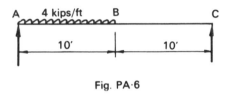

Fig. PA·6

A.7 A semicircular bar is loaded as shown in Fig. PA.7. Write expressions for M, S and N for the segments AC and CD in terms of θ, where θ is the angular distance of the section from OA.

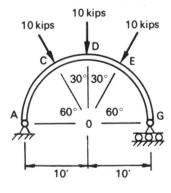

Fig. PA·7

A.8 The beam of Fig. PA.8 has couples C_1 and C_2 applied at B and C ($C_2 > C_1$). Find the reactions at A and D and draw the bending moment and shear force diagrams for the beam.

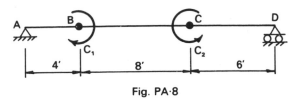

Fig. PA·8

A.9   Draw shear force, axial force and bending moment diagrams for the bar shown in Fig. PA.9. Mark the main values.

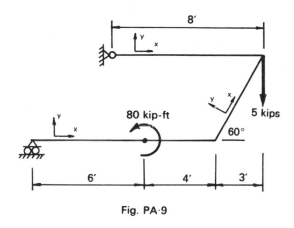

Fig. PA·9

A.10   Draw M and S diagrams for the beam of Fig. PA.10.

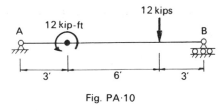

Fig. PA·10

A.11   Draw M and S diagrams for the beam of Fig. PA.11.

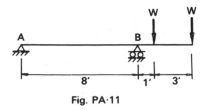

Fig. PA·11

A.12   Draw M and S diagrams for the beam of Fig. PA.12.

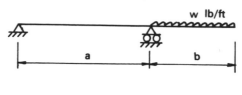

Fig. PA·12

A.13   Draw M and S diagrams for the beam of Fig. PA.13.

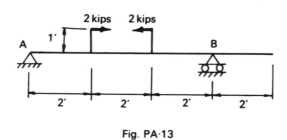

Fig. PA·13

A.14   Draw M and S diagrams for the beam of Fig. PA.14.

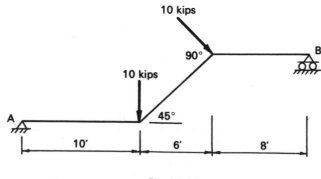

Fig. PA·14

A.15   Draw M, S and N diagrams for the beam of Fig. PA.15.

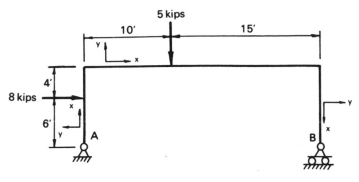

Fig. PA·15

A.16   The beam of Fig. PA.16 carries a distributed load the intensity of which varies from zero at A, to 3 kips/ft at D. For this beam find:

(a) The reactions at A and B.

(b) The equations of S and M using A as origin.

(c) The position of zero shear.

(d) The position and magnitude of the maximum bending moment.

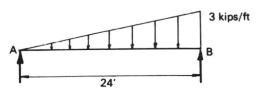

Fig. PA·16

A.17   A beam 12 ft long is cantilevered from one end and carries a uniformly distributed load of 2 kips per foot. By integration, derive expressions for the bending moment and shear force at a section of the beam distance x ft from the free end.

A.18   A simply supported beam of length L supports a distributed load the intensity of which is given by $w = w_0 \sin \pi x/L$ where w_0 is the intensity at the centre of the beam. Obtain expressions for the shear force and bending moment.

A.19   Fig. PA.19 shows the bending moment diagram for a beam 23 ft long. Sketch the shear force diagram and a diagram showing the loading of the beam.

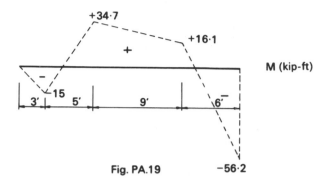

Fig. PA.19

A.20   Fig. PA.20 shows the bending moment diagram for a certain beam. Draw the shear force diagram and the loading diagram.

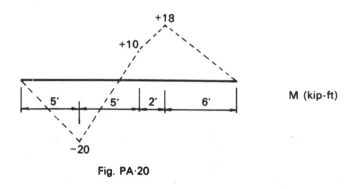

Fig. PA·20

A.21   Fig. PA.21 shows the bending moment graph for a horizontal beam carrying vertical loads. Obtain the load diagram and the shear force diagram.

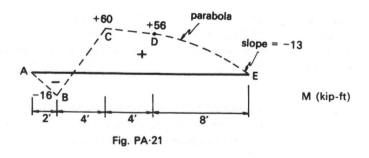

Fig. PA·21

A.22 A beam 17 ft long supports a loading which produces the bending moment diagram shown in Fig. PA.22. Draw the diagrams of shear force and loading and give numerical values at the salient points.

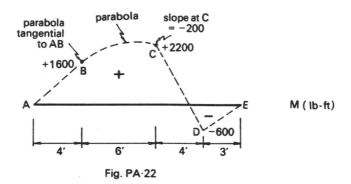

Fig. PA-22

A.23 Derive expressions for S and M for the beam of Fig. PA.23. Find the maximum value of M and the location of the point of contraflexure, i.e., the point where M is zero.

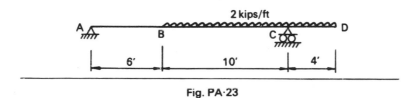

Fig. PA-23

A.24 In Fig. PA.24, the bars ABC and BD both lie in the horizontal plane and carry vertical loads as shown. Draw graphs of bending moment, twisting moment and shear force for each section of the structure.

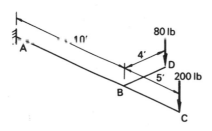

Fig. PA-24

A.25 The bar *ABC* lies in a horizontal plane (Fig. PA.25). The angle *ABC* is 120°. At *C* a couple of 6 kip-ft is applied whose axis is parallel to *AB*. Write expressions for *M*, *S* and *T* along *AB* and *BC*.

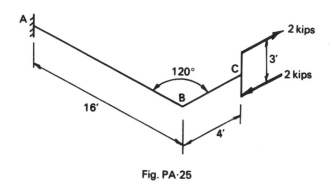

Fig. PA·25

A.26 The bar *ABC* is bent so that *BC* makes an angle of 30° with *AB* (Fig. PA.26). At *C* a couple of moment 10 kip-ft is applied, the axis of the couple being *BC*. Write expressions for *M* and *T* in *AB* and in *BC*.

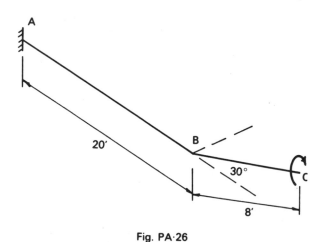

Fig. PA·26

A.27 Draw graphs of *M*, *S* and *T* for each section of the horizontal bar *ABCD* of Fig. PA.27.

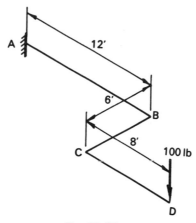

Fig. PA·27

A.28 In the bar of Fig. PA.28, each part is parallel to one of the axes, which are indicated at *A*. At any cross-section, the internal force can be expressed in terms of six components—axial force, shear forces parallel to *y* and *z*, bending moments about axes parallel to *y* and *z* and a twisting moment. Draw diagrams for each of these six components for the bar.

   $AB = 10$ ft; $BC = 8$ ft; $CD = 6$ ft; $DE = 12$ ft.

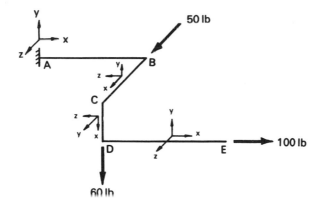

Fig. PA·28

A.29 The semicircular arch of Fig. PA.29 is pinned at *A* and *E* and has a hinge at *C*. For each of the portions *AB* and *BC* write equations for *M*, *S* and N in terms of θ. Find the positions and magnitudes of the maximum and minimum values of each quantity.

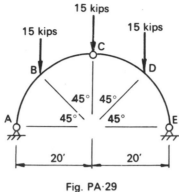

Fig. PA·29

A.30   Draw graphs of M and S for the beam of Fig. PA.30.

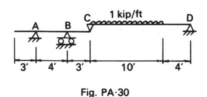

Fig. PA·30

A.31   The structure of Fig. PA.31 is direction fixed at A and G. It has hinges at C, D and E. Draw graphs of M, S and N for the portions AB, BC and CD.

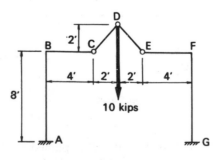

Fig. PA·31

Geometrical Properties of Plane Figures

The stresses in a bar depend upon the shape and size of the cross-section. Any such stress determination will require a knowledge of a certain geometrical property of the cross-section. Some geometrical properties are special to a particular class of problems in stress analysis—the so-called "torsion constant", for instance, required for the calculation of torsion stresses in bars of non-circular section. Other properties are of more general application, not only within the sphere of stress calculations but elsewhere.

These latter properties are reviewed in this Appendix, and the method of calculating them is briefly outlined. A table at the end of the Appendix gives some commonly used values.

The present work is related specifically to bar cross-sections. In the main chapters of the book, the plane of the bar cross-section is always taken as the yz plane. For consistency we shall therefore consider here figures lying in the yz plane.

B.1 First Moment of Area—Centroid

Consider a small element of area, dA, with co-ordinates y', z' (Fig. B.1). The *first moment* of this element about axis Oy' is defined as $dA \cdot z'$. The first moment about Oy' of the whole figure is denoted by $Q_{y'}$. Then

$$Q_{y'} = \int z'\, dA \tag{B.1}$$

Similarly, the first moment about axis Oz' is

$$Q_{z'} = \int y'\, dA \tag{B.2}$$

For regular figures, $Q_{y'}$ and $Q_{z'}$ may be found by simple integration.

Example B.1. For a semicircle of radius r (Fig. B.2) find Q about the diameter.

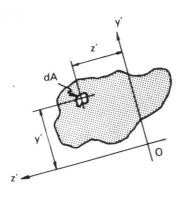

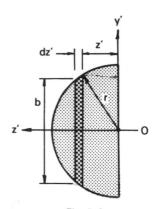

Fig. B·1

Fig. B·2

Solution. The element dA is taken as a strip parallel to the axis of moments

$$dA = b \cdot dz' = 2\sqrt{r^2 - z'^2}\, dz'$$

$$Q_{y'} = \int z'\, dA = \int_0^r z'(2\sqrt{r^2 - z'^2})\, dz'$$

$$= 2r^3/3$$

The first moment $z'\, dA$ will be positive or negative according to the sign of z'. Consequently, there will be some axes parallel to Oy' about which the first moment will be zero. Similarly, there will be an axis parallel to Oz' about which the first moment is also zero. These two axes intersect at the *centroid* (Fig. B.3).

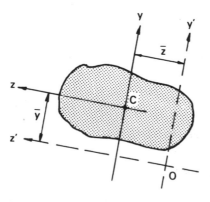

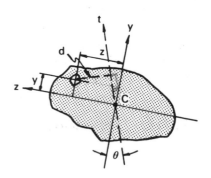

Fig. B·3

Fig. B·4

The centroid, C, of an area is the point such that the first moment of the area about any axis (in the plane of the figure) through C is zero.

It is clear that if the first moments Q_y and Q_z about two axes through C are zero, then the first moment about *any* axis through C is zero. Let Ct be any axis at θ to Cy (Fig. B.4). The element dA (at y, z) is d from Ct. Then

$$d = -y \sin \theta + z \cos \theta$$

$$Q_t = \int dA \cdot d = -\sin \theta \int y\, dA + \cos \theta \int z\, dA$$

$$= -Q_z \sin \theta + Q_y \cos \theta$$

$$= 0$$

About an axis Oy' parallel to Cy (Fig. B.3), the first moment is equal to $A\bar{z}$, where $\bar{z}$ is the distance of C from the axis Oy' (Fig. B.3). This is so because the distance of any element from Oy' can be expressed as z (the distance from Oy) plus $\bar{z}$. Hence

$$Q_{y'} = \int \bar{z}\, dA + \int z\, dA$$

$$= A\bar{z} + 0 \qquad\qquad\qquad (B.3)$$

This enables us to find the centroid of a given area. We select an arbitrary axis Oy' and find $Q_{y'}$ and A. Then from equation (B.3)

$$\bar{z} - Q_{y'}/A \qquad\qquad\qquad (B.4)$$

If necessary, the distance, $\bar{y}$, of C from an axis Oz' can be found in a similar way.

For instance, for the semicircle of Fig. B.2 we found that $Q_{y'} = 2r^3/3$. Now $A = \pi r^2/2$. Hence

$$\bar{z} = \frac{2r^3/3}{\pi r^2/2} = \frac{4r}{3\pi}$$

Any axis of symmetry passes through the centroid, therefore the axis Oz' in Fig. B.2 passes through the centroid and might be called Oz. So C is on the axis of symmetry at $4r/3\pi$ from the centre of the circle.

Equation (B.3) also enables us to locate the centroid of composite figures, such as that of Fig. B.5. This is, in fact, the usual problem, since for simple shapes the position of the centroid is given either by symmetry or by Table B.1

(p. 388). Shapes such as that of Fig. B.5 frequently occur as the cross-sections of bars which are extruded, built up from plates welded together, or cast (concrete or plastics).

The composite figure is divided into elements each of whose centroids is known. Suppose, then, that the ith element has an area A_i and has its centroid at y'_i from some arbitrary axis Oz'. Then

$$\text{the total moment about } Oz' = \sum A_i y'_i$$

$$\text{and the total area} = \sum A_i$$

Hence

$$\bar{y} = \frac{\sum A_i y'_i}{\sum A_i} \tag{B.5}$$

Similarly

$$\bar{z} = \frac{\sum A_i z'_i}{\sum A_i} \tag{B.6}$$

It is convenient to tabulate the values of A_i, $A_i y'_i$ and $A_i z'_i$ for ease of summation.

Example B.2. Locate the centroid of the composite figure of Fig. B.5.

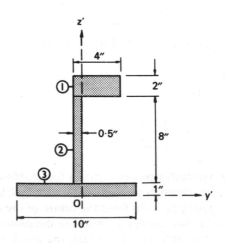

Fig. B·5

Solution. Arbitrary axes Oy' and Oz' are chosen. The figure is decomposed into elements which, in this case, are all rectangles. For each element is tabulated the area and the coordinates of the centroid.

Element	A_i	y_i	z_i	$A_i y_i$	$A_i z_i$
1	8	1.5	10	12	80
2	4	−0.25	5	−1	20
3	10	−0.25	0.5	−2.5	5
Sum	22			8.5	105

Then

$$\bar{y} = \frac{\Sigma A_i y_i}{\Sigma A_i} = \frac{+8.5}{22} = +0.386 \text{ in}$$

$$\bar{z} = \frac{\Sigma A_i z_i}{\Sigma A_i} = \frac{+105}{22} = +4.773 \text{ in}$$

That is to say, the centroid is at the point (+0.386, +4.773) relative to the arbitrary axes Oy', Oz'.

B.2   Second Moments of Area

Consider again the elemental area dA at position (y,z) relative to axes which, for the present, are taken as passing through the centroid C (Fig. B.6).

It is possible to define several second moments. The second moment about Cy is $dA z^2$. The second moment about Cz is $dA y^2$. The second moment about the x axis through C (the polar moment) is $dA r^2$. The product second moment is $dA yz$. The first three of these quantities are positive for all positions of dA. The last quantity is positive if y and z have the same sign, i.e., for elements dA in the first and third quadrants.

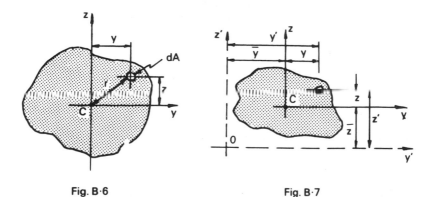

Fig. B·6   Fig. B·7

By analogy with similar properties of solid bodies, these *second moments of area* are commonly called *moments of inertia* and this term will be used. For the whole figure we have:

(a) The moment of inertia about the y axis, I_{yy}, is

$$I_{yy} = \int z^2 \, dA \tag{B.7}$$

(b) The moment of inertia about the z axis, I_{zz}, is

$$I_{zz} = \int y^2 \, dA \tag{B.8}$$

(c) The polar moment of inertia, I_p,* is

$$I_p = \int r^2 \, dA \tag{B.9}$$

This is a second moment about an axis normal to the plane of the figure. Since $r^2 = y^2 + z^2$ we have

$$I_p = \int y^2 \, dA + \int z^2 \, dA$$
$$= I_{zz} + I_{yy} \tag{B.10}$$

(d) The product of inertia about the y and z axes, I_{yz}, is

$$I_{yz} = \int yz \, dA \tag{B.11}$$

Similar definitions will apply to quantities taken with respect to axes Oy' and Oz' distant by $\bar{y}$ and $\bar{z}$ from the centroid, and parallel to Cy and Cz (Fig. B.7). A convenient relationship exists between the quantities taken about axes Oy' and Oz' and the corresponding quantities taken about Cy and Cz. This relation is expressed as the *theorem* of *parallel axes*.

$I_{y'y'}$ is defined as $\int (z')^2 \, dA$.
Since $z' = z + \bar{z}$

$$I_{y'y'} = \int z^2 \, dA + 2 \int z\bar{z} \, dA + \int \bar{z}^2 \, dA$$
$$= I_{yy} + 2\bar{z}Q_y + \bar{z}^2 A$$

Since Cy passes through the centroid, $Q_y = 0$.

$$I_{y'y'} = I_{yy} + A\bar{z}^2 \tag{B.12}$$

---

* I_p is sometimes denoted by I_{xx} since the x axis is normal to the plane of the figure.

Similarly

$$I_{z'z'} = I_{zz} + A\bar{y}^2 \tag{B.13}$$

$$I_{y'z'} = \int y'z'\,dA$$

$$= \int (y + \bar{y})(z + \bar{z})\,dA$$

$$= \int yz\,dA + \int y\bar{z}\,dA + \int \bar{y}z\,dA + \int \bar{y}\bar{z}\,dA$$

$$= I_{yz} + \bar{z}Q_z + \bar{y}Q_y + \bar{y}\bar{z}A$$

Thus

$$I_{y'z'} = I_{yz} + A\bar{y}\bar{z} \tag{B.14}$$

The form of equation (B.12) shows that the moment of inertia about the axis y through the centroid is always less than that about any parallel axis y', since the term $A\bar{z}^2$ is always positive. Similar remarks apply to equation (B.13), but not to (B.14) since the term $A\bar{y}\bar{z}$ may be either positive or negative.

Example B.3 Find the moment of inertia of a triangle about the base and about an axis through the centroid parallel to the base (Fig. B.8).

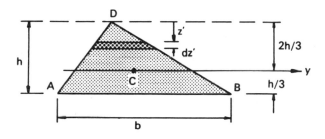

Fig. B·8

Solution. Consider first the moment of inertia about the axis through D parallel to the base.

$$dA = \frac{bz'}{h}\,dz'$$

$$\therefore \quad I_D = \int_0^h \left(\frac{b}{h}z'\right)z'^2\,dz' = \frac{bh^3}{4}$$

I_{yy} is less than this by $A\left(\dfrac{2h}{3}\right)^2$

$$I_{yy} = \frac{bh^3}{4} - \frac{bh}{2} \times \frac{4h^2}{9} = \frac{bh^3}{36}$$

Table B.1   —CENTROIDS OF FIGURES

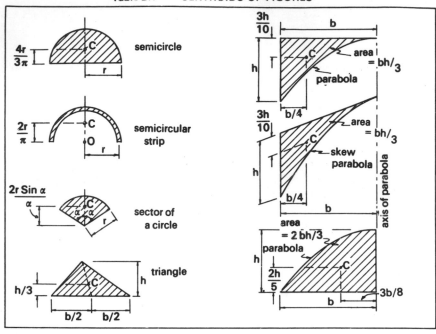

Table B.2   —MOMENTS OF INERTIA;   PRODUCTS OF INERTIA

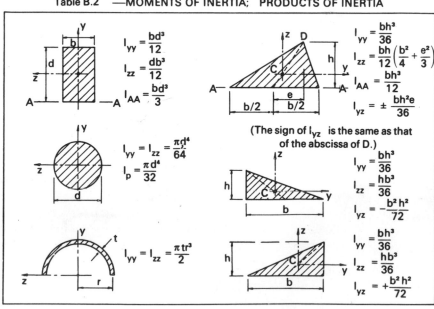

I_{AB} is greater than this by $A \left(\dfrac{h}{3}\right)^2$

$$I_{AB} = \frac{bh^3}{36} + \frac{bh}{2} \times \frac{h^2}{9} = \frac{bh^3}{12}$$

The moments of inertia and products of inertia of some simple figures are given in Table B.2 (p. 388).

B.3   Principal Axes

It was noted above that, unlike moments of inertia, the product of inertia is not necessarily positive. The product of inertia about axes yz of the figure of Fig. B.9 is clearly positive, since the majority of its area lies either in the first or in the third quadrants. If we consider the axes $y_1 z_1$, however, the product of inertia is negative. In fact, it is easy to show that $I_{y_1 z_1} = -I_{yz}$.

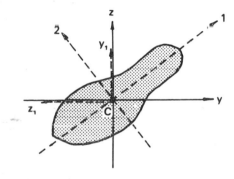

Fig. B·9

Thus, rotation of the axes by 90° has reversed the sign of I_{yz}. The change of I_{yz} as the axes rotate is a gradual one, hence, for some intermediate position of the axes, $I_{yz} = 0$. These axes are called principal axes of the figure.

The principal axes of a figure are those axes through the centroid for which the product of inertia is zero. These axes will be denoted by $C1$ and $C2$.

Suppose that I_{yy}, I_{zz} and I_{yz} are known. The corresponding properties with respect to any other axes y' and z' through C can be found. Suppose that Cy' makes an angle θ with Cy. The co-ordinates y', z' of any point are given by

$$y' = y \cos \theta + z \sin \theta$$

$$z' = -y \sin \theta + z \cos \theta$$

From the basic definition of the quantities concerned, it then follows that

$$I_{y'y'} = I_{yy} \cos^2 \theta + I_{zz} \sin^2 \theta - 2I_{yz} \cos \theta \sin \theta$$

or

$$I_{y'y'} = \tfrac{1}{2}(I_{yy} + I_{zz}) + \tfrac{1}{2}(I_{yy} - I_{zz}) \cos 2\theta - I_{yz} \sin 2\theta \qquad (B.15)$$

Similarly

$$I_{z'z'} = \tfrac{1}{2}(I_{yy} + I_{zz}) - \tfrac{1}{2}(I_{yy} - I_{zz}) \cos 2\theta + I_{yz} \sin 2\theta \qquad (B.16)$$

and

$$I_{y'z'} = \tfrac{1}{2}(I_{yy} - I_{zz}) \sin 2\theta + I_{yz} \cos 2\theta \qquad (B.17)$$

For $I_{y'z'}$ to be zero we require (equation B.17) that

$$\tan 2\theta = \frac{-I_{yz}}{\tfrac{1}{2}(I_{yy} - I_{zz})} \qquad (B.18)$$

and this equation determines the orientation of the principal axes $C1$ and $C2$ relative to the axes Cy and Cz.

Substituting the value of θ from equation (B.18) into equations (B.15) and (B.16) we obtain the values of the *principal moments of inertia*, I_1 and I_2.

$$\left.\begin{array}{c}I_1\\I_2\end{array}\right\} = \tfrac{1}{2}(I_{yy} + I_{zz}) \pm \sqrt{\left(\frac{I_{yy} - I_{zz}}{2}\right)^2 + I_{yz}^2}$$

If we define I_1 as the larger principal moment of inertia, we obtain specifically

$$I_1 = \tfrac{1}{2}(I_{yy} + I_{zz}) + \sqrt{\left(\frac{I_{yy} - I_{zz}}{2}\right)^2 + I_{yz}^2} \qquad (B.19)$$

$$I_2 = \tfrac{1}{2}(I_{yy} + I_{zz}) - \sqrt{\left(\frac{I_{yy} - I_{zz}}{2}\right)^2 + I_{yz}^2} \qquad (B.20)$$

The axis $C1$ is at an angle θ with Cy. The value of θ is obtained from equation (B.18), and to ensure that $C1$ is the axis about which I is greater, we take $0 < 2\theta < 180°$ if I_{yz} is negative, and $180° < 2\theta < 360°$ (or $-180° < 2\theta < 0$) if I_{yz} is positive.

The relationship between $I_{yy}, I_{zz}, I_{yz}, I_1$ and I_2 is exactly the same as that which exists between $\sigma_x, \sigma_y, \tau_{xy}, \sigma_1$ and σ_2 (see equations 7.7 and 7.8 of Chapter 7). Having determined I_{yy}, I_{zz} and I_{yz} for a given figure, it is therefore possible to use the Mohr's circle construction to determine the position of the principal axes and the magnitude of the principal moments of inertia.

If this construction is used (Fig. B.10), the point Y is plotted with co-ordinates (I_{yy}, I_{yz}) and the point Z is plotted with co-ordinates $(I_{zz}, -I_{yz})$. The circle is drawn on diameter YZ. The direction of the principal axes and the magnitudes of the principal moments of inertia are then obtained as described in Chapter 7.

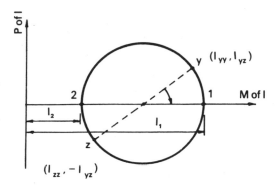

Fig. B·10

Example B.4. Determine the principal axes and the principal moments of inertia of the triangle shown in Fig. B.11 (a).

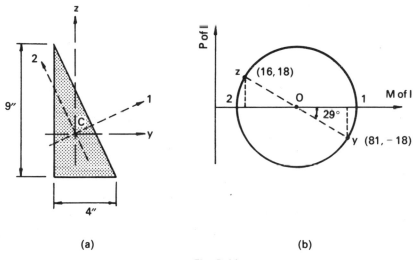

(a)   (b)

Fig. B·11

Solution. By reference to Table B.2 we find that

$$I_{yy} = \frac{bh^3}{36} = \frac{4 \times 9^3}{36} = 81 \text{ in}^4$$

$$I_{zz} = \frac{hb^3}{36} = \frac{9 \times 4^3}{36} = 16 \text{ in}^4$$

$$I_{yz} = \frac{-b^2 h^2}{72} = \frac{-4^2 \times 9^2}{72} = -18 \text{ in}^4$$

From equation (B.18)

$$\tan 2\theta = \frac{-(-18)}{\frac{1}{2}(81-16)} = \frac{18}{32.5} = +0.5538$$

Since I_{yz} is negative,

$$2\theta = 28° 59'$$

$$\theta = 14° 30'$$

Thus axis $C1$ is $14° 30'$ from Cy and $C2$ is $104° 30'$ from Cy.

$$\frac{I_{yy} + I_{zz}}{2} = \frac{81+16}{2} = 48.5 \text{ in}^4$$

Then from equation (B.19)

$$I_1 = 48.5 + \sqrt{32.5^2 + 18^2} = 48.5 + 37.2 = 85.7 \text{ in}^4$$

and from equation (B.20)

$$I_2 = 48.5 - 37.2 = 11.3 \text{ in}^4$$

Alternatively, once I_{yy}, I_{zz} and I_{yz} have been determined, the principal axes and principal moments of inertia can be found by Mohr's circle (Fig. B.11 (b)).

Point Y is plotted with co-ordinates $(81, -18)$ and point Z is plotted with co-ordinates $(16, 18)$. When the circle is drawn on diameter YZ we find that

$$I_1 = 86 \text{ in}^4 \quad \text{and} \quad I_2 = 11 \text{ in}^4$$

The angle $Y01$ is $29°$, from which we see that axis $C1$ is $14° 30'$ from Cy.

B.4   Composite Figures

The most common problem is the calculation of I_{yy} and I_{zz} for composite figures possessing at least one axis of symmetry.

If there are two axes of symmetry, the centroid position is known. Otherwise it is located by the method of section B.1. If there is even one axis of symmetry, the direction of the principal axes are known since an axis of symmetry is a principal axis and the two principal axes are at right-angles.

The composite figure is divided into simple elements. Then I_{yy} for the whole figure is the sum of I_{yy} for all the components. For any one component,

$$I_{yy} = I_{yy}^0 + Az^2$$

where I_{yy}^0 is the moment of inertia of the element about an axis through its own centroid parallel to the y axis of the whole figure, and z is the distance of the element centroid from the y axis.

A similar calculation gives I_{zz}.

Example B.5. For the figure shown in Fig. B.12, find the position of the centroid and the values of I_{yy} and I_{zz}. (This section was used in example 4.2.)

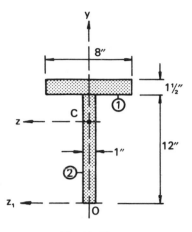

Fig. B·12

Solution. The centroid, C, lies on the axis of symmetry but the vertical position is not known. The temporary axis Oz' is therefore adopted, and the position of C found as described in section B.1. This calculation occupies the first part of the tabulation below.

Element	Calculation of centroid					Calculation of I_{yy}		Calculation of I_{zz}	
	A	y'	Ay'	y	z	I^0_{yy}	Az^2	I^0_{zz}	Ay^2
1	12	12.75	153	3.375	0	64	0	2.25	136.69
2	12	6	72	−3.375	0	1	0	144	136.69
Sum	24		225			65	0	146.25	273.38

$$\bar{y} = \frac{\Sigma\,Ay'}{\Sigma\,A} = \frac{225}{24} = 9.375 \text{ in} \qquad I_{yy} = 65 \text{ in}^4 \qquad I_{zz} = 419.63 \text{ in}^4$$

When C has been located the origin is shifted to C and the co-ordinates y, z of the centroid of each element are noted. Then I_{yy} and I_{zz} for each element are found. For instance, for element 1,

$$I_{zz} = I^0_{zz} + Ay^4$$

$$= \frac{8 \times 1.5^3}{12} + 12 \times 3.375^2$$

$$= 2.25 + 136.69 \text{ in}^4$$

These calculations occupy the latter part of the tabulation. The I_{yy} and I_{zz} for the whole figure are then obtained by summation.

When the figure has no axis of symmetry it is necessary to find the direction of the principal axes and principal moments of inertia. The steps in this procedure are:

(a) Adopt axes Oy' and Oz' and locate the centroid.

(b) Shift the origin to the centroid and with axes Cy, Cz compute I_{yy}, I_{zz} and I_{yz} (as in example B.5).

(c) From these values locate the direction of the principal axes $C1$ and $C2$, and compute I_1 and I_2 as in example B.4.

Example B.6. Find the principal axes and the principal moments of inertia of the figure shown in Fig. B.13 (a).

Solution. The calculations of steps (a) and (b) are shown in the following tabulation.

			Centroid						I_{yy}		I_{zz}		I_{yz}	
Element	A	y'	Ay'	z'	Az'	y	z	I^0_{yy}	Az^2	I^0_{zz}	Ay^2	I^0_{yz}	Ayz
1	4	2	8	12	48	+2.736	+7.153	0.33	204.66	5.33	29.94	0	+78.4
2	5	0.25	1.25	6.5	32.5	+0.986	+1.653	41.67	13.66	0.01	4.86	0	+8.2
3	9	−2.5	−22.5	0.75	6.75	−1.764	−4.097	1.69	151.05	27.00	28.01	0	+65.0
Sum	18		−13.25		87.25			43.69	369.37	32.34	62.81	0	+151.6

$$\bar{y} = \frac{-13.25}{18} \qquad \bar{z} = \frac{87.25}{18}$$

$$= -0.736 \text{ in} \qquad = +4.847 \text{ in}$$

$$\underbrace{413 \text{ in}^4}_{= I_{yy}} \quad \underbrace{95 \text{ in}^4}_{= I_{zz}} \quad \underbrace{+152 \text{ in}^4}_{= I_{yz}}$$

The principal axes and moments of inertia can now be determined by drawing a Mohr's circle (Fig. B.13 (b)).

Alternatively, by calculation, we have

$$\tan 2\theta = \frac{-152}{\frac{1}{2}(413 - 95)} = \frac{-152}{156} = -0.975$$

Since I_{yz} is positive, 2θ lies between 180° and 360°

Hence

$$2\theta = 315° 42'$$

$$\theta = 157° 50'$$

Thus axis $C1$ is 157° 50′ anticlockwise from Cy.

$$\left. \begin{matrix} I_1 \\ I_2 \end{matrix} \right\} = \frac{413 + 95}{2} \pm \sqrt{156^2 + 152^2}$$

$$= 254 \pm 218$$

$$I_1 = 472 \text{ in}^4$$

$$I_2 = 36 \text{ in}^4$$

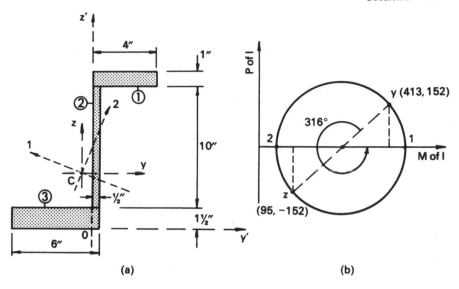

Fig. B·13

Problems

B.1 For the figure shown in Fig. PB.1 find the moment of inertia about the z axis. (This beam section is used in example 4.1.)

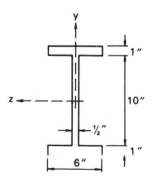

Fig. PB·1

B.2 Find I_{yy} and I_{zz} for the figure of Fig. PB.2.

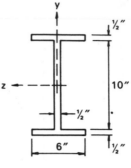

Fig. PB·2

B.3   For the figure of Fig. PB.3 find I_{yy} and I_{zz}. (This beam section is used in example 8.3.)

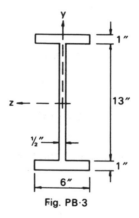

Fig. PB·3

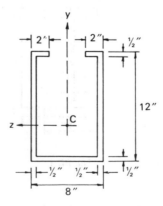

Fig. PB·4

B.4   For the figure of Fig. PB.4 calculate the position of the centroid and the value of I_{zz}. (This beam section is used in example 5.4.)

B.5   For the figure of Fig. PB.5 find the position of the centroid and the values of I_{yy} and I_{zz}.

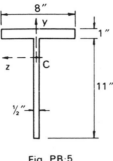

Fig. PB·5

B.6   Find I_{zz} for the figure shown in Fig. PB.6.

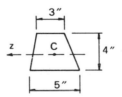

Fig. PB·6

B.7   Find the position of the centroid, the direction of the principal axes, and the values of the principal moments of inertia for the figure of Fig. PB.7.

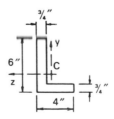

Fig. PB·7

B.8   Find the position of the centroid, the direction of the principal axes, and the values of the principal moments of inertia for the figure of Fig. PB.8.

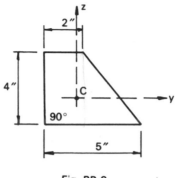

Fig. PB·8

B.9   Find the position of the centroid, the direction of the principal axes, and the values of the principal moments of inertia for the figure of Fig. PB.9.

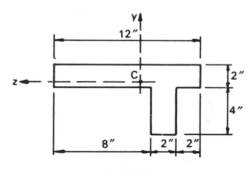

Fig. PB·9

Answers to Problems

Chapter 1

1.1 23,900 psi; normal stress

1.2 5310 lb

1.3 (a) 22,500 lb at 0.833 in below the top on the vertical centreline
(b) $N = 22,500$ lb
$M = 9375$ lb-ln

1.4 $833y^2 \delta y$; 60,000 lb; 4.5 in from C

1.5 72 kips

1.6 253.5 kips at 5.22 in from the top

1.7 48 kips

1.8 (a) 10,000 lb in the y direction cutting the z axis at $+5\pi$ in from O
(b) $S = 10,000$ lb; $T = 50,000\pi$ lb-in

1.9 −0.0025

1.10 $\epsilon = 7.2 \times 10^{-4}$; $e = 0.144$ in

1.11 (a) 3.6×10^{-4}
(b) 0.0384 in

1.12 0.0405 in³; 0.0015

1.13 −0.1077; −0.0405

1.14 0.001 in; $0.001/\sqrt{2}$ in

Chapter 2

2.1 60.8 kips

2.2 (a) 90,000 psi
(b) 3.575×10^{-3}
(c) 0.179 in

2.3 6.20 kips

2.4 (a) 31×10^6 psi
(b) 20.4×10^6 psi
(c) 24.6×10^6 psi

2.5 0.15 in
2.6 1:2
2.7 $-0.05\sqrt{2}$ in
2.8 (a) $e_{AB} = 0.2$ in; $e_{BC} = -0.1$ in
      (b) 0.2 in downward; 0.34 in to the right
2.9 -0.0015; 0.1 radian

Chapter 3

3.1 $\sigma = 8500$ psi
      $E = 29.4 \times 10^6$ psi
      $U = 469$ in-lb
      $f = 4.17 \times 10^{-6}$ in/lb
      $k = 2.4 \times 10^5$ lb/in
3.2 5100 psi; 1.53 in
3.3 $U = 868$ ft-lb
3.4 10,560 in-lb; 8450 in-lb
3.5 $k = 1.48 \times 10^5$ lb/in
      $f = 6.75 \times 10^{-6}$ in/lb
      At $\sigma = 10^5$ psi,
      $k_t = 7.95 \times 10^4$ lb/in
      $f_t = 1.26 \times 10^{-5}$ in/lb
3.6 (a) 14,140 lb
      (b) 0.0256 in
3.7 (a) 149,000 psi
      (b) 0.22 in
      (c) 13,700 lb/in
3.8 2.49 in
3.9 (a) 0.065 in
      (b) More than twice
3.10 (a) $U_{AB} = 2$ in-lb; $U_{BC} = 1$ in-lb
      (b) To increase
3.11 $f = 0.175$ in/kip
      $k = 5.714$ kips/in
3.12 (a) $\sigma_c = 2190$ psi
          $\sigma_s = 13,150$ psi
          $U = 9200$ in-lb
      (b) $\sigma_c = 1770$ psi
          $\sigma_s = 31,900$ psi
3.13 (a) 7.7 in from the bottom
      (b) 0.104 in
      (c) 2170 psi

3.14 (a) $\sigma_A = 14.1$ ksi; $\sigma_B = 33.9$ ksi
   (b) $k = 353$ kips/in
3.15 -2400 lb; -4800 psi; -0.0096 in
3.16 $-17,600$ psi
3.17 9.9 ksi
3.18 (a) $N = 144x - 1.5x^2$ lb
   (b) 2.54×10^{-3} in
   (c) 2080 rpm

Chapter 4

4.1 (a) 5 in
   (b) 1600 psi
   (c) 4800 psi
   (d) 400 kip-in
4.2 1815 kip-in; 17.33 ksi
4.3 16/9 in
4.4 21.7 kip-in
4.5 $b = 6.7$ in; $d = 13.4$ in
4.6 (a)

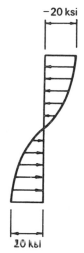

-20 ksi

20 ksi

   (b) 200 kips at 3.125 in from the N.A.
   (c) 1250 kip-in
   (d) 5×10^{-3}
   (e) 10^{-3} in^{-1}
4.7 2860 kip-in
4.8 5.5 in; 5×10^{-3}; 4.09×10^{-3}; 9.09×10^{-4} in^{-1}; no

4.9   (a) 4×10^{-4}
(b) 3.33×10^{-4}
(c)

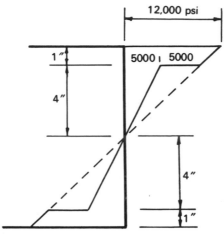

(d) 819 kip-in

4.10   11.67 in; $\sigma_1 = 3.64$ ksi; $\sigma_2 = 2.55$ ksi

4.11   (a) 946.7 kip-in
(b) 5.6%
(c) 4.17×10^{-4} in^{-1}

4.12   530 kip-in

4.13   1560 kip-in; 2.12×10^{-4} in^{-1}

4.14

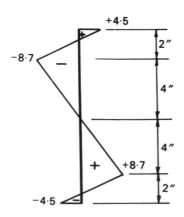

4.15 $-0.22\,bh^2\,\sigma_y$

4.17 0.69 in; 0.0092 radian

4.18 -0.25 in

4.19 -0.36 in at 10.39 ft from A

4.20 $\dfrac{Wa}{EI}\left(\dfrac{L^2}{8}-\dfrac{aL}{2}-\dfrac{a^2}{3}\right)$

4.21 $\dfrac{dv}{dx}=\dfrac{L^3\,w_0}{\pi^3\,EI}\cos\dfrac{\pi x}{L};\ v=\dfrac{L^4\,w_0}{\pi^4\,EI}\sin\dfrac{\pi x}{L}$

4.23 For AC, $EIv=-20.8x^4+365x^3-31{,}900x$
For CB, $EIv=52.1\bar{x}^3-20{,}200\bar{x}$
   (ft, lb units; $\bar{x}$ is measured from B)
-0.096 in at 11.37 ft from B

4.24 $-\dfrac{2233}{EI}$ ft (EI in feet, kip units)

4.25 -1.16 in

4.26 $0<x<a\quad v=D(x^3+2L^2x-6aLx+3a^2x)$
$a<x<L\quad v=D(x^3-3Lx^2+2L^2x+3a^2x-3a^2L)$
$\theta_A=D(2L^2-6aL+3a^2)$
$\theta_B=D(3a^2-L^2)$
$\theta_E=D(2L^2-6aL+6a^2)$
$v_E=D(4a^3-6a^2L+2aL^2)$

   where $D=\dfrac{C}{6EIL}$

4.27 $v=\dfrac{w}{24EI}\{-x^4+(2L-2a^3/L)x^3-(L^3-2a^3L)x\}$

   $\theta_B=\dfrac{w}{24EI}(L^3-4a^2L)$

4.28 $v=\dfrac{C}{EI}\left(\dfrac{x^3}{6b}-\dfrac{x^2}{2}+\dfrac{bx}{3}\right)$

4.29 $C_A = 33.6$ kip-ft; $C_B = -33.6$ kip-ft

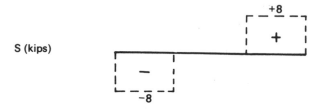

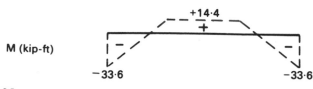

4.30 $M = \dfrac{25}{3} x^3 - 750x^2 + 18,000x - 90,000$ lb-ft

4.31 -0.58 in

4.32

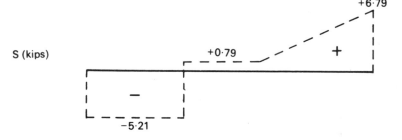

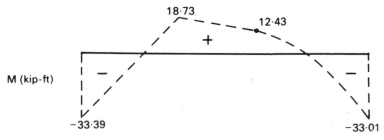

4.33 $R_A = 10.13$ kips
$R_B = 3.87$ kips
$C_A = 49.6$ kip-ft

4.34   3.12 kips

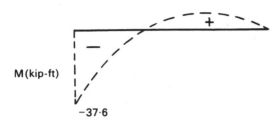

M(kip-ft)

$-37 \cdot 6$

4.35 $C_A = \dfrac{5}{192} wL^2$

$C_B = -\dfrac{11}{192} wL^2$

4.36   117 ft-lb

4.37 $L^5 w^2 / 4EI\pi^4$

4.38 $\dot{C}^2 L/6EI$; $\dot{C}L/3EI$

4.39   354 lb-ft; 0.708 in

4.40   As for problem 4.17

4.41   As for problem 4.22

4.42   As for problem 4.24

4.43 $\theta_A = -200/EI$ radian
$v_D = -917/EI$ ft

4.44 $W_2/W_1 = 4$

4.45

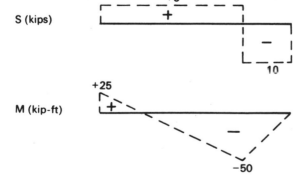

S (kips)

+5

+

−

10

+25

M (kip-ft)

+

−

−50

4.46 $\theta_A = -360/EI$ radian
$v_C = -3360/EI$ ft

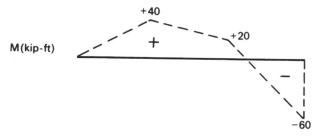

4.47 0.923 kips
4.48 $32W/EI$ ft (EI in kip-ft^2)
4.49 $R_A = 6EI/L^2$
$C_A = 4EI/L$
$R_B = -6EI/L^2$
$C_B = 2EI/L$
4.50 (a) 6.068×10^{-4} in/lb
6.32×10^{-6} rad/lb
(b) 6.32×10^{-6} in/lb-in
1.37×10^{-7} rad/lb-in
4.51 $R_P = -2.68 \times 10^3$ lb/in; $R_Q = 2.68 \times 10^3$ lb/in;
$C_P = -5.36 \times 10^5$ lb-in/in; $C_Q = -2.68 \times 10^5$ lb-in/in
4.52 $0 < \epsilon < \epsilon_y$ $M = EI\rho$

$\epsilon_y < \epsilon$ $M = EI\rho_y\left(\dfrac{3}{2} - \dfrac{\lambda^2}{2}\right)$

where $\lambda = \dfrac{2\epsilon_y}{d\rho}$

Chapter 5

5.1 (a) 113 lb/in
(b) 417 lb/in
5.2 (a) 522 lb/in
(b) 1711 lb/in
(c) 5184 psi
5.3 1.25
5.4 654 lb/in
5.5 375 lb/in
5.6 469 lb/in; 625 lb/in
5.7 (a) 5.184 kips/in; 1.728 ksi
(b) 2.67 ksi

5.8 $q_A = 113$ lb/in; $q_B = 95.3y - 10.66y^2$; 948 lb; 104 lb

5.9 $q = \dfrac{S}{a^4}(8y^3 - 6\sqrt{2}ay^2 + \sqrt{2}a^3)$; no

5.10 102 lb/in

5.11 431 lb/in

5.12 $S/209.6$ psi

5.13 On the axis of symmetry, 12.75 in from the centre of the circle

5.14 $h^2/1.2L^2$; $h/L = 0.35$

5.15 18.2 ft

5.16 $3S^2 dx/5Ght$ where h is the total depth and t is the web thickness

5.17 $\theta_A = \dfrac{L}{3EI}\left(1 + \dfrac{\alpha}{2}\right)$; $\theta_B = \dfrac{-L}{6EI}(1 - \alpha)$

where $\alpha = \dfrac{1}{GA} \times \dfrac{6EI}{L^2}$

5.18 $R_A = \dfrac{6EI}{L^2}\left(\dfrac{1}{1 + 2\alpha}\right)$; $C_A = \dfrac{2EI}{L}\left(\dfrac{1 - \alpha}{1 + 2\alpha}\right)$

$R_B = \dfrac{-6EI}{L^2}\left(\dfrac{1}{1 + 2\alpha}\right)$; $C_B = \dfrac{4EI}{L}\left(\dfrac{1 + \alpha/2}{1 + 2\alpha}\right)$

where $\alpha = \dfrac{1}{GA} \times \dfrac{6EI}{L^2}$

Chapter 6

6.1 5300 psi; 1.05°

6.2 $d = 3.63$ in; $\phi = 2°47'$; $k = 1.67 \times 10^6$ lb-in/radian

6.4 6.32 in; 4.76 in

6.5 $\frac{1}{2}$ inch bar: $k = 1840$ lb-in/rad; $f = 5.44 \times 10^{-4}$ rad/lb-in
   $\frac{3}{4}$ inch bar: $k = 9320$ lb-in/rad; $f = 1.07 \times 10^{-4}$ rad/lb-in
   Assembly: $k = 1530$ lb-in/rad; $f = 6.51 \times 10^{-4}$ rad/lb-in

6.6 Aluminium: $k = 1230$ lb-in/rad; $f = 8.15 \times 10^{-4}$ rad/lb-in
   Brass: $k = 17,400$ lb-in/rad; $f = 5.75 \times 10^{-5}$ rad/lb-in
   Assembly; $k = 18,630$ lb-in/rad; $f = 5.37 \times 10^{-5}$ rad/lb-in
   $\tau_{aluminium} = 2690$ psi; $\tau_{brass} = 8060$ psi

6.7 $10.33T^2 L/\pi Gd^4$; $20.67TL/\pi Gd^4$ radian

6.8 $d = 0.53$ in; $D = 5.3$ in; $n = 12$; $k = 66$ lb/in

6.9 (a) 0.8 in; (b) 29.3 lb/in; (c) 0.375 in; (d) 2.06 in-lb

6.10 (a) $\dfrac{32P^2 R^3 n\cos\alpha}{Gd^4}$; (b) $\dfrac{64P^2 R^3 n\sin^2\alpha}{2.5Gd^4\cos\alpha}$; (c) 1.04

6.11   270 kip-in; 0.687°

6.12 $D = 4$ in; $d = 2$ in

6.13   9130 lb-in; 1500 psi; 0.517×10^{-3} rad/in; 30

6.14   665 lb/in; 5540 psi. The corners will remain in a plane normal to the axis due to symmetry

6.15   7150 psi; 13.32 in-lb

6.16   833 lb-in; 2.47°

6.17   9%

6.18   31,200 lb-in

Chapter 7

7.1 $\sigma_x = +850$ psi; $\tau_{Xy} = -780$ psi

7.2 $\sigma_x = +850$ psi; $\tau_{Xy} = -780$ psi
$\sigma_y = -50$ psi; $\tau_{Yx} = -780$ psi
$+28.2°$ and $-88.2°$ to the X plane
$\tau_{max} = 900$ psi on planes at $+15°$ and $-75°$ to the X plane

7.3   (a) -8000 psi; $+3464$ psi
(b) -4000 psi; $+3464$ psi

7.4   As for problem 7.3

7.5   (a) $\sigma_1 = 2270$ psi at $-10.27°$ to Ox
$\sigma_2 = -6270$ psi at $+79.73°$ to Ox
(b) $+20.78°$ and $-41.32°$ with plane X: 3750 psi

7.6   As for problem 7.5

7.7   (a) $\sigma_1 = +8480$ psi at $+31.7°$ to Ox
$\sigma_2 = -480$ psi at $+31.7°$ to Oy
(b) $\sigma_{\bar{X}\bar{x}} = +8000$ psi; $\sigma_{\bar{Y}\bar{y}} = 0$; $\tau_{\bar{X}\bar{y}} = -2000$ psi
(c) $\tau_1 = 240$ psi; $\tau_2 = 4240$ psi; $\tau_3 = 4480$ psi
(d) $\tau_{max} = 4480$ psi

7.8   As for problem 7.7

7.10   0.247; the plane at $54°\,44'$ from P_1

7.11   Between 3333 psi and 18,000 psi

7.12 -600 psi; 800 psi

7.13   3500 psi

7.14 $63°\,26'$

7.15   (a) $\tau = +7000$ psi; $\sigma_1 = +12,610$ psi; $\sigma_2 = -2610$ psi
(b) $\tau = -13,000$ psi; $\sigma_1 = +12,340$ psi; $\sigma_2 = -14,340$ psi

7.16 $\tau = +4920$ psi; $\sigma_1 = +2280$ psi; $\sigma_2 = 1 -11,960$ psi

7.17 $+8200$ psi; $+1140$ psi

7.18 $\epsilon_x = 4.83 \times 10^{-4}$ $\alpha_x = -3.214 \times 10^{-4}$
$\epsilon_y = -2.83 \times 10^{-4}$ $\alpha_y = +3.214 \times 10^{-4}$
$\gamma_{xy} = +6.428 \times 10^{-4}$

7.19 As for problem 7.18

7.20 (a) At 30° to ϵ_1, $\epsilon = 5.5 \times 10^{-4}$, $\alpha = 4.34 \times 10^{-4}$ radian
At 60° to ϵ_1, $\epsilon = 0.5 \times 10^{-4}$, $\alpha = 4.34 \times 10^{-4}$ radian
(b) $\pm 63.42°$ to ϵ_1
(c) 5×10^{-4} radian

7.21 $\epsilon_1 = 9.9 \times 10^{-4}$ at $-16.8°$ to Ox
$\epsilon_2 = 0.9 \times 10^{-4}$ at $-16.8°$ to Oy
-6×10^{-4} radian; 1.08×10^{-3} radian

7.22 As for problem 7.21

7.23 $+0.6567 \times 10^{-3}$; $+0.0067 \times 10^{-3}$; -0.3833×10^{-3}
7.56×10^{-3} in^3

7.24 -1000 psi; 3.75×10^{-4}

7.25 0.021%

7.26 $\dfrac{100}{E}(\sigma_1 + \sigma_2 + \sigma_3)(1 - 2\mu)$; $\dfrac{300\sigma}{E}(1 - 2\mu)$

7.29 -1.0×10^{-4}

7.30 $\sigma_1 = -1625$ psi, $\sigma_2 = +375$ psi

7.31 4.12×10^{-4}; 8240 psi

7.32 6×10^6 psi

7.33 $2\sqrt{2} \times 10^{-4}$ radian

7.35 8.20×10^{-4}; 1.14×10^{-4}; 9050 psi; 3400 psi

7.36 $\tau_{max} = 6000$ psi; $\tau_{oct} = 4990$ psi

7.37 (a) $34,500$ psi; 0; 0
(b) 0; $17,250$ psi; $17,250$ psi
(c) $16,270$ psi

Chapter 8

8.1 -2280 lb
-326 psi

8.2 72 kips; 3.4 in above the bottom

8.3 $D/8$

8.4 $N = -4000$ lb; $M = -13,860$ lb-ft; $+61,400$ psi; $-63,400$ psi

8.5 $N = +116,000$ lb
$M_z = 200,000$ lb-in (compression AB)
$M_y = 56,000$ lb-in (compression AD)

8.6 (a) At top, $\sigma_1 = -2318$ psi; $\sigma_2 = +1146$ psi; $\tau_{max} = 1732$ psi
At bottom, $\sigma_1 = 2318$ psi; $\sigma_2 = -1146$ psi; $\tau_{max} = 1732$ psi
(b) At one end $\sigma_1 = -\sigma_2 = \tau_{max} = 1712$ psi
At the other end $\sigma_1 = -\sigma_2 = \tau_{max} = 1549$ psi

8.7 7620 psi

8.8   (a) −4860 psi and 0
     (b) −5809 psi and +1433 psi
     (c) ±3870 psi

8.9 $U = 9.30$ in-lb; deflection $= 0.093$ in

8.10 At A: $\sigma = +7800$ psi; $\tau = 3190$ psi
     At B: $\sigma = +7800$ psi; $\tau = 970$ psi

8.11 $\sigma = 407$ psi; $\tau = 444$ psi
     $\sigma_1 = +691$ psi; $\sigma_2 = -285$ psi

8.12 $U_T = 8.00$ in-lb; $U_M = 9.63$ in-lb; deflection $= 0.035$ in

8.13 4.9 in-lb; 2.44 in-lb; 0.147 in

8.14 +210 psi and −8890 psi

8.15 At A: +3410 psi; −1980 psi
     At B: +3760 psi; −2330 psi
     At C: +1150 psi; −7580 psi
     At D: +920 psi; −7350 psi

8.16 (a) At mid-point of top $\sigma_1 = -6150$ psi; $\sigma_2 = +1460$ psi; $\sigma_3 = 0$
       At mid-point of bottom $\sigma_1 = +6150$ psi; $\sigma_2 = -1460$ psi; $\sigma_3 = 0$
       At one end of N.A. $\sigma_1 = +4875$ psi; $\sigma_2 = -4875$ psi; $\sigma_3 = 0$
       At other end of N.A. $\sigma_1 = +1125$ psi; $\sigma_2 = -1125$ psi; $\sigma_3 = 0$
     (b) At top corners $\sigma_1 = -4690$ psi; $\sigma_2 = \sigma_3 = 0$
       At bottom corners $\sigma_1 = +4690$ psi; $\sigma_2 = \sigma_3 = 0$

8.17 7.3 in-lb

8.18 $\sigma = +4890$ psi
     $\tau = +890$ psi

8.19 650 lb

8.20 2200 psi

8.21 $N = -120$ kips
     $M = +4421$ kip-in

8.22 (a) 648 kip-ft
     (b) 216 kip-ft
     (c) 1134 kips

8.23 (a) $8.8\sigma_y$; (b) $9.75\sigma_y$; (c) $10.22\sigma_y$

8.24 (a) 57 kips compression, 0.27 in below the top
     (b) $N = -57$ kips; $M = +241$ kip-in

8.25 (a) 96 kips compression, 21.15 in above the top of the beam
     (b) $N = -96$ kips
       $M = +2990$ kip-in

8.26 (a) 159 kip-in
     (b) 159 kip-in

8.27 (a) 13.9 kip-in
     (b) 16.1 kip-in

8.28 47,500 lb-in

Chapter 9

9.1 $e = 0.05$ in for each bar
$v_A = -0.10$ in
$v_D = -0.15$ in

9.2 $e_{AB} = +0.111$ in
$e_{AC} = -0.055$ in
$u_A = +0.055$ in
$v_A = -0.160$ in

9.3 $u_C - -0.09$ in
$v_C = -0.756$ in

9.4 As for problem 9.3

9.5 $u_C = -0.39$ in
$v_C = -6.44$ in

9.6 $u_D = -0.256$ in
$v_D = -1.008$ in
$u_C = +0.256$ in
$v_C = -1.080$ in
$u_E = -0.384$ in
$v_E = -2.600$ in

9.7 $+16$ kips

9.8 $v_D = -1.008$ in
$v_E = -2.600$ in

9.9 $v_D = -1.712$ in
$v_E = -4.179$ in
$u_E = -0.512$ in

9.10 $u_B = 0.624$ in
$v_B = 0.083$ in
$u_C = 0.072$ in
$v_C = 0.125$ in

9.11 (a) $v_A = -0.141$ in
$v_D = -0.212$ in
(b) $v_A = -0.355$ in
$v_D = -0.532$ in

9.12 $+0.09$ in

9.13 -0.163 in

9.14 (a) $+0.26$ in; (b) $+0.26$ in; (c) no

9.15 1.57 in

9.16 $+8$ kips

9.17 16.98 kips

9.18 2.6 kips upward

9.19 $AB = 2.22$ kips
$AC = 2.22$ kips
$AD = 5$ kips
$AE = 2.78$ kips
$v_A = -0.044$ in

9.20 As for problem 9.19

9.21 $N_1 = 6.5$ kips
$N_2 = 15.4$ kips
$N_3 = 4.6$ kips
$u_A = -0.069$ in
$v_A = -0.115$ in

9.22 As for problem 9.21

9.23 $N_1 = 11.3$ kips
$N_2 = 12.0$ kips
$N_3 = 8.0$ kips
$u_A = -0.12$ in
$v_A = -0.20$ in

9.24 $N_{AB} = N_{BC} = 10$ kips
$N_{CD} = -4.33$ kips
$N_{CE} = +4.33$ kips
$N_{CF} = +5.67$ kips
$u_B = -0.134$ in
$v_B = +1.383$ in

9.25 $AE = BE = +\sqrt{2}$ kips
$CE = DE = -2\sqrt{2}$ kips

9.26 $AE = BE = +\sqrt{2}$ kips
$CE = DE = -2\sqrt{2}$ kips

9.27 $\sigma_{AB} = 10,000$ psi
$\sigma_{CD} = 4800$ psi
9.48 in from B

9.28 (a) 12.65 in from B
(b) $\sigma_{AB} = 1412$ psi
$\sigma_{CD} = 1324$ psi
$\sigma_{EF} = 1765$ psi

9.29 $\sigma_T = -15,400$ psi
$\sigma_P = -7120$ psi

9.30 $N_c = -12,020$ lb
$N_s = +12,020$ lb
$U = 376$ in-lb

9.31 $N_N = 6688$ lb; $\sigma_N = 16{,}720$ psi
$N_M = 2964$ lb; $\sigma_M = 4940$ psi
$N_T = 10{,}336$ lb; $\sigma_T = 12{,}920$ psi

9.32 (a) $\sigma_C = 2190$ psi
$\sigma_S = 13{,}150$ psi
(b) $\sigma_C = 1770$ psi
$\sigma_S = 31{,}900$ psi

9.33 (a) $\sigma_{MS} = 40{,}000$ psi
$\sigma_{HTS} = 63{,}000$ psi
(b) $\sigma_{MS} = -17{,}300$ psi
$\sigma_{HTS} = +5700$ psi

9.34 48.98 kips

9.35 (a) $N_A = 24$ kips; $\sigma_A = 40$ ksi
$N_B = 24$ kips; $\sigma_B = 30$ ksi
(b) $N_A = -1.4$ kips; $\sigma_A = -2.33$ ksi
$N_B = +1.4$ kips; $\sigma_B = +1.75$ ksi

9.36 (a) 25,000 psi; 15,000 psi
(b) 52,500 psi; 24,000 psi

9.37 (a) $\sigma_A = 3000$ psi
$\sigma_B = 4000$ psi
(b) $\sigma_A = 15{,}000$ psi
$\sigma_B = 13{,}000$ psi

9.38 $v_D = 0.173$ in
$u_D = -0.199$ in

9.39 $N_1 = N_2 = +2087$ lb
$N_3 = N_4 = -3479$ lb
$N_5 = +2783$ lb

9.40 $\sigma_{AB} = \sigma_{AD} = -3000$ psi
$\sigma_{AC} = +3850$ psi

9.41 $\sigma_S = 12{,}000$ psi
$\sigma_Z = -15{,}000$ psi

9.42 3650 lb; 0.0102 in towards A

9.43 $\sigma_A = \sigma_B = -15{,}000$ psi; $+1.48 \times 10^{-3}$ in

9.44 (a) $\sigma_s = -9230$ psi; $\sigma_c = +370$ psi; (b) $+160$ kips

9.45 (a) $N_1 = N_3 = 20$ kips; $N_2 = 24$ kips
(b) $N_1 = N_3 = 5.33$ kips; $N_2 = -5.33$ kips

9.46 (a) $N_1 = 47.1$ kips; $N_2 = 65.2$ kips; $N_3 = 8.2$ kips
(b) $N_1 = 7.1$ kips; $N_2 = -4.1$ kips; $N_3 = 8.2$ kips

9.47 -0.01 ft; -0.059 ft

9.48 3.69 kips; -13.1 kip·ft

9.49 $u_B = 0$; $v_B = 0$; $\phi_B = 0.045$ rad (counterclockwise)
$u_C = 0.36$ ft; $v_C = 0$; $\phi_C = 0.045$ rad

9.50 $v_B = -0.51$ ft; $M_A = -46.8$ kip-ft
$$M_C = +24.6 \text{ kip-ft}$$

9.51 10 kips; 100 kip-ft

9.52 In AB, $M_E = 13.5$ kip-ft
In CD, $M_E = 22.5$ kip-ft

Chapter 10

10.1 37.3°

10.2 15,400 lb

10.3 $\dfrac{kl}{2}$

10.4 $\dfrac{kL}{9}$; $\dfrac{kL}{3}$

10.5 0.423 lk; 1.577 lk

10.6 54 kips

10.7 No

10.8 (*a*) 54 kips; (*b*) 49.4 kips; (*c*) 51.6 kips

10.9 1780 lb

10.10 6000 lb

10.11 $1.355EI/L^2$

10.12 $4\pi^2 EI/L^2$

10.13 8000 lb; 32,000 lb; a deformation with $v_B = -v_D$

10.14 8.8%; 5.0%

Chapter 11

11.1 12.1 cycles/sec; 0.0825 sec

11.2 2.71 cycles/sec; 0.369 sec

11.3 3.62 cycles/sec; 0.28 sec

11.4 (*a*) 4.80 cycles/sec; 0.21 sec
(*b*) 3.39 cycles/sec; 0.295 sec
(*c*) 2.77 cycles/sec; 0.361 sec

11.5 2.11 cycles/sec; 0.474 sec

11.6 2.08 cycles/sec; 0.481 sec

11.7 3.43 cycles/sec

11.8 1.79 cycles/sec

11.9 12,700 psi; 0.025 in

11.10 (*a*) 0.027 in
(*b*) 18,000 psi

11.11 34,500 psi; 0.014 in

11.12 16.2 in

11.13 (a) 0.295 in
  (b) 1.66 in

11.14 1000 ft

11.15 60 lb/in

11.16 (a) 1.44 cycles/sec; 0.69 sec; $a_1/a_2 = 0.82$
  (b) 9.12 cycles/sec; 0.11 sec; $a_1/a_2 = -6.07$

11.17 (a) 4.31 cycles/sec; 0.232 sec; $a_1/a_2 = 0.317$
  (b) 6.34 cycles/sec; 0.158 sec; $a_1/a_2 = -6.317$

11.18 (a) 0.485 cycles/sec; 2.06 secs; $a_1/a_2 = 0.327$
  (b) 2.51 cycles/sec; 0.40 sec; $a_1/a_2 = -1.527$

11.19 (a) 2.71 cycles/sec; 0.37 sec; $a_1/a_2 = 1.0$
  (b) 10.5 cycles/sec; 0.095 sec; $a_1/a_2 = -1.0$

11.20 (a) 0.96 cycles/sec; 1.04 secs; $a_1/a_2 = -0.28$
  (b) 3.17 cycles/sec; 0.315 sec; $a_1/a_2 = +3.61$

11.21 (a) 1.79 cycles/sec; 0.559 sec; $a_1/a_2 = 1.0$
  (b) 2.31 cycles/sec; 0.433 sec; $a_1/a_2 = -1.0$

11.22 (a) 3.36 cycles/sec; 0.30 sec
  (b) 4.81 cycles/sec; 0.21 sec
  (c) 6.13 cycles/sec; 0.16 sec

11.23 (a) 2.71 cycles/sec
  (b) 2.71 cycles/sec
  (c) 2.70 cycles/sec

Appendix A

A.1 $N = -5$ kips; $S = +0.45$ kips; $M = +93.3$ kip-ft

A.2 At E: $N = 0$; $S = -2.75$ kips; $M = +12.375$ kip-ft
  At F: $N = +3.71$ kips; $S = +3.71$ kips; $M = +14.25$ kip-ft
  At G: $N = +7.95$ kips; $S = +0.53$ kips; $M = +1.875$ kip-ft

A.3 At D: $N = +10$ kips, $S = 0$, $M = +43.3$ kip ft
  At E: $N = +5$ kips; $S = +8.66$ kips; $M = +21.65$ kip-ft

A.4 $M_A = 0$; $M_B = +51.0$ kip-ft; $M_C = +36.6$ kip-ft; $M_D = +68.8$ kip-ft
  $S_A = -25.2$ kips; $S_B = +4.8$ kips; left of C, $S = +4.8$ kips;
  above C, $S = -8.0$ kips; left of E, $S = +14.8$ kips;
  right of E, $S = -5.0$ kips; $N_{AB} = +8.0$ kips; $N_{BC} = +8.0$ kips;
  $N_{CD} = +14.8$ kips

A.5

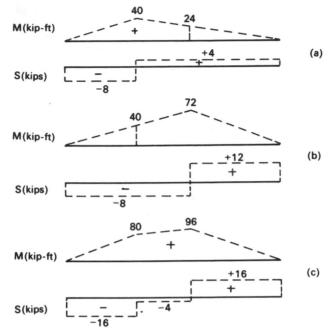

A.6   Segment AB: $S = -30 + 4x$ kips
$\qquad\qquad\qquad M = 30x - 2x^2$ kip-ft
$\qquad$ Segment BC: $S = +10$ kips
$\qquad\qquad\qquad M = 200 - 10x$ kip-ft

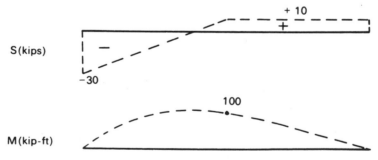

A.7   Segment AC: $N = -13.66 \cos \theta$ kips
$\qquad\qquad\qquad S = -13.66 \sin \theta$ kips
$\qquad\qquad\qquad M = +136.6(1 - \cos \theta)$ kip-ft
$\qquad$ Segment CD: $N = -13.66 \cos \theta - 10 \sin(\theta - 60°)$ kips
$\qquad\qquad\qquad S = -13.66 \sin \theta + 10 \cos(\theta - 60°)$ kips
$\qquad\qquad\qquad M = +136.6(1 - \cos \theta) - 100 \sin(\theta - 60°)$ kip-ft

A.8 $R_A = (C_2 - C_1)/18$; $R_D = -(C_2 - C_1)/18$

S

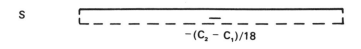

$-(C_2 - C_1)/18$

M

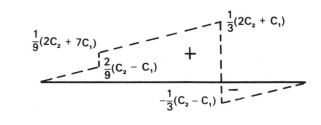

$\frac{1}{9}(2C_2 + 7C_1)$

$\frac{2}{9}(C_2 - C_1)$

$+$

$\frac{1}{3}(2C_2 + C_1)$

$-\frac{1}{3}(C_2 - C_1)$

A.9

N(kips)

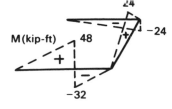

O

O

-6.93

S(kips)

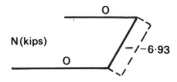

$+3$

$+$

-4

-8

M(kip-ft)

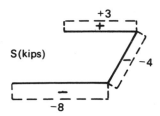

24

-24

$+$

48

$+$

-32

A.10

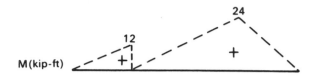

M(kip-ft)

S(kips)

A.11

S

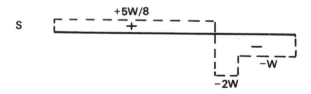

M

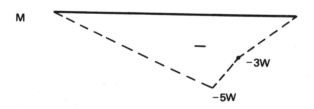

A.12

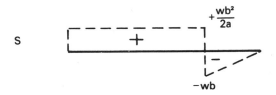

S

$+\dfrac{wb^2}{2a}$

$-wb$

M

$-\dfrac{wb^2}{2}$

A.13

S = O

M (kip-ft)

$+2$

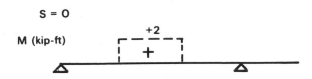

A.14

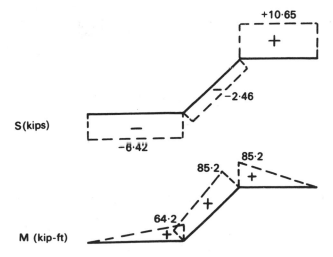

$+10\cdot65$

$-2\cdot46$

S(kips)

$-6\cdot42$

$85\cdot2$

$85\cdot2$

$64\cdot2$

M (kip-ft)

A.15

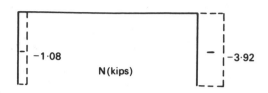

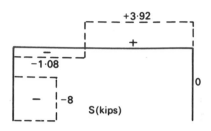

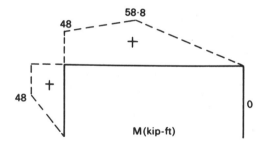

A.16 (a) $R_A = +12$ kips; $R_B = +24$ kips
   (b) $S = -12 + x^2/16$ kips; $M = +12x - x^3/48$ kip-ft
   (c) $S = 0$ at 13.86 ft from A
   (d) $M_{max} = 111$ kip-ft at 13.86 ft from A

A.17 $S = +2x$ kips
   $M = -x^2$ kip-ft

A.18 $S = +\dfrac{L}{\pi} w_0 \cos \dfrac{\pi x}{L}$

   $M = -\dfrac{L^2}{\pi^2} w_0 \cos \dfrac{\pi x}{L}$

A.19

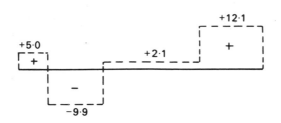

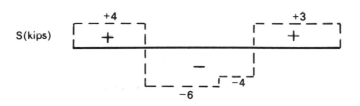

A.20

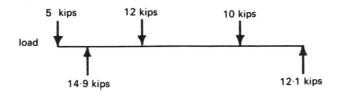

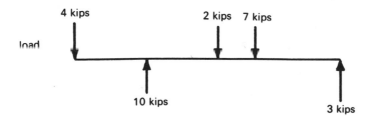

A.21

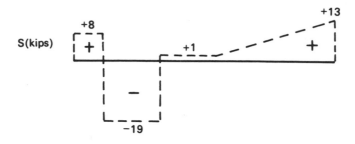

S(kips)

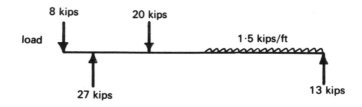

load

A.22

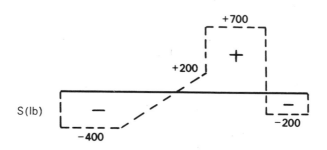

S(lb)

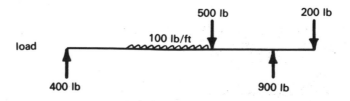

load

A.23 Segment AB: $S = -5.25$ kips; $M = +5.25x$ kip-ft
Segment BC: $S = +2x - 17.25$ kips; $M = -x^2 + 17.25x - 36$ kip-ft
Segment CD: $S = +2x - 40$ kips; $M = -x^2 + 40x - 400$ kip-ft
$M_{max} = +38.39$ kip-ft; $M = 0$ at 14.82 ft from A

A.24

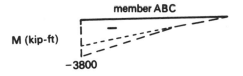

M (kip-ft)   member ABC   member BD -320 -3800

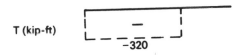

S (kips) -280 -200 -80

T (kip-ft) -320 $T = 0$

A.25 Segment AB: $S = 0$; $M = 0$; $T = -6$ kip-ft
Segment BC: $S = 0$; $M = -5.2$ kip-ft; $T = -3$ kip-ft

A.26 Segment AB: $M = +5$ kip-ft; $T = -8.66$ kip-ft
Segment BC: $M = 0$; $T = -10$ kip-ft

A.27

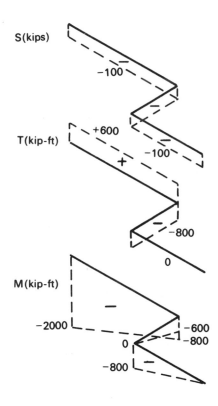

A.28

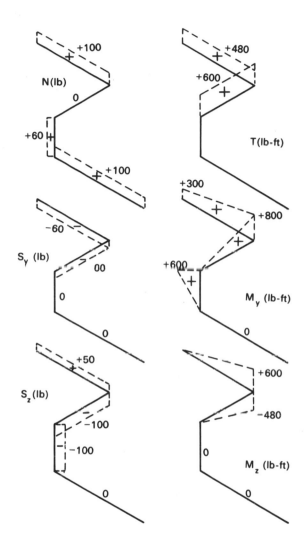

A.29 Segment AB: $N = -22\cos\theta - 11.9\sin\theta$ kips

$S = -22\sin\theta + 11.9\cos\theta$ kips

$M = 450 - 450\cos\theta - 238\sin\theta$ kip-ft

Segment BC: $N = -7.5\cos\theta - 11.9\sin\theta$ kips
$$S = -7.5\sin\theta + 11.9\cos\theta \text{ kips}$$
$$M = 238 - 150\cos\theta - 238\sin\theta \text{ kip-ft}$$
$N_{max} = -11.9$ kips at $\theta = 90°$
$N_{min} = -25.6$ kips at $\theta = 27° 50'$
$S_{max} = +11.9$ kips at $\theta = 0°$
$S_{min} = -7.5$ kips at $\theta = 45°$ and $90°$
$M_{max} = 0$ at $\theta = 0°$ and $90°$
$M_{min} = -59$ kip-ft at $\theta = -27° 50'$

A.30

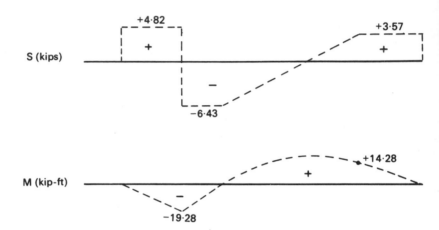

A.31

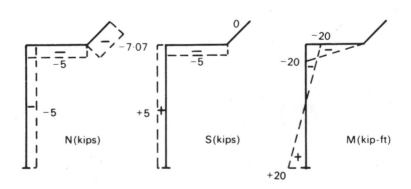

Appendix B

B.1 $I_{zz} = 405.67$ in^4

B.2 $I_{yy} = 18.12$ in^4; $I_{zz} = 254$ in^4

B.3 $I_{yy} = 36.13$ in^4; $I_{zz} = 680.5$ in^4

B.4 5.32 in from the bottom; $I_{zz} = 301.6$ in^4

B.5 9.06 in from the bottom; $I_{yy} = 42.8$ in^4; $I_{zz} = 173.6$ in^4

B.6 $I_{zz} = 20.90$ in^4

B.7 1.08, 2.08 in from the back of the angle;
axis 1 is $+23°\,12'$ from Cy; $I_1 = 28.03$ in^4; $I_2 = 5.13$ in^4

B.8 1.86 in from the left, 1.71 in from the bottom;
axis 1 is $+49°\,01'$ from Cy; $I_1 = 25.09$ in^4; $I_2 = 11.82$ in^4

B.9 $y = 1.75$ in from top, $z = 6.75$ in from left; axis 1 is $-10°\,50'$ from Cy;
$I_1 = 355.0$ in^4; $I_2 = 62.33$ in^4

Index